FLORA OF THE GUIANAS

Edited by

M.J. JANSEN-JACOBS

Series C: Bryophytes
Fascicle 2

MUSCI IV

(J. Florschütz-de Waard
with H.R. Zielman
M.A. Bruggeman-Nannenga)

2011
Royal Botanic Gardens, Kew

The Flora of the Guianas

is a modern, critical and illustrated Flora of Guyana, Suriname, and French Guiana designed to treat Phanerogams as well as Cryptogams of the area.

Contents: Publication takes place in fascicles, each treating a single family, or a group of related families, in the following series: A: Phanerogams; B: Ferns and Fern allies; C: Bryophytes; D: Algae; and E: Fungi and Lichens. A list of numbered families in taxonomic order has been established for the Phanerogams. Publication of fascicles will take place when available.
In the Supplementary series other relevant information concerning the plant collections from the Guianas appears, like indexes of plant collectors.

The Flora in general, follows the format of other, modern Floras such as the *Flora of Ecuador* and *Flora Neotropica*. The treatment provides fundamental and applied information; it covers, when possible, wood anatomy, chemical analysis, economic uses, vernacular names, and data on endangered species.

ORGANISATION: The Flora is a co-operative project of: Botanischer Garten und Botanisches Museum Berlin-Dahlem, *Berlin*; Institut de Recherche pour le Développement, IRD, Centre de Cayenne, *Cayenne*; Department of Biology, University of Guyana, *Georgetown*; Herbarium, Royal Botanic Gardens, *Kew*; Nationaal Herbarium Nederland, Leiden University branch, *Leiden;* New York Botanical Garden, *New York*; Nationaal Herbarium Suriname, *Paramaribo*; Muséum National d'Histoire Naturelle, *Paris*, and Department of Botany, Smithsonian Institution, *Washington, D.C.*

The Flora is edited by the Advisory Board: Executive Editor: M.J. JANSEN-JACOBS, Leiden. Members: H. SIPMAN, Berlin; P.G. DELPRETE, Cayenne; P. DA SILVA, Georgetown; E. LUCAS, Kew; T.R. VAN ANDEL, Leiden; B. TORKE, New York; D. TRAAG, Paramaribo; O. PONCY, Paris, and P. ACEVEDO RODR., Washington, D.C.

PUBLICATION: The *Flora of the Guianas* is a publication of The Royal Botanic Gardens, Kew. The prices of the fascicles are determined by their size. Authors are requested to submit a hard copy of their manuscript as well as an electronic version; Word and other standard word processing packages are acceptable.

INFORMATION: http://www.nationaalherbarium.nl/FoGWebsite/index.htm

Editorial office for correspondence on contributions, etc.:

S. Mota de Oliveira and M.J. Jansen-Jacobs
Nationaal Herbarium Nederland
Leiden University branch
P.O. Box 9514
2300 RA Leiden
The Netherlands

Publisher
Royal Botanic Gardens
Kew
Richmond
Surrey, TW9 3AB
U.K.

Email: s.motadeoliveira@gmail.com; M.J.Jansen-Jacobs@uu.nl

Printed in the USA by The University of Chicago Press

Contents

MUSCI IV
by
Jeanne Florschütz-de Waard [1,2]

family SPHAGNACEAE
by
H.R. (Rudi) Zielman [1]

family FISSIDENTACEAE
by
M.A. (Ida) Bruggeman-Nannenga [1]

INTRODUCTION

This final part of the Moss Flora of the Guianas is a compilation of the preceding publications on this subject. The start was made in 1964 by the thesis of my late husband (P.A. Florschütz), published as a part of the Flora of Suriname (Vol. VI, part I, Musci). This study was mainly based on collections we made during a half year stay in Suriname (1950-1951). At that time only few moss collections were available in herbaria; collections from Guyana (BM) and French Guiana (PC) were studied, but were not fully described and illustrated.

Thus far only the acrocarpous families were treated. With a lot of work to do on the pleurocarpous families we decided to continue this work together. After the sudden decease of my husband in 1976 I continued the project on my own. This resulted in a second part of the Moss Flora, published in 1986 in the Flora of Suriname (Vol. VI, part I, Musci II), containing NECKERACEAE, HOOKERIACEAE, and PLAGIOTHECIACEAE. This time all collections available from the Guianas were fully described and illustrated.

The remaining families of pleurocarpous mosses (LEUCOMIACEAE, THUIDIACEAE, SEMATOPHYLLACEAE, HYPNACEAE) were published in 1996 as a part of the Flora of the Guianas (Series C, Musci III).

[1] National Herbarium of the Netherlands, Utrecht University branch. Since 2009 housed in NHN-Leiden: P.O. Box 9514, 2300 RA Leiden, The Netherlands.

[2] To Ida Bruggeman-Nannenga and Rudi Zielman I am grateful for their share in the accomplishment of this publication. Thanks are due to William Buck and Bruce Allen for the loan of their moss collections from the Guianas. I wish to thank Rob Gradstein for correcting the English text and William Buck for critically reading the manuscript. Gea Zijlstra was very helpful with assistance in nomenclatural problems.

All drawings, except for Sphagnum and Fissidens, are made by Jeanne Florschütz-de Waard.

The publication of the Flora of Suriname had stopped and was replaced by participation in the international project Flora of the Guianas.
It seems useful to make a compilation of these discontinuous publications. During the last decennia many collecting trips in divers habitats of the Guianas have notably enlarged the number of known species. In total 83 species and 4 new varieties could be added, most of them in the acrocarpous families. These are all described and included in new keys. For the species already treated in preceding parts is referred to the descriptions in those treatments, amplified with new data about ecology and distribution.
The sequence of the families is mainly based on the classification in orders as used in the moss flora of Mexico (Sharp et al., 1994). In each order a key is given to find the family in question. A general key to the genera directly leads to the concerning genus.

GENERAL LITERATURE

Allen, B. 1994. Moss Flora of Central America. Part 1. Sphagnaceae-Calymperaceae. Monogr. Syst. Bot. Missouri Bot. Gard. 49: 1-242.

Allen, B. 2002. Moss flora of Central America. Part 2. Encalyptaceae-Orthotrichaceae. Monogr. Syst. Bot. Missouri Bot. Gard. 90: 1-699.

Britton, E.G. 1918. Musci. In N.L. Britton, Flora of Bermuda, pp. 430-448.

Buck, W.R. 1990. Contributions to the Moss Flora of Guyana. Mem. New York Bot. Gard. 64: 184-196.

Buck, W.R. 2003. Guide to the Plants of Central French Guiana. Part 3. Mosses. Mem. New York Bot. Gard. 76(3): 1-167.

Florschütz, P.A. 1964. Musci. Part I. In: J. Lanjouw (ed.), Flora of Suriname 6(1): 1-271.

Florschütz-de Waard, J. 1986. Musci. Part II. In: A.L. Stoffers & J.C. Lindeman (eds.), Flora of Suriname 6(1): 273-361.

Florschütz-de Waard, J. 1990. A catalogue of the bryophytes of the Guianas II, Musci. Trop. Bryol. 3: 89-104.

Florschütz-de Waard, J. 1996. Musci III. In: A.R.A. Görts-van Rijn (ed.), Flora of the Guianas, ser. C, 1: 363-490.

Gradstein, S.R., S.P. Churchill & N. Salazar-Allen. 2001. Guide to the Bryophytes of Tropical America. Mem. New York Bot. Gard. 86: 1-577.

Sharp, A.J., H. Crum & P.M. Eckel (eds.). 1994. The Moss Flora of Mexico (in 2 parts). Mem. New York Bot. Gard. 69: 1-1113, I-XVII.

CLASSIFICATION
(According to Sharp *et al.*, 1994)

Class SPHAGNOPSIDA

Branches spirally arranged in fascicles. Plants usually in wet habitats. Capsules without seta, globose, peristome lacking.

Order **Sphagnales**
  1. Sphagnaceae

Class BRYOPSIDA

Branches not arranged in fascicles. Plants erect, simple or sparsely branched, producing sporophytes on end of stem (acrocarpous) or plants prostrate, freely branched, producing sporophytes laterally (pleurocarpous). Capsules on long or short setae, with single or double peristome.

Subclass Bryidae
  Order **Archidiales**
    2. Archidiaceae
  Order **Fissidentales**
    3. Fissidentaceae
  Order **Dicranales**
    4. Dicranaceae
    5. Ditrichaceae
    6. Leucobryaceae
    7. Leucophanaceae
  Order **Pottiales**
    8. Calymperaceae
    9. Pottiaceae
  Order **Funariales**
    10. Ephemeraceae
    11. Funariaceae
    12. Splachnobryaceae
  Order **Bryales**
    13. Bartramiaceae
    14. Bryaceae
    15. Phyllodrepaniaceae
    16. Rhizogoniaceae
  Order **Orthotrichales**
    17. Macromitriaceae
  Order **Leucodontales**
    18. Leptodontaceae

    19. Meteoriaceae
    20. Neckeraceae
    21. Phyllogoniaceae
    22. Pterobryaceae
    23. Racopilaceae
    24. Rhacocarpaceae
  Order **Hookeriales**
    25. Daltoniaceae
    26. Leucomiaceae
    27. Pilotrichaceae
  Order **Hypnales**
    28. Fabroniaceae
    29. Hydropogonaceae
    30. Hypnaceae
    31. Sematophyllaceae
    32. Stereophyllaceae
    33. Thamnobryaceae
    34. Thuidiaceae

Subclass Buxbaumiidae
  Order **Buxbaumiales**
    35. Diphysiaceae

Subclass Polytrichidae
  Order **Polytrichales**
    36. Polytrichaceae

KEY TO THE GENERA OF MOSSES OF THE GUIANAS
(This key is based on the characters of the taxa occurring in the Guianas and does not take account of characters of the species outside this area.)

1 Branches arranged in fascicles on the stem. Laminal cells forming a regular network of wide hyaline cells enclosed by narrow chlorophyllose cells . . .
. . . . *1-1. Sphagnum*
Branches not arranged in fascicles. Laminal cells not forming a regular network of two cell types 2

2 Leaves with duplicate, sheathing bases, distichous . . . . *3-1. Fissidens*
Leaves different 3

3 Leaves with broad, sheathing bases and upper laminae with numerous longitudinal lamellae on the costa . . . . *36-1. Polytrichum*
Leaves different 4

4 Minute plants, less than 1 mm high, growing scattered from a persistent, algae-like protonema . . . . *10-1. Micromitrium*
Plants larger, protonema not persistent 5

5 Plants whitish green. Leaf for the greater part composed of the costa, lamina reduced to the basal part of the leaf; costa in cross-section with two or more layers of large hyaline cells (leucocysts) and a central layer of small chlorophyllose cells (chlorocysts) 6
Plants seldom whitish green (except *Bryum argenteum*). Leaf lamina well developed, costa various, if wide not occupying the whole leaf width . . 10

6 Chlorocysts in cross-section triangular; leaf apex flat, lingulate. . . . .
. . . . *6-4. Octoblepharum*
Chlorocysts in cross-section four-angled; leaf apex tubulose or cucullate . . . 7

7 Costa with a central bundle of stereids from base to apex . . . *7-1. Leucophanes*
Costa without a central bundle of stereids 8

8 Small plants, often bearing budlike propagules. Capsule globose, immersed, peristome absent . . . . *6-3. Ochrobryum*
Plants larger. Capsule cylindrical, exserted, peristome present 9

9 Leaf apex blunt, cucullate, hyaline lamina conspicuous to the apex. Capsule erect, smooth, peristome teeth papillose, not striate. . . . *6-1. Holomitriopsis*
Leaf apex acute, hyaline lamina indistinct or absent in upper leaf. Capsule arcuate, furrowed when dry, peristome teeth vertically striate. . . . .
. . . . *6-2. Leucobryum*

10 Leaves without costa or with very short, indistinct, double costa. . . . Group C
Leaves with a distinct costa, extending at least 1/4 of leaf length 11

11 Costa single...Group A
 Costa double..Group B

GROUP A. Genera with unicostate leaves.

1 Very small plants; stems not over 5 mm long. Upper laminal cells bistratose.
 Perichaetium terminal, conspicuous; capsule immersed..............
 ....*35-1. Diphyscium*
 Plants small to medium sized. Laminal cells predominantly unistratose ... 2

2 Plants silvery white, upper part of leaves hyaline .... *14-2. Bryum* p.p.
 Plants green..3

3 Costa strong, percurrent or excurrent............................4
 Costa well developed but not reaching apex 37

4 Leaf base with a conspicuous central group of wide, hyaline cells
 (cancellinae), usually clearly distinct from the smaller green upper
 lamina cells .... *8. CALYMPERACEAE*
 Cells of leaf base, if wide and hyaline, not forming a conspicuous central
 group...5

5 Alar cells differentiated, forming a distinct group...................6
 Alar cells, if differentiated, not in a distinct group.................13

6 Alar cells coloured and/or inflated, often forming auricles 7
 Alar cells small, quadrate, not coloured 11

7 Costa wide, occupying 1/3 of leaf base or more....................8
 Costa narrower 9

8 Stems interruptedly foliate; leaves over 10 mm long. Seta straight or
 slightly twisted............................. *4-1. Bryohumbertia*
 Stems evenly foliate or comose at apex, if interruptedly foliate leaves shorter
 than 10 mm. Seta cygneous .... *4-2. Campylopus*

9 Leaves with a hyaline border of narrow, elongate cells, at least in the
 basal part .... *4-6. Leucoloma*
 Leaves without hyaline border.................................10

10 Laminal cells linear and strongly incrassate throughout the leaf; leaf apex
 hyaline or long-tubulose.................... *4-4. Eucamptodontopsis*
 Only basal cells elongate and sometimes incrassate, upper laminal cells more
 or less quadrate; leaf apex acute, serrate .... *4-5. Holomitrium*

11 Leaves narrowly lanceolate, folded lengthwise. Capsules immersed.......
 ....*2-1. Archidium*
 Leaves ovate or oblong, concave. Capsules exserted................12

12 Plants often with microphyllous branches. Lamina cells fusiform, minutely papillose at both cell ends. Endostome reduced. . . . *18-1. Pseudocryphaea*
No microphyllous branches. Lamina cells linear, flexuose. Peristome complete . . . . *22-6. Pireella*

13 Plants with erect stems, simple or sparingly branched 14
Plants with creeping primary stems (stolons) and erect or prostrate secondary stems or branches. 29

14 Upper laminal cells lax, elongate-hexagonal, over 10 μm wide. 15
Upper laminal cells quadrate, rectangular or linear, less than 10 μm wide 20

15 Marginal laminal cells differentiated 16
Marginal laminal cells not differentiated 19

16 Marginal laminal cells shorter, more or less quadrate in several rows; apex obtuse . . . . *12-1. Splachnobryum*
Marginal laminal cells elongate, forming a narrow border; apex acute or rounded-acute and apiculate. 17

17 Costa percurrent or excurrent 18
Costa ending below apex . . . . *11-1. Entosthodon*

18 Capsule erect, endostome shorter than exostome . . . . . *14-1. Brachymenium*
Capsule inclined, endostome as long as exostome . . . . *14-2. Bryum* p.p.

19 Leaves oblong or obovate-spathulate, 3-4 mm long. . . . . *11-2. Funaria*
Leaves lanceolate, not over 3 mm long . . . . *14-2. Bryum* p.p.

20 Upper laminal cells isodiametric, quadrate or rounded hexagonal. 21
Upper laminal cells elongate, rectangular to linear 25

21 Leaf margin thickened, coarsely double-serrate . . . . *16-1. Pyrrhobryum*
Leaf margin not thickened 22

22 Laminal cells smooth or indistinctly papillose 23
Laminal cells pluri-papillose 24

23 Leaves 1.5-3 mm long, apex obtuse to broad-acute, mucronate; upper laminal cells obscure, sometimes minutely papillose, less than 10 μm in largest diameter . . . . *9-2. Hyophila*
Leaves 1-2 mm long, apex acute-acuminate; upper laminal cells pellucid, over 10 μm in largest diameter. . . . . *9-3. Hyophiladelphus*

24 Leaves not over 1.5 mm long; inner basal laminal cells hardly wider, 2-3 times as long as upper laminal cells . . . . *9-1. Barbula*
Leaves 2 mm or more long; inner basal laminal cells wider and 4-6 times as long as upper laminal cells . . . . *9-4. Trichostomum*

7

25 Laminal cells smooth 26
    Laminal cells prorulose.. 27

26 Laminal cells irregularly rectangular. Capsule exserted.... *4-3. Dicranella*
    Laminal cells linear. Capsule immersed.................. *5-1. Garckea*

27 Plants small. Leaves less than 2 mm long .... *13-3. Philonotis*
    Plants medium-sized. Leaves over 2 mm long...................... 28

28 Leaves lanceolate with gradually acuminate apex; all cells linear, prorulose
    at both cell ends .... *13-1. Breutelia*
    Leaves linear-subulate with long-excurrent costa; cells in basal part of the
    leaf linear, smooth, in upper part oblong, strongly prorulose by projecting
    apical cell ends .... *13-2. Leiomela*

29 Secondary stems erect... 30
    Secondary stems prostrate 34

30 Plants with distant, simple or branched, secondary stems often frondose or
    dendroid .... *22-6. Pireella* p.p.
    Plants with closely spaced, mostly simple branches, forming dense mats ... 31

31 Inner basal laminal cells quadrate or transversely elongate (oblate), at margin
    bordered with elongate cells........................ *17-1. Groutiella*
    All basal laminal cells elongate (or all isodiametric), no border
    differentiated.. 32

32 Leaves crispate or spirally twisted when dry ..... *17-2. Macromitrium* p.p.
    Leaves erect, appressed when dry, only at end of branch sometimes
    twisted 33

33 Calyptra cucullate. Midlaminal cells elongate to linear ....
    .... *17-2. Macromitrium* p.p.
    Calyptra campanulate, lobed at base. Midlaminal cells rounded or transversely
    elongate .... *17-3. Schlotheimia*

34 Leaves dimorphous, dorsal leaves much smaller than lateral leaves, costa
    long-excurrent .... *23-1. Racopilum*
    Leaves uniform... 35

35 Leaves spreading in all directions, symmetric or slightly falcate..........
    .... *28-1. Anacamptodon*
    Leaves complanate-spreading, curved-asymmetric 36

36 Laminal cells rounded-quadrate, pluri-papillose .... *15-1. Mniomalia*
    Laminal cells elongate-rhomboidal, smooth....... *15-2. Phyllodrepanium*

37 Alar cells strongly differentiated in a distinct group 38
    Alar cells, if differentiated, not forming a distinct group............. 43

8

53 Leaves 5 mm or more long with long piliform acumen ....*22-8. Spiridentopsis* Leaves not over 2 mm long. 54

54 Stem and branch leaves little differentiated, spreading from the insertion ... . *19-3. Meteoridium* Stem leaves long-acuminate, laxly appressed with a clasping base, branch leaves acute, often squarrose-recurved. .... *19-7. Zelometeorium*

GROUP B. Genera with bicostate leaves (costae extending 1/4 of leaf length or more).

1 Plants with erect secondary stems, stipitate at base, pinnately branched .... *27-10. Pilotrichum* Plants with prostrate or ascending branches. 2

2 Laminal cells papillose. .... 3 Laminal cells smooth 5

3 Laminal cells pluri-papillose (sometimes indistinct) .... *27-7. Hypnella* Laminal cells uni-papillose or prorulate. 4

4 Leaves ovate-oblong with rounded apex .... *27-3. Callicostella* p.p. Leaves lanceolate with a long, flexuose apex .... *27-12. Trachyxiphium*

5 Leaves linear, longitudinally plicate. .... *27-6. Hemiragis* Leaves ovate, oblong or lanceolate, not longitudinally plicate. 6

6 Leaves transversely undulate in upper part. .... 7 Leaves not transversely undulate 8

7 Leaves oblong-lingulate, apex short-acute .... *27-2. Brymela* Leaves lanceolate, apex acuminate. .... *27-11. Thamniopsis* p.p.

8 Leaf margins narrowly recurved from base nearly to apex .... *27-1. Actinodontium* Leaf margins flat or partly inflexed. 9

9 Laminal cells large and lax, over 15 μm wide, along the margin a distinct border of narrow cells. .... 10 Laminal cells narrower, border not distinct 11

10 Costae slender, less than 40 μm wide at base; leaf margin subentire. .... *27-5. Cyclodictyon* Costae firm, at least 40 μm wide at base; leaf margin dentate in upper part .... *27-9. Lepidopilum* p.p.

11 Costae extending ca. 3/4 of leaf length 12
     Costae shorter, extending 1/2-3/4 of leaf length 13

12 Leaves ovate or oblong; upper laminal cells isodiametric . . . .
     . . . . *27-3. Callicostella* p.p.
     Leaves lanceolate; upper laminal cells oblong-linear . . . .
     . . . . *27-11. Thamniopsis* p.p.

13 Seta smooth; capsule with transversely striate and furrowed exostome teeth.
     Leaves with abruptly acuminate apex. Plants often reddish . . . .
     . . . . *27-8. Lepidopilidium*
     Seta papillose; capsule with papillose exostome teeth without median furrow.
     Leaves with acute or gradually acuminate apex . . . . *27-9. Lepidopilum* p.p.

GROUP C. Genera with ecostate leaves (or with short and indistinct, double costa).

1 Laminal cells papillose, prorulose or reticulate 2
    Laminal cells smooth 9

2 Laminal cells covered with a reticulum of many fine pits, appearing papillose;
    marginal cells smooth, forming a distinct border . . . . . . *24-1. Rhacocarpus*
    Laminal cells papillose or prorulose; leaf not bordered 3

3 Laminal cells prorulose by projecting cell ends 4
    Laminal cells with papillae over the lumen 5

4 Laminal cells prorulose at both ends . . . . *30-1. Chryso-hypnum*
    Laminal cells prorulose only at distal ends . . . . *30-4. Mittenothamnium*

5 Laminal cells uni-papillose 6
    Laminal cells pluri-papillose 7

6 Leaf apex pungent, with involute margins; inflated alar cells large (70-170 μm
    long), curved to the insertion . . . . *31-1. Acroporium* p.p.
    Leaf apex flat or with partly revolute margins, not pungent; inflated alar cells
    not over 100 μm long . . . . *31-8. Trichosteleum*

7 Papillae spiny, irregularly scattered over the cell lumen frequently at cell ends;
    leaf apex truncate with coarsely dentate margin . . . . *30-5. Phyllodon*
    Papillae arranged in rows over the cell lumen; margin smooth or slightly
    serrulate 8

8 Papillae multifid, stalked . . . . *27-7. Hypnella*
    Papillae low and rounded . . . . *31-7. Taxithelium*

9 Plants with elongate, often pendulous secondary stems 10
    Plants with creeping or floating stems, usually freely branched and forming
    dense mats 15

10 Stem and branches strongly flattened, often attenuate. Leaves arranged in 2
   or 4 opposite rows 11
   Stem and branches not flattened 12

11 Leaves falcate to sickle-shaped in 4 rows; apex flat, acute . . . .
   . . . . 20-1. Isodrepanium
   Leaves concave-cymbiform, closely imbricated in 2 rows; apex boat-shaped,
   apiculate . . . . 21-1. Phyllogonium

12 Branch leaves lanceolate, flat . . . . 19-2. Lepyrodontopsis
   Branch leaves ovate-oblong, strongly concave, with broadly inflexed margins
   in upper part 13

13 Leaves not conspicuously ranked; apex with long, flexuose acumen; leaf
   base strongly auriculate, with a small, round group of coloured alar cells
   . . . . 22-7. Renauldia
   Leaves ranked in spiral rows; apex apiculate; leaf base not auriculate, alar
   cells few in a poorly defined group 14

14 Slender plants, branch leaves not over 1 mm long . . . . . 19-4. Orthostichella
   Medium-sized plants, branch leaves 1-2 mm long . . . 22-3. Hildebrandtiella

15 Plants growing periodically submerged. Stems and branches often elongate,
   sometimes floating. Leaves usually with rounded apex 16
   Plants not typically growing submerged. Stems creeping with branches erect
   or prostrate 19

16 Laminal cells uniform, elongate-hexagonal, alar cells not differentiated.
   Capsule immersed 17
   Laminal cells at midleaf at least twice as long as in apex, alar cells
   differentiated, the basal row usually inflated. Capsule exserted 18

17 Leaves closely imbricate particularly towards ends of branches. Capsule
   with single peristome . . . . 29-1. Hydropogon
   Leaves distant, flaccid. Peristome absent . . . . 29-2. Hydropogonella

18 Leaves oval or semi-circular, to 1.4 mm long. Capsule with slender, fragile
   peristome; exostome teeth not transversely striolate on outer surface,
   endostome reduced to filiform segments, often rudimentary . . . .
   . . . . 31-2. Colobodontium
   Most leaves longer than 1.4 mm. Capsule with firm peristome; exostome
   teeth transversely striolate on outer surface, endostome well-developed,
   with a high basal membrane and broad, keeled segments . . . .
   . . . . 31-6. Sematophyllum p.p.

19 Alar cells conspicuously differentiated in a well-defined group, specially in
   lateral leaves 20
   Alar cells not or scarcely differentiated, not forming a conspicuous group . . 27

20 Alar cells quadrate or rectangular, in the basal row sometimes oval, but not inflated 21
Alar cells in the basal row distinctly inflated 23

21 Leaves asymmetric, oval-oblong with obtuse or broad-acute apex; alar cells in lateral leaves differentiated in one leaf edge only . . . . . . *32-3. Pilosium*
Leaves symmetric, ovate-lanceolate with acute apex; alar cells always differentiated in both leaf edges 22

22 Laminal cells incrassate with fusiform lumen; leaf margin entire . . . .
. . . . *31-3. Donnellia*
Laminal cells thin walled; leaf margin serrulate. . . . . . *31-5. Pterogonidium*

23 Leaves on complanate branches dimorphous, at dorsal side broad-ovate, usually falcate, at ventral side lanceolate or triangular, subsymmetric . . .
. . . . *30-5. Rhacopilopsis*
Leaves uniform, branches complanate or not 24

24 Leaf apex with involute margins, pungent; inflated alar cells large, 70-170 µm long. . . . . *31-1. Acroporium*
Leaf apex plane or partly reflexed; inflated alar cells not over 100 µm long
. . . . 25

25 Plants pinnately or bipinnately branched, branches attenuate, often curved. Branch leaves smaller than stem leaves . . . . *31-9. Wijkia*
Plants irregularly branched, branches erect or prostrate. Branch leaves usually equal to stem leaves, sometimes larger 26

26 Leaf margins narrowly reflexed from just below apex to base. Peristome single, exostome teeth slender, pale, widely spaced . . . . *31-4. Meiothecium*
Leaf margins flat or partly reflexed. Peristome double, exostome teeth thick, brown, endostome well-developed with a high basal membrane and broad, keeled segments . . . . *31-6. Sematophyllum* p.p.

27 Branches / secondary stems erect or ascending. Leaves squarrose or erect-spreading 28
Branches prostrate. Leaves complanate 29

28 Leaves erect-spreading, narrowly lanceolate with slender, serrulate apex . . .
. . . . *19-2. Lepyrodontopsis*
Leaves squarrose-spreading, broad-ovate with acute apex (costa often variable, short and double or single) . . . . *22-4. Jaegerina*

29 Midlaminal cells lax, elongate-rhomboidal, over 10 µm wide. 30
Midlaminal cells linear, less than 10 µm wide 31

30 Midlaminal cells 70-240 µm long. Peristome teeth with a median furrow . .
. . . . *26-1. Leucomium*
Midlaminal cells to 70 µm long. Peristome teeth not furrowed . . . .
. . . . *30-6. Vesicularia*

31 Branch leaves apparently in 4 rows 32
Leaves in more than 4 rows 33

32 Leaves inserted at dorsal side of the stem, at ventral side dense clusters of
rhizoids. Epiphyllous plants . . . . *27-4. Crossomitrium*
Leaves on complanate branches dimorphous, at dorsal side broad-ovate,
usually falcate, at ventral side lanceolate or triangular, subsymmetric . . . .
. . . . *30-5. Rhacopilopsis*

33 Plants pinnately branched. Leaves usually falcate and strongly homomallous.
Perichaetia conspicuous, inner perichaetial leaves erect, to 4.5 mm long . .
. . . . *30-2. Ectropothecium*
Plants irregularly branched. Leaves subsymmetric to slightly falcate, more or
less complanate-spreading. Perichaetia not conspicuous, inner perichaetial
leaves patent, to 1.5 mm long. . . . . *30-3. Isopterygium*

Order **Sphagnales**

1. **SPHAGNACEAE** [3]

Erect stems without rhizomes and with continuous apical growth forming
cushions, often submerged; branches in fascicles, crowded at the stem
apex in a capitulum. Stem with a central cylinder of wide thin-walled
cells, outwards fading into smaller thick-walled, often coloured cells,
surrounded by one or more layers of hyaline cells (hyalodermis). Leaves
unistratose with alternating chlorocysts (narrow living green cells) and
hyalocysts (inflated hyaline cells, often reinforced with fibrils and porose),
without costa. Dioicous or monoicous. Antheridia and archegonia on
separate branches, with differentiated leaves. Sporophytes black, globose
on a short hyaline stalk, opening explosively to release the spores.
Protonema thalloid.

Distribution: Worldwide; 1 genus.

---

[3] by H.R. (Rudi) Zielman.
All drawings of *Sphagnum* are made by Rudi Zielman.

14

LITERATURE

Crum, H., 1990. A new look at Sphagnum sect. Acutifolia in South America. Contr. Univ. Michigan Herb. 17: 83-91.

Daniels, R.E. & A. Eddy. 1990. Handbook of European Sphagna. HMSO, London.

Dirkse, G.M. & P. Isoviita. 1986. Sphagnum denticulatum, an older name for S. auriculatum. J. Bryol. 14: 2.

Gradstein, S.R. 1989. Results of a botanical expedition to Mount Roraima, Guyana. 1. Bryophytes. Trop. Bryol. 1: 25-54.

McQueen, C.B. & R.E. Andrus. 2007. Sphagnaceae. In: Flora of North America Editorial Committee (eds.), Flora of North America, north of Mexico 27(1): 45-101.

1. **SPHAGNUM** L., Sp. Pl. 2: 1106. 1753.
   Type: S. palustre L.

N o t e s : Stem characters are taken from parts just below the capitulum. Leaves of pending branches are always more elongate than those of spreading branches. Leaves at the base of branches are longer than at the middle of the branch. Chlorocyst descriptions are based on branch leaves. No distinction has been made between true pores (just holes) and pseudopores (holes with a thickened ring).

Cortical fibrils are best visible in pending branches and distal parts of spreading branches. Pores can be situated in various parts of (fibrillose) hyalocysts: along the commissures (walls between hyalocysts and chlorocysts), in trios (section Sphagnum) at the edge where three hyalocysts meet, or just at midcell.

KEY TO THE SECTIONS

1   Branch leaves broad, cucullate-concave and rough at the apex, cortical cells of branches usually spirally fibrillose . . . . Section Sphagnum
     Branch leaves narrower, apex not cucullate, often involute, cortical cells of branches not fibrillose 2

2   Chlorocysts of branch leaves in cross-section equally exposed at both sides, pores serial along commissures . . . . Section Subsecunda
     Chlorocysts of branch leaves in cross-section trapezoid, more exposed at one leaf surface, pores not serial 3

3   Chlorocysts of branch leaves in cross-section more exposed at ventral side
     . . . . Section Acutifolia

Chlorocysts of branch leaves in cross-section broadly exposed at dorsal side, less exposed at ventral side .... Section Cuspidata

Section Acutifolia Wilson, Bryol. Brit. 20. 1855.

Slender to medium-sized plants, often tinged with various red to purplish colors, hyalodermis of stem and branches without spiral fibrils, with or without pores. Branch leaves ovate to lanceolate, apex often more or less abruptly narrowed with inflexed margin in upper half, hyalocysts at dorsal surface with large ringed pores, chlorocysts in most species trapezoid to triangular in cross-section, widest exposure at ventral leaf surface. Dioicous.

KEY TO THE SPECIES OF Section Acutifolia

1   Stem leaves triangular-lingulate, with broad border, apex cuspidate; branch leaves narrowly lanceolate, with involute-cuspidate apex, 5-ranked ......
....5. *S. oxyphyllum*
Stem leaves narrowly ovate to ovate-triangular, with narrow border, apex rounded; branch leaves lanceolate, apex not involute ...... 8. *S. tenerum*

Section Cuspidata Lindb., Öfvers. Kongl. Vetensk.-Akad. Förh. 19: 134. 1862.

Small to medium-sized plants. Hyalodermis of stem and branches without spiral fibrils, without pores. Branch leaves lanceolate to linear, chlorocysts trapezoid to triangular, widest exposure at dorsal leaf surface. Dioicous, occasionally monoicous.

One species in the Guianas: *9. S. trinitense*.

Section Sphagnum

Plants usually robust, epidermis of stems and branches spirally fibrillose, most conspicuous in pending branches. Branch leaves ovate, cucullate with a border of narrow cells, rough at dorsal side of apex by resorption of hyaline cells, ventral pores usually in groups of three at the conjuncture of three cells, chlorocysts in section various. Dioicous.

KEY TO THE SPECIES OF Section Sphagnum

1   Branches club shaped, upper leaves larger than basal leaves. Cortical cells of branches funnel-like and nested together . . . . *7. S. portoricense*
Branches not club shaped, upper leaves not conspicuously larger. Cortical cells of branches not funnel-like and nested together 2

2   Chlorocysts exposed at ventral and dorsal side of leaf, trapezoid in cross-section . . . . *6. S. perichaetiale*
Chlorocysts of branch leaves immersed, lenticular in cross-section 3

3   Branch leaves 5-ranked, hyalocysts with large pores in trios and small commissural pores . . . . *4. S. ornatum*
Branch leaves not ranked in 5 distinct rows, commissural pores lacking . . .
. . . . *3. S. magellanicum*

Section Subsecunda Lindb., Öfvers. Kongl. Vetensk.-Akad. Förh. 19: 135. 1862.

Robust plants, hyalodermis of stem and branches without spiral fibrils. Stem and branch leaves unequal in size, mostly ovate but varying from triangular to lingulate, bordered with a narrow margin of linear cells. Branch leaves ovate or elliptical, sometimes slightly cucullate, hyalocysts dorsally with numerous, mostly small, serial pores along commissural walls, pores at ventral surface lacking or like those of dorsal surface, chlorocysts of branch leaves equally exposed at both surfaces. Dioicous.

N o t e s : Species in this section show considerable morphological plasticity, as a result of moisture conditions and nutrient levels, sometimes giving rise to "obese" growth forms.
A relatively new species in this section is: *Sphagnum boomii* H.A. Crum, Mem. New York Bot. Gard. 64: 185. 1990. Type: Guyana, Upper Potaro R. Region, Mt. Wokomung, 1530 m, Boom & Samuels 9216 (holotype MICH, isotypes BRG, NY). Unfortunately no material was seen by the author.

KEY TO THE SPECIES OF Section Subsecunda

1   Stems unbranched or branches few, without capitulum . . *1. S. cyclophyllum*
Stems with numerous branches in distinct capitulum . . . . *2. S. lescurii*

1. **Sphagnum cyclophyllum** Sull. & Lesq. in Sull. in A. Gray, Manual ed. 2: 611. 1856. Type: U.S.A., 'in arenosis humidis per montes Alabamae', sin coll., in hb. Mitt. (NY).   – Fig. 1

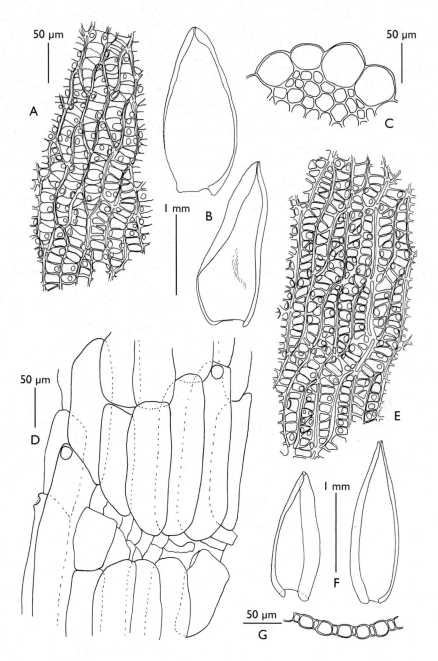

Fig. 1. *Sphagnum cyclophyllum* Sull. & Lesq.: A, stem leaf cells, dorsal view; B, stem leaves; C, stem section; D, branch cortex; E, branch leaf cells, dorsal view; F, branch leaves; G, branch leaf section (A-G, Newton 3453).

Moderate sized plants without capitulum, growing in lax mats. Stems tumid, branches lacking, seldom few and single, cylinder distinct, brownish, surrounded by a single layered hyalodermis. Branch hyalodermis with some pores. Stem leaves ovate, 1.9-2.1 mm long, 0.6-0.9 mm wide, apex rounded to obtuse, strongly inflexed, occasionally cucullate, margin with a continuous border of about 3 rows of linear cells, hyalocysts fibrillose throughout the leaf, 140-180 μm long, 20-30 μm wide, pores dorsally numerous 8-12 per cell, 6-10 μm, serial along commissures, ventrally scarce, one per cell or lacking, 6-8 μm. Branch leaves ovate, 1.6-2.1 mm long, ca. 0.7 mm wide, apex obtuse, margin with a border of about 2 rows of linear cells, hyalocysts 210-270 μm long, 20 μm wide, pores dorsally 8-15 per cell, 8-10 μm, serial along commissures, ventral pores few or none, chlorocysts in cross-section rectangular to barrel shaped.

D i s t r i b u t i o n : Southern N America, C and tropical S America.

E c o l o g y : On rock in river, temporarily dry; apparently rare, not seen from Suriname or French Guiana.

S p e c i m e n e x a m i n e d : Guyana: Kaieteur Falls, Newton 3453 (L).

N o t e s : This peculiar species is easily recognized by the (nearly) absence of branches. Stem leaves are fully developed, without any degradation of fibrils or cell walls (often seen in other species). When present, branches and their leaves are not different from stems and stem leaves. The hyalodermis is single-layered in the Guyanan collection, however collections from other S American countries showed considerable variation, the hyalodermis was often partly bi-stratose or occasionally tri-stratose. Stem leaves are generally larger than reported here (to 4 mm long and 2 mm wide) with up to 25 pores per cell.
A photograph of the type, showing the typical outline of the species is available at the web-site of the New York Botanical Garden (http://sciweb.nybg.org/science2/VirtualHerbarium.asp): http://sweetgum.nybg.org/vh/specimen.php?irn=708072.

2. **Sphagnum lescurii** Sull. in A. Gray, Manual ed. 2: 611. 1856. Type: U.S.A., 'in palude', Virginia. sin coll. (not seen).

Usually growing in dense mats, with branches just beneath capitulum curved like the horns of 'long-horn' cows. Stem with a distinct brown cylinder, surrounded by a single layered hyalodermis. Branches ca. 3 per fascicle. Stem leaves ovate, 1.6-2.0 mm long, 0.9-1.3 mm wide, apex rounded, margin with a continuous border of about 2 rows of linear cells, hyalocysts fibrillose throughout the leaf, 110-150 μm long,

20-30 μm wide, occasionally septate, pores dorsally 5-7 per cell, 8-10 μm, ringed, along commissures, ventrally scarce, occasionally 1 ringed pore of 12-15 μm per cell. Branch leaves triangular-ovate, ca. 2 mm long, 1.0-1.2 mm wide, apex rounded to truncate, denticulate, margin with a border of about 2 rows of linear cells ending just below apex, hyalocysts 130-150 μm long, 20-30 μm wide, dorsal pores 11-15 per cell, ringed or unringed, 6-10 μm, serial along commissures, most cells without ventral pores or 1 ringed pore of 10-15 μm, chlorcysts in cross-section barrel shaped to lens shaped.

Distribution: Probably worldwide.

Ecology: Collected once in Guyana along minerotrophic water; apparently rare, not known from Suriname and French Guiana.

Specimen examined: Guyana: Jenman 1486 (K), s.n. (NY).

Note: This collection is reported in Musci I in Fl. Suriname 6: 25. 1964, as *Sphagnum auriculatum* Schimp. var. *ovatum* Warnst.. *S. auriculatum* (discussed by Dirkse & Isoviita, 1986) is reputed for its great morphological plasticity. The nomenclature is rather dubious. The collection probably should correctly be named *S. lescurii* Sull., as this is the name for the large species in section Subsecunda in the Western Hemisphere (R. Andrus, pers. comm.).

3. **Sphagnum magellanicum** Brid., Muscol. Recent. 2(1): 24. 1798.
   Type: Chili, Straits of Magellan, Commerson s.n. (BM, PC). – Fig. 2

Ochre, red or purple tinged robust plants. Stems with dark cylinder, rather stiff, hyalodermis in 2-3 layers, without or with one pore, uniform, short rectangular, 40-160 μm long. Fascicles with 2-4 branches, 1-2 spreading and 1-2(-3) pending. Stem leaves bright brown, usually shorter than branch leaves, lingulate to rectangular, 0.9-1.5 mm long and 0.5-0.7(-0.9) mm wide, apex obtuse, often cucullate, margin resorbed, dorsal surface with large irregular pores or resorbed, hyalocysts linear in basal part, quadrate in upper part, without fibrils to fibrillose throughout the leaf length, chlorocysts with strongly thickened walls. Branch leaves ovate to ovate-lanceolate, 1.1-2.1 mm long and 0.6-1.1 mm wide, apex cucullate, hyalocysts 100-200 μm long, 30-50 μm wide, plane at both surfaces, with large (10-20 μm) ringed pores in trio's; chlorocysts in cross-section lenticular, immersed.

Distribution: Worldwide.

Ecology: Wet soil and dripping rocks at higher altitudes; not common, not seen from French Guiana.

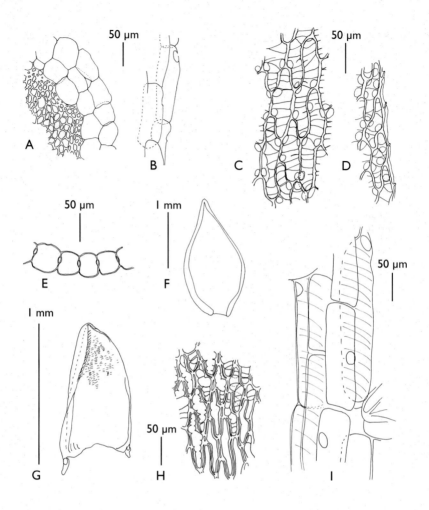

Fig. 2. *Sphagnum magellanicum* Brid.: A, stem section; B, stem cortex; C, branch leaf midleaf cells, dorsal view; D, branch leaf border, dorsal view; E, branch leaf section; F, branch leaf; G, stem leaf; H, stem leaf midleaf cells, dorsal view; I, branch cortex (A-D, G-I, Gradstein 5412A; E-F, Newton 3387).

Specimens examined: Guyana: Mt. Latipu, Maas *et al.* 2666 (L); Mt. Roraima, Gradstein 5238, 5248, 5412A (L); Kaieteur Falls, Korume Cr., Newton 3387 (L); summit Mt. Makarapan, Maas *et al.* 7465, p.p. (L). Suriname: Wilhelmina Mts., BW 7139, 7140 (L) (as *S. palustre* L. in Musci I in Fl. Suriname 6: 27. 1964).

Note: The hyalocysts are remarkably flat in cross-section.

4. **Sphagnum ornatum** H.A. Crum, Cryptog. Bryol. Lichénol. 6: 181. 1985. Type: Venezuela, Cerro de la Neblina, Venezuelan-Brazilian frontier, Planicie de Zuloaga, Rio Titirico, 2300 m, Steyermark 103890 (holotype MICH, isotype NY).　　　　　　　　 – Fig. 3

Plants stiff, mostly brownish red. Stems with dark cylinder, hyalodermis in 3 layers, cells without fibrils, with no or one distal pore, uniform, short rectangular, 60-120 μm long and 40-70 μm wide. Fascicles with 2-3 branches, 1(-2) spreading, 1-2 pending. Stem leaves bright brown, due to pigmented cell walls, lingulate to rectangular, 0.7-1.0(-1.3) mm long and 0.4-0.8 mm wide, apex obtuse, often cucullate, margin resorbed. Basal hyalocysts linear, in upper part quadrate, sometimes septate in distal part, fibrils lacking or only in upper quart, dorsal surface with large irregular pores or resorbed, chlorocysts with strongly thickened walls. Branch hyalodermis without fibrils, with distal pores. Branch leaves in upper part of plants ranked in five rows (most obvious when moist), ovate to ovate-lanceolate, 1.0-1.6(-2.0) mm long and 0.5-0.7 mm wide, apex cucullate, hyalocysts 100-150 μm long and (25-)30-40 μm wide, large ringed pores in trios and smaller, ringed, commissural pores at dorsal and ventral surface, chlorocysts lenticular in cross-section, immersed with thickened cell walls reaching both leaf surfaces.

Distribution: Endemic on the Guayana Shield.

Ecology: On soil and rocks (on rotten wood), in shrubs and open forest, at altitudes above 700 m; uncommon, not known from Suriname or French Guiana.

Specimens examined: Guyana: Mt. Latipu, Maas *et al.* 2660, 2665, 4214 (L), Gradstein 5609 (L); Mt. Roraima, Gradstein 5372 (L), Aptroot 17099, 17099A (L).

Note: This species is recognized by its five-ranked branch leaves and commissural pores. It resembles *S. magellanicum* which occasionally shows a single pore at the commissural walls in addition to the larger trio-pores. There is some variation in the extent to which the chlorocysts are immersed even within one leaf. Examination of the type has confirmed the variabilty of this character. *S. ornatum* is one of the few species in section Sphagnum with efibrillose hyalodermis.

5. **Sphagnum oxyphyllum** Warnst., Hedwigia 29: 192. 1890. Type collection: Brazil, Tubarão, 'am Rande der Serra Geral', Ule 1102, 1890 (B).　　　　　　　　　　　　　　 – Fig. 4

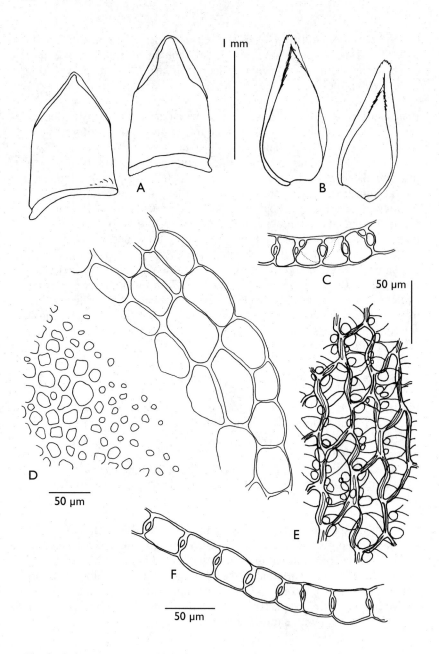

Fig. 3. *Sphagnum ornatum* H.A. Crum: A, stem leaves; B, branch leaves; C, branch leaf section; D, stem section; E, branch leaf midleaf cells, dorsal view; F, branch leaf section (A, D, F, Aptroot 17099A; B-C, E, Aptroot 17099).

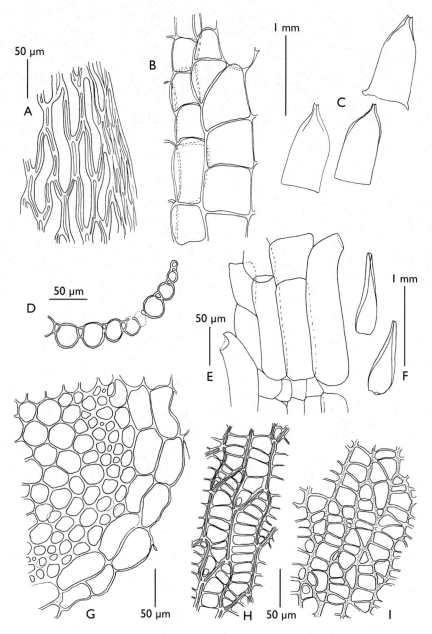

Fig. 4. *Sphagnum oxyphyllum* Warnst.: A, stem leaf border, dorsal view; B, stem cortex; C, stem leaves, dorsal view left, ventral view right; D, branch leaf section; E, branch cortex; F, branch leaves; G, stem section; H, branch leaf midleaf cells, ventral view; I, branch leaf midleaf cells, dorsal view (A-I, Maas & Westra 4322).

Medium sized plants with purplish tinges. Stems with purplish cylinder, hyalodermis of 2-3 layers, without fibrils and without pores. Spreading branches 2, pending 1, branch epidermis with conspicuous retort cells. Stem leaves ovate-triangular, 1.0-1.2(-1.4) mm long and 0.4-0.7 mm wide, apex acuminate, cuspidate, margin entire, border conspicious with very narrow cells, broader towards base, hyalocysts without fibrils, and occasionally septate. Branch leaves ranked in 5 rows when moist, secund when dry, narrowly ovate to linear-lanceolate, 1.1-1.4 mm long and 0.3-0.5 mm wide, apex involute, truncate, dentate, margin entire, narrowly bordered, hyalocysts 130-190 µm long and 25-40 µm wide, slightly bulging at ventral leaf side, more distinctly at dorsal leaf side, dorsal pores large and ringed, 4-10 per cell, ventral pores less numerous, chlorocysts in cross-section rectangular to trapezoid, exposed at both leaf sides but more conspicuous at the ventral leaf side.

Distribution: Brazil, Guyana.

Ecology: In spray zone at vertical rock surface along streams in low forest, at higher altitudes; apparently rare, not known from Suriname and French Guiana.

Specimens examined: Guyana: Pakaraima Mts., Maas *et al.* 4322, 5730 (L), Hoffman 2931 (L).

Note: The combination of characters is important to separate this species from *S. tenerum* as both species are variable. Apart from the characters in the key the shorter stem leaves without fibrils can be indicative for this species. When present, the leaves ranked in five rows form a striking feature but this is not always distinct.

6. **Sphagnum perichaetiale** Hampe, Linnaea 20: 66. 1847. Type: Brazil, Rio de Janeiro, Beyrich s.n. (BM).

6a. var. **perichaetiale**   – Fig. 5

Large plants, growing in rather dense pale to reddish cushions. Stems with dark cylinder, hyalodermis in 2-3 rings, cells without fibrils or faintly fibrillose, 70-140 µm long, with one large (25-30 µm) median or distal pore. Fascicles with 2-3 branches, 1-2 spreading, 1-2 pending. Stem leaves lingulate, ovate to rounded-rectangular, 1.3-3.0 mm long and 0.5-1.8 mm wide, apex rounded, often cucullate, margin with resorption furrow, hyalocysts with fibrils throughout in large leaves, to completely lacking fibrils in short leaves; hyalocysts of fibrillose leaves with large

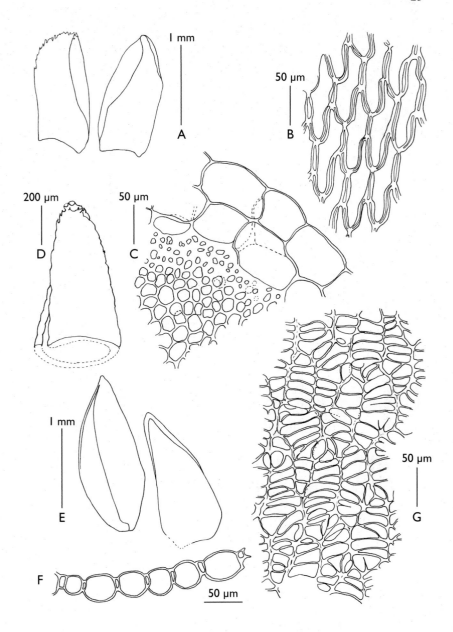

Fig. 5. *Sphagnum perichaetiale* Hampe var. *perichaetiale:* A, stem leaves; B, stem leaf midleaf cells, dorsal view; C, stem section; D, branch leaf apex; E, branch leaves; F, branch leaf section; G, branch leaf midleaf cells, dorsal view (A-G, Florschütz 841).

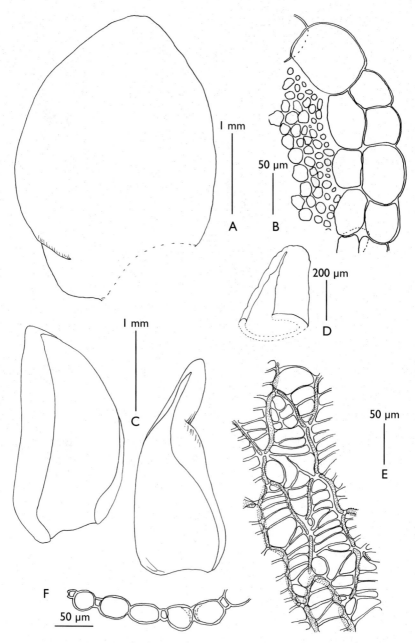

Fig. 6. *Sphagnum perichaetiale* Hampe var. *tabuleirense* (Yano & H.A. Crum) H.R. Zielman: A, stem leaf; B, stem section; C, branch leaves; D, branch leaf apex; E, branch leaf midleaf cells, dorsal view; F, branch leaf section (A-F, Gradstein 5009).

27

(ca. 20 μm) ringed pores. Branch hyalodermis fibrillose, with large distal pores. Branch leaves ovate to ovate-lanceolate, sometimes contracted in distal part, (1.5-)1.8-2.5(-2.8) mm long and 0.6-1.3 mm wide, apex cucullate, margin with narrow resorption furrow, hyalocysts 100-190 μm long and (30-)40-50(-60) μm wide, strongly bulging at dorsal surface, pores large, 12-25 μm, in trios and commissural, ringed or unringed, ventrally less numerous and frequently unringed, chlorocysts trapezoid in cross-section and exposed at both leaf surfaces.

Distribution: Tropical and subtropical lowlands of the Americas, Asia and Australia.

Ecology: Wet sand savannas with shrub or light forest, creeks in lowland forest and shrubs and humid rocks at higher altitudes, occasionally at streaming water; common, not yet seen from French Guiana.

Specimens examined: Guyana: Soesdyke, Cooper 345 (L); Pakaraima Mts., Maas et al. 4325 (L); Mt. Makarapan, Maas et al. 7465 p.p. (L); Mazaruni Distr., Makwaima, Gradstein 4981A (L); Mt. Latipu, Gradstein 5625 (L); Kanuku Mts., Jansen-Jacobs et al. 880 (L); Essequibo R., Wai-Wai-area, Jansen-Jacobs et al. 1894 (L); Kaieteur Falls, Newton 3358 (L). Suriname: Brinckheuvel, Teunissen & Wildschut 11925 (L); Zanderij, Florschütz 690, 4776 (L).

Note: All collections mentioned in Musci I in Fl. Suriname 6: 27. 1964 under Sphagnum palustre L. belong to this species, with exception of BW 7139 and BW 7140 (both belong to S. magellanicum). It is the most widespread species of the genus in the Guianas, and the only lowland member of the section Sphagnum. The variability of the stem leaves is great: some collections have large stem leaves which are strongly fibrillose whereas other collections have only tiny, almost quadratic stem leaves without any fibrils. However, several collections show both expressions at the same stem which may indicate that these differences are seasonal variations. In cross-section the hyalocysts are strongly bulging, especially at the dorsal surface.

6b. var. **tabuleirense** (Yano & H.A. Crum) H.R. Zielman, comb. nov.
– Sphagnum tabuleirense Yano & H.A. Crum, Bryologist 95: 343. 1992. Type: Brazil, Baraiba, Municipio de Santa Rita, Xavier Filha s.n. (holotype SP, isotype MICH). – Fig. 6

Yellowish robust plants with a greasy appearance. Stems with light brown cylinder. Hyalodermis in 2 broad layers, with weakly developed fibrils, quadratic, 70-90 μm long, larger cells with a distal pore. Branches when present single, mostly pending, hyalodermis with spiral fibrils, without pores. Stem leaves almost circular to broadly ovate or oblong, 2.3-3.5 mm long and 1.5-2.5 mm wide, apex obtuse, cucullate, margin bordered, with a resorption furrow, hyalocysts with fibrils throughout, not septated, dorsal pores in trio's, ca. 20 µm wide, ventral pores rare, in trios. Branch leaves ovate to oblong, 1.7-2.1 mm long and 0.7-1.5 mm wide, larger towards branch tip, apex cucullate, leaf margin bordered, with a small resorption furrow, hyalocysts 110-180 µm long and 30-50 µm wide, bulging at both surfaces, most conspicuous at the dorsal leaf side, pores at dorsal side in trios, ringed, ca. 20-25 µm, at ventral side rare, chlorocysts trapezoid in cross-section, broadly exposed at ventral leaf side, reaching dorsal leaf side.

Distribution: Insufficiently known, northern Brazil and Guyana.

Ecology: Sub-aquatic, river side; rare, not known from Suriname or French Guiana.

Specimens examined: Guyana: Mazaruni Distr., Waruma R., Gradstein 5009, 5078 (L).

Notes: Although the overall stature of this variety is quite distinct, examination of the type showed that the differential characters mentioned by Yano & Crum are gradual. The type and the collections from Guyana show a slightly roughened dorsal leaf profile. The type shows some fascicles of 2 branches, the Guianan collections are more extreme with no or single branches; typical *S. perichaetiale* as described above sometimes also shows fascicles with only two 2 branches.
This variety resembles *S. cyclophyllum* in general appearance, due to the general lack of branch fascicles and greasy appearance, the presence of fibrils in the cortices of stem (weak) and branches (conspiciously) will guide proper identification.

7. **Sphagnum portoricense** Hampe, Linnaea 25: 359. 1853. Type: Porto Rico, Schwanecke s.n. (BM). – Fig. 7

Stems with brown cylinder. Hyalodermis in 2-3 layers, fibrils weakly developed, cells with one or two central pores, quadratic, 40-60 µm long. Fascicles with 3 branches, 2 spreading, 1 pending. Stem leaves broadly ovate to lingulate, 1.0-1.4 mm long and 0.8-1.0 mm wide, apex obtuse, often cucullate, narrow border with resorption furrow, hyalocysts with fibrils throughout the leaf or in larger leaves lacking in the lower third, not septated. Hyalodermis cells of spreading branches with spiral

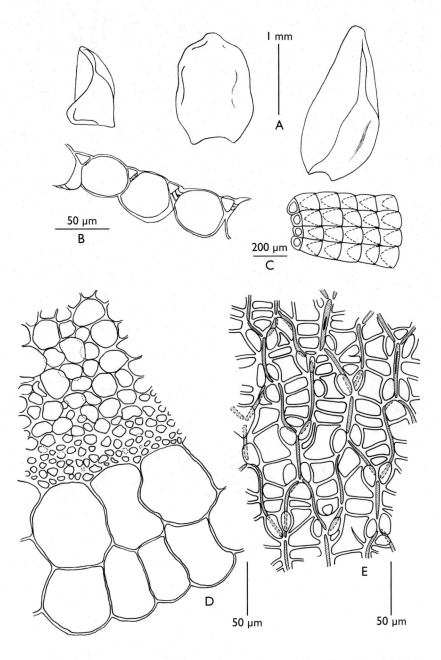

Fig. 7. *Sphagnum portoricense* Hampe: A, stem leaves, left and mid; branch leaf right; B, branch leaf section; C, branch cortex (schematic); D, stem section; E, branch leaf midleaf cells, dorsal view (A-E, Hoffman 2862).

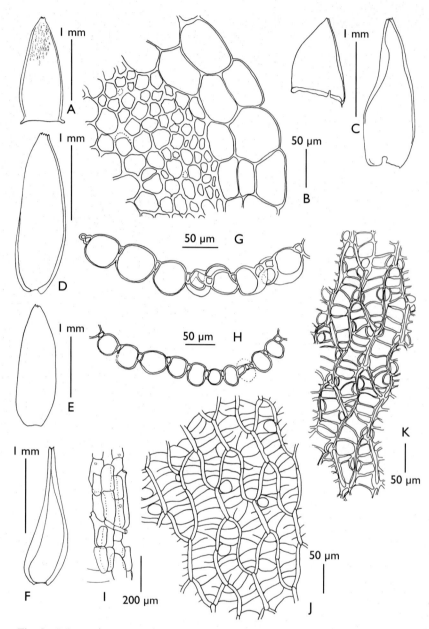

Fig. 8. *Sphagnum tenerum* Sull. & Lesq.: A, stem leaf; B, stem section; C, stem leaves; D, branch leaf; E, branch leaf; F, branch leaf; G, branch leaf section; H, branch leaf section; I, branch cortex; J, branch leaf midleaf cells, ventral view; K, branch leaf midleaf cells, dorsal view (A, D-E, Gradstein 4981; B-C, F-K, Maas & Westra 4264).

fibrils, basal part funnel-shaped nested in proximal cell. Branch leaves ovate, 1.7-2.1 mm long and 1.3-1.5 mm wide, larger towards branch tip, apex cucullate, hyalocysts 130-180 µm long and 30-50 µm wide, often with comb-like lamellae where adjacent to chlorocysts, dorsal pores in trios, ringed, 14-20 µm, at ventral side smaller, 6-8 µm, chlorocysts in cross-section equilateral-triangular, broadly exposed at ventral surface, sometimes slightly trapezoid.

Distribution: Southern U.S.A., Mexico, West Indies, C and tropical S America.

Ecology: Rock crevices in woodland.

Specimen examined: Guyana: Pakaraima Mts., Hoffman 2862 (L, NY).

Note: Under the dissection microscope this species can easily be recognized by the oblique airfilled funnel-shaped cells of the branch cortex when dried specimens are remoistened.

8. **Sphagnum tenerum** Sull. & Lesq. in Sull. in A. Gray, Manual ed. 2: 611. 1856. Type: U.S.A., 'ad margines rivulorum per montes Raccoon Alabamea', Lesquereux s.n.? (NY, not seen). – Fig. 8

Medium sized, soft textured, yellowish plants with purplish tinges. Stems with purplish cylinder, hyalodermis in 2-3 layers, without fibrils, without pores. Spreading branches 2, pending 1, branch epidermis with low retort cells. Stem leaves ovate to ovate-triangular or lingulate, apex rounded to truncate, dentate, 1.5-2.1 mm long and 0.6-0.9 mm wide, margin entire, with narrow border, ca. 2 cells wide, hyalocysts in proximal part with or without fibrils, occasionally septate, distal part always fibrillose, fibrillose hyalocysts dorsally with many large ringed pores, ventral pores less numerous or lacking, with or without ring, non fibrillose hyalocysts occasionally with dorsal membrane gaps, pores more numerous towards apex and decreasing in size. Branch leaves ovate-lanceolate, apex truncate, dentate, 2.0-2.1 mm long and 0.6-0.7 mm wide, margin entire, narrowly bordered, hyalocysts ca. 180-300 µm long, 40-60 µm wide, moderately bulging at ventral leaf side, more extremely so at dorsal leaf side, often dorsally without or with a few ringed or unringed large pores, ventrally with less pores. Chlorocysts in cross-section rectangular to trapezoid exposed at both leaf sides, widest at the ventral leaf side.

Distribution: Southern U.S.A., C America, northern S America.

32

Ecology: On soil and rock near streaming water; at higher altitudes; uncommon, not known from Suriname or French Guiana.

Specimens examined: Guyana: Pakaraima Mts., Mt. Membaru, Maas *et al*. 4264 (L); Mazaruni Distr., Makwaima Savanna, Gradstein 4981 (L); Potaro-Siparuni, Kaieteur Falls, Newton 3357 (L).

Notes: The delimitation of this species has been discussed e.g. by Crum (1990). *Sphagnum tenerum* is said to be isophyllous, but although stem and branch leaves are almost equal in size, branch leaves are narrowly ovate to lanceolate whereas stem leaves are more variable in outline.
A photo image of the isotype is available at the web-site of the New York Botanical Garden (http://sciweb.nybg.org/science2/VirtualHerbarium. asp): http://sweetgum.nybg.org/vh/specimen.php?irn=739514.

9. **Sphagnum trinitense** Müll. Hal., Syn. Musc. Frond. 1: 102. 1848.
Type: Trinidad, Crüger 14 (BM).   – Fig. 9

Sphagnum cuspidatum Ehrh. ex Hoffm. var. *serrulatum* (Schlieph.) Schlieph., Irmischia 2: 67. 1882. Type: not indicated.

Slender whitish green, often submerged plants with a plumose appearance. Stems with distant fascicles of branches, without clearly defined hyalodermis. Fasicles with 2-4 branches, 1-2 spreading and 0-2 pending. Stem leaves ovate-triangular, apex obtuse, denticulate, 1.0-1.6 mm long and 0.4-0.9 mm wide, margin entire or slightly dentate, bordered, hyalocysts fibrillose, sometimes less conspicious in lower third, occasionally septate, without pores. Lower branch leaves ovate-lanceolate, upper branch leaves slenderly linear-lanceolate to linear, undulate, 3-7 mm long and 0.3-0.5(-0.8) mm wide, apex truncate and dentate, with a narrow serrulate border, hyalocysts 100-250 μm long and 10-15(-25) μm wide, pores rare, chlorocysts trapezoid, broadly exposed at dorsal surface, less so at ventral surface, nearly as wide as hyalocysts. Monoicous?

Distribution: Eastern N America, Mexico?, West Indies, C America, northern S America.

Ecology: Submerged in swamps and along creeks in savannas, at low altitudes; not known from French Guiana.

Selected specimens: Guyana: Santa Mission, Werda Cr., Florschütz-de Waard 6131 (L); North-West Distr., Moruca R., van Andel *et al*. 1963 (L). Suriname: Rijsdijkweg, 25 km S of Paramaribo, Lindeman 4620 (L);

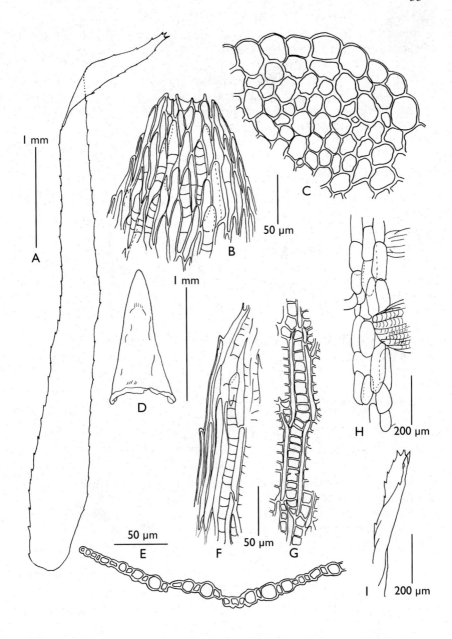

Fig. 9. *Sphagnum trinitense* Müll. Hal.: A, branch leaf; B, stem leaf apex; C, stem section; D, stem leaf; E, branch leaf section; F, branch leaf border; G, branch leaf midleaf cells, dorsal view; H, branch cortex and leaf insertion; I, branch leaf apex (A-I, Teunissen & Moonen 16300).

Upper Nanni R., Teunissen 14938 (L); Wana Cr., E of Prati, Teunissen & Moonen 16300 (L); Powakka, Teunissen s.n. (2x) (L).

N o t e : This remarkable plumose species is usually easy to recognize without sectioning the leaves. This taxon is often described as a variety of *Sphagnum cuspidatum* Ehrh. ex Hoffm. (in Musci I in Fl. Suriname 6: 22. 1964, as var. *serrulatum* (Schlieph.) Schlieph.).

Order **Archidiales**

N o t e s : Some authors put this order in a separate subclass based on the unique development of the spores in which the quadrant phase of sporogenesis is passed over, resulting in only a few large spores.
This order consists of only one family, the ARCHIDIACEAE.

2. ARCHIDIACEAE

Stems erect, simple or forked. Leaves larger and more crowded at end of stem, costa single, strong. Capsules immersed, globose, rupturing at maturity, operculum and peristome absent; spores few, large.

D i s t r i b u t i o n : Worldwide; 1 genus.

N o t e : Snider (1975) described the sporophyte development in *Archidium* which in this family is fundamentally different from that in other families of the Bryoideae. The most important differences are:
– the seta is not developed; the capsule is connected with the gametophyte by a large globose foot.
– the endothecium is not differentiated into a sterile columella and sporogenous archesporium; any cell of the endothecium can differentiate into a spore mother cell, 1-44 per capsule.

LITERATURE

Frahm, J.-P. & S. Porembski. 1994. Moose von Inselbergen aus Westafrika. Trop. Bryol. 9: 59-68.
Snider, J.A. 1975. Sporophyte development in the genus Archidium. J. Hattori Bot. Lab. 39: 85-104.
Snider, J.A. 1975a. A revision of the genus Archidium. J. Hattori Bot. Lab. 39: 105-201.

1. **ARCHIDIUM** Brid., Bryol. Univ. 1: 747. 1827.
   Type: A. phascoides Brid., nom. illeg. [= A. alternifolium (Dicks. ex
   Hedw.) Schimp. (Phascum alternifolium Dicks. ex Hedw.)]

Ephemeral or perennial, terrestrial plants; stems erect, simple or branched. Leaves ovate to linear-lanceolate, acute, frequently channeled, clasping at base, costa single, percurrent to excurrent into a hairpoint; laminal cells prosenchymatous, uniform in upper leaf, shorter towards base, quadrate in alar region. Capsule sessile, globose, without columella, operculum or peristome; spores few, large, released by rupturing of the single-layered capsule wall. Calyptra undeveloped.

Distribution: Worldwide; in the Guianas 1 species.

1. **Archidium globiferum** (Brid.) Frahm, Trop. Bryol. 9: 62. 1994. –
   Pleuridium globiferum Brid., Muscol. Recent. Suppl. 4: 10. 1819.
   Type: Mascarene Islands, Mauritius, sin. coll. (B).   – Fig. 10

   Archidium ohioense Schimp. ex Müll. Hal., Syn. Musc. Frond. 2: 517. 1851.
   Lectotype (Snider 1975a): U.S.A., Sullivant, Musci Allegh. 213 (FH).

Tiny, perennial, green or yellow-green, glossy plants growing in thin mats or low cushions. Stems simple or little branched, erect or flexuose, to 2 cm high. Leaves distant, lanceolate to narrow-triangular, 0.7-1.2 mm long, complicate-carinate, at base clasping; margin entire or faintly serrulate towards apex; costa firm, 30-35 μm wide at base, percurrent or excurrent in a short point; laminal cells elliptic to linear, 50-110 μm long and 7-12 μm wide, towards base rectangular or irregular, in alar region forming 3-4 rows of short-rectangular to quadrate cells, 10-34 μm long and ca. 18 μm wide. Autoicous. Perichaetia and perigonia sessile or on short branches in leaf axils; perichaetia ca. 0.5 mm high, with ovate, short-acuminate leaves; perigonia budlike, ca. 0.3 mm high. Capsule sessile, globose with single-layered, semi-transparent wall, 0.3-0.4 mm in diameter; spores yellow, 12-16 per capsule, rounded-triangular, 175 μm in greatest diam., smooth or finely papillose.

Distribution: Africa, SE Asia, N America, Mexico, West Indies, the Guianas.

Ecology: In the Guianas collected in crevices of rock pavements and on clay of riverbank; rare.

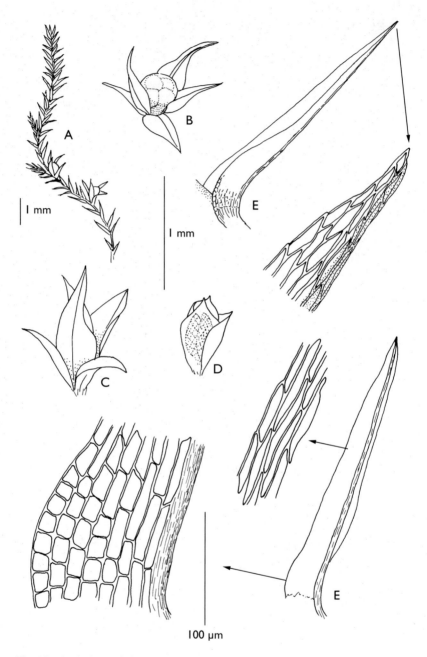

Fig. 10. *Archidium globiferum* (Brid.) Frahm: A, plant with capsules; B, mature capsule; C, perichaetium; D, perigonium; E, leaves (A-E, Florschütz-de Waard & Zielman 5608).

Selected specimens: Guyana: Rupununi savanna near Dadanawa, Florschütz-de Waard 6017 (L). Suriname: area of Kabalebo Dam project, road to Amatopo, Florschütz-de Waard & Zielman 5608 (L). French Guiana: confluence of Sinnamary R. and Marouina Cr., Cremers 5441 (CAY, L).

Note: This species was published in 1851 as *A. ohioense*, new for America, but it was already known from Africa as *Pleuridium globiferum* Brid. (Frahm & Porembski 1994).

Order **Fissidentales**

3. FISSIDENTACEAE[4,5]

Plants of varying size, less than 1 mm to several centimeters tall, with or without axillary nodules, simple or branched; stems and branches with basal rhizoids. Persistent protonemata infrequent. Leaves distichous, pinnate or frondiform, consisting of 2 vaginant laminae clasping the stem, a dorsal lamina and an apical lamina, lanceolate, elliptical, oblong or lingulate, limbate on all laminae, limbate on vaginant laminae of some or all leaves or elimbate; vaginant laminae equal, subequal or unequal; dorsal laminae tapering or abruptly narrowed below, reaching the insertion, or ending above insertion, infrequently decurrent; costa bryoides-, taxifolius- or oblongifolius-type, ending below apex to long-excurrent, rarely lacking, sometimes distally branched; laminal cells parenchymatous, rarely prosenchymatous, flat, convex, or conical, smooth, uni- or pluripapillose, guttulate or eguttulate, small to large, 3-165 µm, mostly less than 25 µm long. Gemmae and tubers infrequent. Dioicous, autoicous (cladautoicous, gonioautoicous, rhizautoicous) or synoicous, often polyoicous. Perigonia terminal or axillary, bud-shaped or not. Perichaetia mostly terminal; perichaetial leaves often longer than stem leaves. Peristome mostly taxifolius-, similiretis-, scariosus-, bryoides- or zippelianus-type, 16 teeth, each divided into 2 filaments, infrequently undivided; spores prolate, papillose or smooth.

Distribution:Worldwide; in the Guianas represented by 32 species of which 2 are endemic.

[4] by M.A. (Ida) Bruggeman-Nannenga.
  All drawings of *Fissidens* are made by Ida Bruggeman-Nannenga.
[5] I am obliged to R.A. Pursell for letting me consult the manuscript of his Neotropical treatment of the Fissidentacae that has now been published as Flora Neotropica Monograph 101 and for critically reading the manuscript.

38

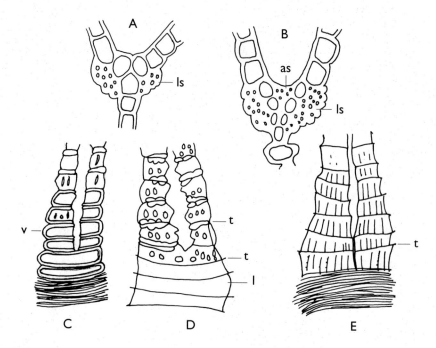

Fig. 11. *Fissidens*. Schematic drawings costa types and peristome types. A, **bryoides**-type: costa with 2 lateral bands of stereid cells, 2 large adaxial cells and 1 large central cell; B, **oblongifolius**-type: costa with 2 lateral and 1 adaxial band of stereid cells and 7 large central cells; C, outer surface bifurcation **scariosus**-type peristome; D, outer surface of bifurcation of **bryoides**-type; E, outer surface bifurcation **similiretis**-type peristome (ls = lateral band of stereids; as = adaxial band of stereids; t = trabecula; l = lamella; v = vertical wall).

Notes: New, important characters used in the infrageneric classification are the structure of the costa as seen in cross-section (Bruggeman-Nannenga 1990; Pursell & Bruggeman-Nannenga 2004) and peristome-types (Bruggeman-Nannenga & Berendsen 1990; Pursell & Bruggeman-Nannenga 2004). Three main costa types are recognized, 2 of which are known from the Guianas:

– The **bryoides**-type (Fig. 11 A) characterized by 2 lateral bands of stereid cells, 2 large adaxial cells and 1, infrequently more, large central cells.
– The **oblongifolius**-type (Fig. 11 B) characterized by 3 stereid bands, 2 lateral and 1 adaxial, and 4-16 large central cells.
– The types can be obscured and unrecognizable at the proximal and distal ends of the costa as well as in perichaetial leaves.

Three of the 5 main peristome types are found in the Guianas:
- The **similiretis**-type (Fig. 11 E) characterized by trabeculae and horizontal lamellar ridges on outer surface of the undivided proximal parts of the teeth equally high; above bifurcation vertical walls absent, trabeculae thin and well distinct from the delicate vertical, lamellar ridges; distal ends of filaments with declinate squamulae.
- The **scariosus**-type (Fig. 11 C) characterized by trabeculae and lamellar ridges on outer surface of the undivided proximal parts of the teeth equally high (as in the similiretis-type); above bifurcation trabeculae and vertical walls confluent, both regularly thickened, lamellar ornamentation various; distal parts of filaments spirally ornamented.
- The **bryoides**-type (Fig. 11 D) characterized by trabeculae on outer surface of the undivided proximal parts of the teeth distinct from and higher than lamellae; above bifurcation trabeculae incompletely and irregularly thickened, vertical walls not or incompletely thickened, lamellar ornamentation various; distal parts of filaments as in scariosus-type.

Two Guianan species, *F. geijskesii* Florsch. and *F. lagenarius* Mitt., have anomalous peristomes.

Unless stated differently figures and descriptions are from Guianan specimens. All data under the heading E c o l o g y are based exclusively on labels from Guianan specimens. In species with pinnate plants leaf measurements are from mid-stem leaves of sterile plants, in species with frondose plants from leaf pairs just below the apical ones. Unless indicated differently length of plants refers to the length of simple, unbranched plants. Unless stated differently descriptions of laminal cells are from mid dorsal laminal cells.

LITERATURE
(see also general literature)

Bruggeman-Nannenga, M.A. & W. Berendsen. 1987. Notes on Fissidens III and IV. Proc. Kon. Ned. Akad. Wetensch., C, 90: 81-86.

Bruggeman-Nannenga, M.A. & W. Berendsen. 1990. On the peristome types found in the Fissidentaceae and their importance for the classification. J. Hattori Bot. Lab. 68: 193-234.

Bruggeman-Nannenga, M.A. 1990. On the anatomy of the costa in Fissidens. Trop. Bryol. 3: 37-44.

Bruggeman-Nannenga, M.A., R.A. Pursell & Z. Iwatsuki. 1994. A re-evaluation of Fissidens subgenus Serridium section Amblyothallia. J. Hattori Bot. Lab. 77: 255-271.

Bruggeman-Nannenga, M.A. & R.A. Pursell. 1995. Notes on Fissidens V. Lindbergia 20: 49-55.

Grout, A.J. 1943. Bryales. Fissidentaceae. N. Amer. Fl. 15 (3): 167-202.

Pursell, R.A. 1984. A preliminary study of the Fissidens elegans complex in the neotropics. J. Hattori Bot. Lab. 55: 235-252.

Pursell, R.A. 1986. Additions and deletions to the genus Fissidens (Fissidentaceae) in Mexico, including four new species. Cryptog. Bryol. Lichénol. 7: 37-46.

Pursell, R.A. 1989. Notes on Neotropical Fissidens I, II and III. I. The relationship of F. leptophyllus. II. The relationship of Fissidens obtusissimus, stat. nov. III. The identity of F. hornschuchii. Bryologist 92: 523-528.

Pursell, R.A. 1994. Taxonomic notes on Neotropical Fissidens. Bryologist 97: 253-271.

Pursell, R.A. & W.R. Buck, 1996. Fissidens saülensis (Fissidentaceae), a new species from central French Guiana. Brittonia 48: 26-28.

Pursell, R.A. 1999. Taxonomic notes on Neotropical Fissidens. III. Addendum II. Bryologist 102: 125-127.

Pursell, R.A. & M.A. Bruggeman-Nannenga. 2004. A revision of the infrageneric taxa of Fissidens. Bryologist 107: 1-20.

Pursell, R.A. 2007. Fissidentaceae. Flora Neotropica Monograph 101: 1-278.

The family is monogeneric.

1. **FISSIDENS** Hedw., Sp. Musc. Frond. 152. 1801.
   Lectotype (E.G. Britton in N.L. Britton 1918): F. bryoides Hedw.

Subgenera 4, of which 3 are found in the Guianas:
– Subgenus Pachyfissidens 3 sections of which 2 are found in the Guianas
    section *Pachyfissidens* 1 species
    section *Amblyothallia* 3 species
– Subgenus Fissidens 2 sections of which 1 is found in the Guianas
    section *Fissidens* 1 species
– Subgenus Aloma, the remaining 27 Guianan species.
Subgenus Aloma is gametophytically diverse: leaves elimbate, semilimbate, infrequently, completely limbate and costate or, in some species, (not in the Guianas) nearly or completely ecostate; costa bryoides-type; laminal cells papillose or smooth, guttulate or eguttulate, small to large, 3-167 μm, mostly less than 30 μm long; peristome scariosus-type and theca with 32(-40) files of exothecial cells.
Subgenus Pachyfissidens section *Pachyfissidens* and section *Amblyothallia* and subgenus Fissidens section *Fissidens* are discussed under the species involved.

KEY TO THE SPECIES

1 Plants rheophilous; costa ill-defined; leaves elimbate, pluristratose; seta
   short, 0.4 mm long .... *9. F. geijskesii*
   Ecology various, when aquatic costa distinct; leaves limbate or elimbate,
   mostly unistratose; setae longer than 1mm 2

2 Inner laminal cells large, at least some longer than 30 μm .... Key A
   Inner laminal cells smaller, rarely longer than 22.5 μm 3

3 Leaves elimbate or limbate; limbidia restricted to 3/4 or less the vaginant
   laminae .... Key B
   Leaves limbate; limbidia extending at least 3/4 the length of the vaginant
   laminae, often present on all laminae .... Key C

KEY A. Inner laminal cells longer than 30 μm.

1 Mid dorsal laminal cells narrow, 7-20 times as long as wide ....
   .... *27. F. scariosus*
   Mid dorsal laminal cells wider, to 6.5 times as long as wide 2

2 Leaves limbate on all laminae 3
   Limbidia restricted to basal part of vaginant laminae or leaves elimbate 5

3 Limbidium conspicuous, 3-4 layers thick; mid stem leaves 2-4 mm long ...
   .... *8. F. flaccidus*
   Limbidium weak to conspicuous, 1-2 layers thick; mid stem leaves less than
   2.2 mm long 4

4 Mid dorsal laminal cells 52.5-120 μm long, 3.5-6.5 times as long as wide;
   costae shorter than vaginant laminae; limbidium inconspicuous ....
   .... *2. F. amazonicus*
   Mid dorsal laminal cells 22.5-49.5 μm long, 1.5-4 times as long as wide;
   costae extending beyond the vaginant laminae; limbidium conspicuous ....
   .... *21. F. palmatus*

5 Costa ending 1-4(-7) cells below the apex .... *12. F. inaequalis*
   Costa excurrent .... *19. F. ornatus*

KEY B. Leaves elimbate or limbate; limbidia restricted to 3/4 or less the vaginant
laminae; laminal cells less than 30 μm long.

1 Dorsal laminae ending well above insertion or narrow and inconspicuous in
   the basal part; leaves lingulate 2
   Dorsal laminae reaching the insertion, or if not; leaves not lingulate,
   variously shaped 4

42

2 Costae in distal part obscured by chlorophyllose cells; all leaves elimbate ...
.... *5. F. brevipes*
Costa not obscured in distal part; basal part of vaginant laminae of perichaetial
leaves weakly limbate 3

3 Laminal cells pluripapillose or smooth; ca. 32 files of exothecial cells on
theca ....*26. F. ramicola*
Laminal cells smooth or unipapillose; ca. 50 files of exot hecial cells on
theca ....*30. F. subramicola*

4 Laminal cells smooth 5
Laminal cells unipapillose or pluripapillose........................9

5 Laminal cells convex, strongly convex; costa in cross-section with 1 adaxial
and 2 lateral stereid bands (Fig. 11 B); leaves elimbate...............6
Laminal cells flat or lowly convex; costa in cross-section with 2 lateral
(Fig. 11 A) stereid bands; leaves elimbate or with short limbidium 8

6 Leaves caducous; peristome teeth undivided or irregularly divided ....
....*25. F. radicans*
Leaves persistent; peristome teeth divided into 2 filaments.............7

7 Costa ending 6-8 below leaf tip; leaf apex widely acute to obtuse........
.... *6. F. dendrophilus*
Costa ending 2-3 cells below apex; leaf apex narrowly acute ending in a
blunt acumen with rounded tip................... *17. F. oblongifolius*

8 Laminal cells not guttulate; leaves oblong to oblanceolate with widely acute
to obtuse apex.............................. *15. F. leptophyllus*
Laminal cells often guttulate; leaves lanceolate to elliptical with acute apex
.... *22a. F. pellucidus* var. *pellucidus*

9 Laminal cells pluripapillose 10
Laminal cells unipapillose, unipapillose mixed with a few bipapillose ones
or prorate 14

10 Most leaves limbate 11
Leaves elimbate or limbidia restricted to upper leaves of female stems ..13

11 Leaves oblong, apex rounded-obtuse, infrequently widely acute and
rounded....................... *10a. F. guianensis* var. *guianensis*
Leaves lanceolate, apex acute..................................12

12 Leaf apex acute ending in a sharp, clear cell ....*7. F. elegans*
Leaf apex acute-acuminate with rounded tip, not ending in clear cell......
.... *16. F. neglectus*

13 Leaves lanceolate, apex acute ending in a clear cell..........*7. F. elegans*
Leaves oblong, apex obtuse, often rounded, not ending in clear cell.......
....*20. F. pallidinervis*

14   Cells prorate and guttulate . . . . *22b. F. pellucidus* var. *papilliferus*
      Cells unipapillose or unipapillose mixed with a few bipapillose ones, not
      guttulate 15

15   Linear-lanceolate, more than 5 times as long as wide; costa percurrent to
      long excurrent. 16
      Leaves wider, less than 4 times as long as wide; costa mostly ending 2-4 cells
      below apex 17

16   Sporophytes axillary on basal half of the stem; perigonia in axillary buds;
      leaves 6-11 times as long as wide. . . . . *1. F. allionii*
      Sporophytes terminal; perigonial plants at base of female stems; leaves 5-7.5
      times as long as wide . . . . *24. F. prionodes*

17   Limbidia restricted to perichaetial and subtending leaves or all leaves
      elimbate 18
      Limbidia not restricted to upper leaves of female stems 20

18   Plants frondiform; margin of vaginant laminae irregularly serrate; peristome
      teeth divided into 2 filaments . . . . *28. F. serratus*
      Plants pinnate; margin of vaginant laminae crenulate to regularly serrate;
      peristome teeth undivided or irregularly divided 19

19   Laminal cells mammillose . . . . *14a. F. lagenarius* var. *lagenarius*
      Laminal cells with long, ± branched papillae. . . . .
      . . . . *14b. F. lagenarius* var. *muricatulus*

20   Leaves 1-1.4 mm long, apex widely acute to obtuse, often bluntly acuminate
      or rounded-apiculate; not ending in a clear cell; laminal cells unipapillose
      or smooth . . . . *15. F. leptophyllus*
      Leaves to 1 mm long, apex various, often ending in a clear cell; laminal cells
      unipapillose often mixed with a few bipapillose ones 21

21   Leaves oblong-lanceolate to oblong, 3-3.5 times as long as wide . . . .
      . . . . *11. F. hornschuchii*
      Leaves falcate-ovate, 2.5 times as long as wide . . . . *23. F. perfalcatus*

KEY C. Leaves limbate; limbidia extending at least 3/4 the length of the vaginant
laminae, often present on all laminae; laminal cells in mid dorsal lamina not
exceeding 22.5 μm.

1   Limbidia present on all laminae 2
      Limbidia restricted to vaginant laminae or extending slightly onto the apical
      lamina. 5

2   Laminal cells smooth, flat or convex 3
      Laminal cells papillose 4

44

3   Plants to 2 cm tall, pinnate without axillary nodules; theca with 40 or more
      files of exothecial cells . . . . *3. F. anguste-limbatus*
      Plants to 0.5 cm tall, typically frondiform with axillary nodules; theca with
      ca. 32 files of exothecial cells . . . . *32. F. zollingeri*

4   Laminal cells unipapillose . . . . *4. F. angustifolius*
      Laminal cells pluripapillose . . . . *31. F. weirii*

5   Leaves oblong or lingulate with obtuse, rounded-apiculate apex 6
      Leaf shape and apex different 7

6   Laminal cells pluripapillose, 4.5-10.5 μm long . . . .
      . . . . *10b. F. guianensis* var. *pacaas-novosensis*
      Laminal cells smooth, 9-12μm long . . . . *18. F. obtusissimus*

7   Laminal cells pluripapillose . . . . *13. F.intramarginatus*
      Laminal cells unipapillose . . . . *29. F. submarginatus*

1. **Fissidens allionii** Broth., Rev. Bryol. 47: 5. 1920. Lectotype (Pursell
    1994): Ecuador, Morona-Santiago, Gualaquiza, ad terram secus
    semitas silvae in monte Sapotilla, Bryotheca E. Levier 8291, Allioni
    s.n. (H).   – Fig. 12

Plants red-brown to pale green, often growing in dense mats, pinnate,
3.5-4 mm long, 1.2-1.6 mm wide, simple or branched. Leaves imbricate,
to 16 pairs, flat when dry, linear-lanceolate, often in part bistratose,
narrowly acute at apex, 1-1.5 mm long, 0.1-0.15 mm wide, 6 to 11 times
as long as wide, elimbate with crenulate to serrulate margin, infrequently
weakly limbate on proximal part of vaginant laminae of upper leaves;
vaginant laminae ± 1/2 the leaf length, unequal; dorsal lamina tapering
towards insertion, often slightly rounded below; costa strongly developed,
percurrent to excurrent; laminal cells unipapillose, irregularly hexagonal-
rectangular, guttulate or not, 4.5-12 μm long and 6-9 μm wide; basal
juxtacostal laminal cells of vaginant laminae to 22.5 μm long.
Gonioautoicous. Perigonia axillary, bud-shaped, on same stem as
archegonial branches. Perichaetia terminal on short branches in basal part
of stem, perichaetial leaves shorter than stem leaves, 0.4-1 mm long, with
poorly developed to lacking dorsal and apical laminae and long excurrent
costae; calyptra smooth. Sporophyte axillary, 1 per perichaetium, seta
2.2-2.7 mm long, theca 0.4 mm long and 0.25 mm wide, peristome
scariosus-type, teeth 30 μm wide at the base, operculum rostrate, 0.3 mm
long; spores 7.5-15 μm.

Distribution: C America, western and northern S America; the Guianas.

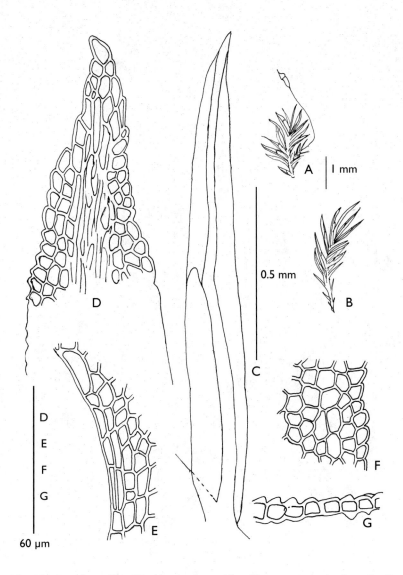

Fig. 12. *Fissidens allionii* Broth.: A, plant with axillary sporophytic plant; B, plant; C, leaf; D, leaf apex; E, margin vaginant lamina at insertion; F, margin mid dorsal lamina; G, cross-section dorsal lamina (A-G, Buck 18425B).

46

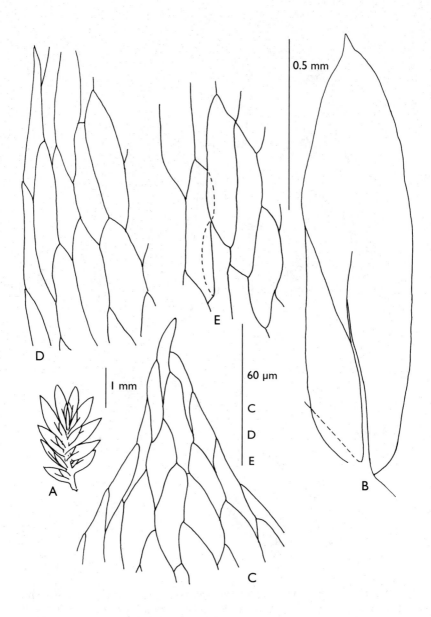

Fig. 13. *Fissidens amazonicus* Pursell: A, plant; B, leaf; C, leaf apex; D, margin mid dorsal lamina with weak limbidium of narrower cells; E, mid dorsal laminal cells (A-E, Gradstein 4968).

47

Ecology: Usually on bare soil, alt. 0-200 m; uncommon.

Selected specimens: Guyana: Junction of Mazaruni and Cuyuni Rs., Graham 371, pp (NY). Suriname: Moengo, virgin forest, Lanjouw 486 C (L). French Guiana: Canton de Maripasoula, 6 km N of Saül, Buck 18425 (NY).

Notes: The linear-lanceolate, narrowly acute leaves with excurrent or percurrent costae, unipapillose laminal cells and axillary sporophytes make recognition of this species easy. Pursell (in Allen 1994) stressed the importance of the width of the costa which in *F. allionii* would occupy 1/2 or more of the width of the leaf. Frequently, however, costae are narrower, 1/4-1/5 the leaf-width (measured in mid leaf). *F. allionii* can be confused with *F. prionodes*. The two differ in the position of the sporophytes, terminal in *F. prionodes* and axillary in *F. allionii*. Rarely a few sporophytes of *F. allionii* can be terminal. Sterile plants cannot always be identified with certainty. Forms of *F. allionii* with almost flat laminal cells (e.g. Aptroot 15194 (L)) are hard to distinguish from *F. pellucidus* var. *pellucidus* which, however, has wider leaves.

2. **Fissidens amazonicus** Pursell, Beiheft Nova Hedwigia 90: 345, fig. 1-7, 21-22, 26. 1988. Type: Brazil, Amazonas, Igarapé Iá-Mirim, no solo, margem da água, Yano 1990 (holotype SP, isotype MO).
   – Fig. 13

Plants pale brown, growing in mats, to 6.5 mm long, 3 mm wide, pinnate, usually unbranched. Leaves distant, to 10 pairs, slightly crisped when dry, lanceolate with acute to acuminate apex ending in a long, pointed cell, leaves 1.5-2 mm long, 0.4-0.6 mm wide, 3.5-6.5 times as long as wide, weakly limbate; vaginant laminae 1/2 or less the leaf length, equal to subequal; dorsal lamina at insertion 3 cells wide, slightly rounded, not decurrent; costa short, indistinct, ending below distal end of vaginant laminae, in perichaetial leaves often slightly longer; laminal cells wide-prosenchymatous, smooth, thin-walled, 52.5-120 μm long, 15-19.5 μm wide and 3.5-6.5 times as long as wide; 1-4 rows of marginal cells narrower, 4.5-7.5 μm wide; inner vaginant laminal cells 7.5-15 μm wide. Sporophyte and fertile parts not seen.

Distribution: Northern S America, Brazil, Colombia, Guyana.

Ecology: On submerged sandstone rocks in river bank, alt. 500 m; rare.

Examined specimen: Guyana: Upper Mazaruni Distr., at confluence of Arobaru and Kako Rs., Gradstein 4968 (L).

Notes: This species is characterized by weakly limbate leaves, costae ending below distal ends of vaginant laminae, equal to subequal vaginant laminae and large laminal cells that are 3.5-6.5 times as long as wide. It can be confused with other species with large laminal cells. *F. palmatus* is distinct by shorter, relatively wider laminal cells (1.5-4 times as long as wide) and costae that extend beyond the vaginant laminae, whereas *F. scariosus* Mitt. has narrower laminal cells (7-20 times as long as wide).

3. **Fissidens anguste-limbatus** Mitt., J. Linn. Soc., Bot. 12: 601. 1869. Type: Brazil, Paraná, in sylva ad rivuli ripas, Weir 19 (holotype NY, isotype H). – Fig. 14

> *Fissidens pennula* Broth., Bih. Kongl. Svenska Vetensk.-Akad. Handl. 26 Afd. 3(7): 13. 1900. Type: Brazil, Matto Grosso, Lindman 535 (holotype H).

Plants bright green to brownish, growing in mats, pinnate, to 20 mm long, 1.5 mm wide, mostly unbranched. Leaves distant to imbricate, to 27 pairs, slightly crisped when dry, ovate-lanceolate to elliptical with acute often slightly acuminate apex, 1.2 mm long, 0.4 mm wide, 3 times as long as wide, limbate; limbidium ending slightly above insertion of dorsal lamina, reaching leaf apex or ending a few cells below, marginal; vaginant laminae 3/4 the leaf length, equal; dorsal lamina slightly rounded at insertion; costa percurrent to short excurrent; laminal cells hexagonal, smooth, 7-15 µm long and 4.5-7.5 µm wide; basal juxtacostal vaginant laminal cells oblong, to 27 µm long. Dioicous. Perigonia and perichaetia terminal, perichaetial leaves 1.5 mm long. Sporophyte 1-2 per perichaetium, seta 4.2-4.4 mm long, theca 0.7 mm long and 0.25 mm wide, peristome bryoides-type, teeth 39-45 µm wide at base, operculum rostrate, 0.5 mm long; spores 12-15 µm.

Distribution: Widespread in the Neotropics; in the Guianas known only from Suriname and Guyana.

Ecology: On river banks at frequently inundated places, usually on stones, also on clay and wood; common in Suriname, rare in Guyana.

Selected specimens: Guyana: S Rupununi savanna near Dadanawa Ranch, Florschütz-de Waard 6016 (L). Suriname: Saramacca, Brokoboto Fall, Florschütz 1218 A (L); Bakhuis Mts., Florschütz & Maas 2354 (L), 2622 (L), 2649 A (L); 10 km below Blanche Marie Falls, Maas F3375 (L); Area of Kabalebo Dam project, Nickerie Distr., Florschütz-de Waard & Zielman 5796 (L).

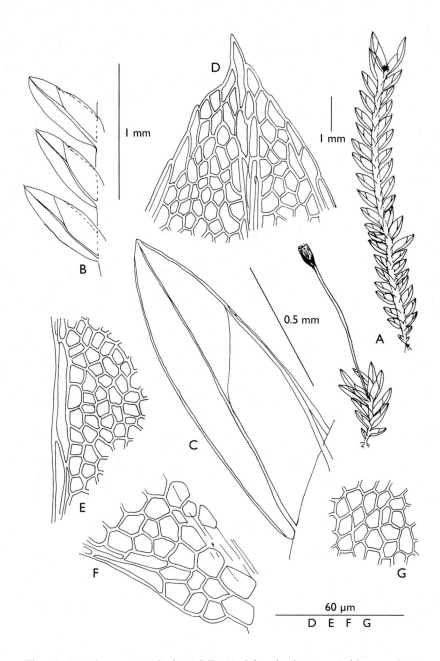

Fig. 14. *Fissidens anguste-limbatus* Mitt.: A, 2 female plants, one with sporophyte; B-C, leaves; D, leaf apex; E, margin mid dorsal lamina; F, base dorsal lamina; G, mid dorsal laminal cells (A-G, Florschütz-de Waard 6016).

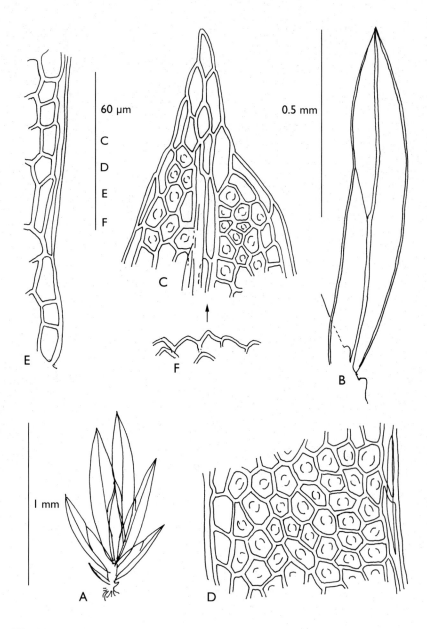

Fig. 15. *Fissidens angustifolius* Sull.: A, plant; B, leaf with 2 axillary nodules; C, leaf apex; D, mid dorsal lamina (costa on the left, limbidium on the right); E, base dorsal lamina; F, side view of mid dorsal laminal cells (leaf fold), arrow indicates the exterior side (A-F, Buck 18797A).

Notes: This species is characterized by limbate, ovate-lanceolate to elliptical leaves and smooth laminal cells. It is the only species of subgenus Fissidens Hedw. section *Fissidens* Hedw. in the Guianas. This section is characterized by limbate leaves, bryoides-type costae, smooth laminal cells, bryoides-type peristomes and theca with 40 or more files of exothecial cells. The only other limbate species in the area with smooth, shorter than 20 μm laminal cells is *F. zollingeri*. This species belongs to subgenus Aloma. It differs from *F. anguste-limbatus* in having frondiform plants, axillary nodules, strongly enlarged vaginant laminal cells, ± 32 files of exothecial cells on the theca and scariosus-type peristomes.

In Guianan specimens of *F. anguste-limbatus* perigonia and perichaetia are terminal, but gonioautoicous specimens have been reported from other areas (Pursell 2007). According to the same source 50% of the collections have sporophytes. In the Guianas only 2 out of 17 collections proved to have sporophytes.

4. **Fissidens angustifolius** Sull., Proc. Amer. Acad. Arts 5: 275. 1861 [1862]. Type: Cuba, wet places on the ground in dense woods, Wright (edited as Sullivant: Musci Cubensis 18) (holotype FH, isotypes BM, NY). – Fig. 15

Plants green, growing scattered, frondiform, 2 mm long, 1.6 mm wide, simple, small axillary nodules present. Leaves imbricate, to 5 pairs, flat when dry, linear-lanceolate to linear-elliptical with narrowly acute apex, 1.1-1.4 mm long, 0.2-0.25 mm wide, 5.5 times as long as wide, limbate; limbidium ± confluent with costa at leaf apex, in upper and middle leaves ± reaching insertion of dorsal lamina, confluent at apex of the vaginant laminae, in mid dorsal lamina 7.5 μm wide; vaginant laminae 1/2 the leaf length, equal; dorsal lamina gradually narrowed towards insertion; costa percurrent to excurrent; laminal cells unipapillose, hexagonal, 7.5-10.5 μm long and 4.5-6 μm wide; basal juxtacostal cells of vaginant laminae lax, to 45 μm long. No fertile plants and sporophytes seen from the Guianas.

Distribution: Pantropical, widespread in the Neotropics; in the Guianas restricted to French Guiana.

Ecology: On rocks in a non-flooded moist forest, alt. 200-300 m; rare.

Examined specimen: French Guiana: Canton de Maripasoula, 6 km N of Saül, vicinity of Eaux Claires, Buck 18797A (NY).

Note: This species is characterized by frondiform plants, small axillary nodules, limbate leaves, unipapillose laminal cells and scariosus-type peristomes. It resembles *F. zollingeri* which differs by smooth laminal cells.

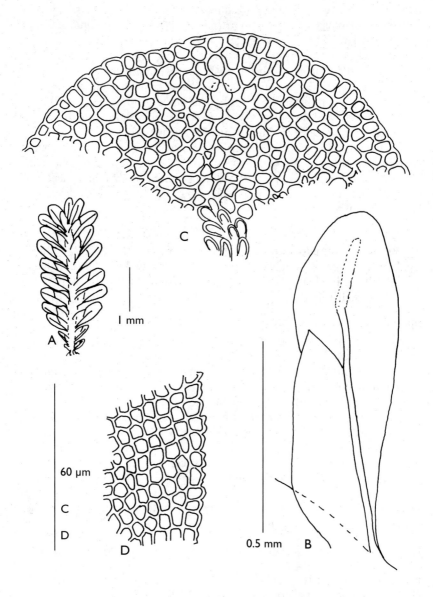

Fig. 16. *Fissidens brevipes* Besch.: A, plant; B, leaf; C, leaf apex; D, margin mid dorsal lamina (A-D, Florschütz-de Waard 1620).

5. **Fissidens brevipes** Besch., J. Bot. (Morot) 5: 252. 1891. Lectotype (Pursell 1994): Paraguay, ad cortices, Balansa 3698 (NY). – Fig. 16

Plants green, scattered, to 3 mm long, 1.5 mm wide, pinnate. Leaves imbricate, almost flat when dry, to 9 pairs, oblong to lingulate, rounded-obtuse, slightly inflexed at apex, 0.8-1 mm long, 0.25 mm wide, 3.5 times as long as wide, elimbate with subentire to crenulate margin; vaginant laminae 1/2-2/3 the leaf length, unequal; dorsal lamina tapering below, ending far above insertion or running down on costa as a single row of cells reaching the insertion; costa ending ± 10 cells below leaf apex, in distal part obscured by chlorophyllose cells; laminal cells finely pluripapillose, 4.5-6 μm long and 4.5 μm wide; basal juxtacostal vaginant laminal cells to 9 μm long. Gonioautoicous and dioicous. Perigonia terminal on plants or bud-shaped in leaf axils; perichaetia terminal, perichaetial leaves 1.4 mm long. Sporophyte 1 per perichaetium, seta 2 mm long, theca 0.7 mm long and 0.3 mm wide, operculum 0.3 mm long (immature); spores not seen.

Distribution: Neotropics.

Ecology: Corticolous; rare.

Examined specimen: Guyana: S Rupununi Savanna near Dadanawa Ranch, Florschütz-de Waard 6020 (L).

Notes: This species is easily recognized by its elimbate, oblong to lingulate leaves with costae distally obscured by small chlorophyllose cells. In Musci I in Fl. Suriname 6: 55. 1964, this species was confused with *F. subramicola* and *F. ramicola* which have similarly shaped leaves, but costae with unobscured distal ends.

6. **Fissidens dendrophilus** Brugg.-Nann. & Pursell, Bryologist 93: 335. 1990. Type: Brazil, Rio Grande do Sul, Portão, on tree, 40 m, Sehnem 300 (holotype NY, isotype FH). – Fig. 17

Plants bright green, scattered or forming mats, pinnate, 2.5-4 mm long, 2 mm wide, simple or not. Leaves loosely imbricate, to 15 pairs, slightly crisped when dry, oblong with widely acute to obtuse and rounded apex, 1.2-1.5 mm long, 0.3-0.4 mm wide, 4 times as long as wide, elimbate with crenulate margin; vaginant laminae 1/2 the leaf length, unequal, minor lamina ending near margin to halfway between margin and

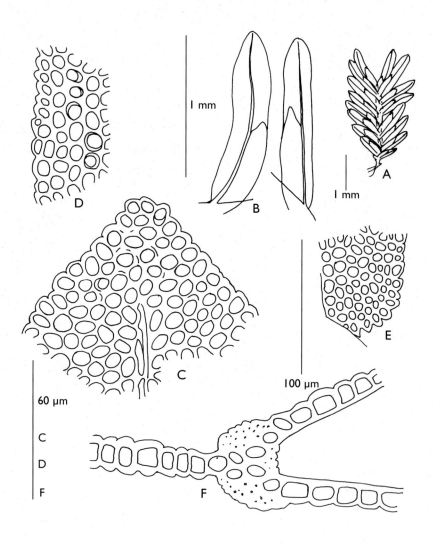

Fig. 17. *Fissidens dendrophilus* Brugg.-Nann. & Pursell: A, plant; B, leaves; C, leaf apex; D, margin mid dorsal lamina; E, base dorsal lamina; F, cross-section costa (A-E, Montfoort & Ek 591; F, Sehnem 300).

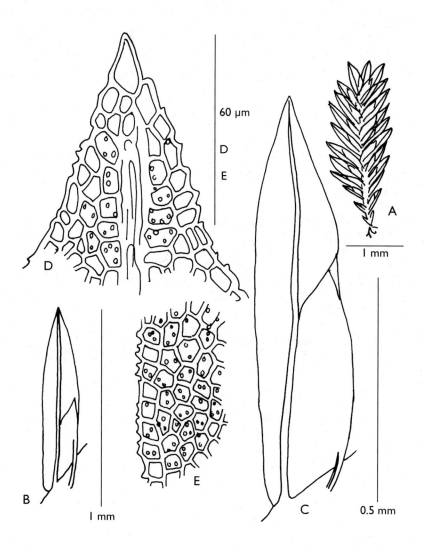

Fig. 18. *Fissidens elegans* Brid.: A, plant; B, leaf; C, leaf; D, leaf apex; E, margin mid dorsal lamina (A-B, D-E, Florschütz & Maas 2649; C, Florschütz 1829).

costa; dorsal lamina slightly rounded at insertion; costa oblongifolius-type, ending 6-8 cells below leaf apex; laminal cells smooth, convex, 4.5-7.5 μm long and 4.5 μm wide; basal juxtacostal vaginant laminal cells flat, thick-walled, to 15 μm long. Dioicous. Perigonia not seen; perichaetia terminal, perichaetial leaves 1.7 mm long. Sporophyte 1-2 per perichaetium, peristome similiretis-type. Sporophyte not seen in Guianan specimen.

Distribution: Brazil, Paraguay and French Guiana.

Ecology: On bark, alt. 180-210 m; rare.

Examined specimen: French Guiana: 2 km SW of Saül, lowland moist forest on lateritic soil, epiphytic on upper trunk of Bagassa guianensis, Montfoort & Ek 591 (L).

Note: This species is characterized by non caducous, oblong, elimbate leaves, smooth, convex laminal cells, oblongifolius-type costa and similiretis-type peristome. The similar *F. radicans* Mont. has caducous leaves and undivided to imperfectly divided peristome teeth. *F. dendrophilus* and *F. radicans* belong to subgenus Pachyfissidens section *Amblyothallia*. See note under *F. radicans*.

7. **Fissidens elegans** Brid., Muscol. Recent. Suppl. 1: 167. 1806. Type: Dominican Republic, Poiteau s.n. (holotype B, isotype PC).
– Fig. 18

Fissidens ravenelii Sull., Mem. Amer. Acad. Arts, n.s. 4: 171. 1849. Type: U.S.A., Ravenel s.n. (FH).
Fissidens leptopodus Cardot, Rev. Bryol. 37: 120, 1910. Type: Mexico, Barnes & Land 558 (NY).
Fissidens densiretis Sull. var. *latifolius* Grout, N. Am. Fl. 15: 181. 1943. Type: Puerto Rico, Steere 6686 (DUKE).

Plants dark green to brownish, usually scattered, pinnate, 2-8.5 mm long, 0.9-2 mm wide, mostly simple. Leaves often imbricate, 8-19 pairs, inflexed when dry, lanceolate with acute often slightly acuminate apex ending in a pointed, clear apical cell, 0.5-1.4 mm long, 0.15-0.3 mm wide, 3-4.5 times as long as wide, often limbate on 1/3-3/4 the vaginant laminae of most leaves, infrequently all leaves of plant elimbate; elimbate margins crenulate, denticulate or serrulate; limbidium typically intramarginal; vaginant laminae 1/2-3/4 the leaf length, subequal; dorsal lamina rounded-truncate at insertion or tapering below; costa ending ± 2 cells below apex to percurrent; laminal cells obscure, pluripapillose with 2-6 papillae per cell, roundish-hexagonal, 4.5-7.5 µm long and 3-6 µm wide; basal juxtacostal vaginant laminal cells to 13.5 µm long. Dioicous. Perigonia terminal on dwarf or longer plants; perichaetia terminal, perichaetial leaves longer than stem leaves; calyptra papillose or smooth. Sporophyte 1 per perichaetium, seta 1.5-3 mm long, theca 0.35-0.5 mm long and 0.2-0.3 mm wide, peristome scariosus-type, teeth 33-39 µm wide at base, operculum not seen; spores 12-13.5 µm, finely papillose.

Distribution: Widespread in the Neotropics, southern U.S.A, West Indies, tropical S America; the Guianas.

Ecology: Typically saxicolous, frequently in places that are regularly inundated; also collected on termitaries, trees and loamy soil, alt. 0-500 m; common in Suriname, rare in Guyana and French Guiana.

Selected specimens: Guyana: Mabura Hill, 180 km SSE of Georgetown, mixed lowland rain forest near Yaya Cr., Cornelissen & ter Steege 147, pp (L). Suriname: Stonemans-hill near km 121 of railroad, on rock, Florschütz 1829 (L); Bakhuis Mts., between Kabalebo and Coppename Rs., on moist bank, Florschütz & Maas 2649 (L). French Guiana: Eaux Claires, 5 km N of Saül, Florschütz-de Waard 5928 (L); village de Saül, on Mt. Galbao, Bekker 2305-2, pp (L).

Notes: *F. elegans* is characterized by narrow, lanceolate leaves terminated by a clear pointed cell contrasting with the dark, pluripapillose laminal cells. Most leaves have short, often intramarginal limbidia extending 1/3-3/4 the length of the vaginant laminae. *F. neglectus* and *F. guianensis* also have pluripapillose laminal cells and short limbidia on the vaginant laminae of most leaves, but only *F. elegans* has leaves ending in a clear apical cell. *F. hornschuchii* another species with leaves ending in a clear apical cell differs by unipapillose laminal cells (though often mixed with a few bipapillose ones).
F. elegans and some closely related species constitute the so-called *F. elegans* complex. This complex is characterized by pluripapillose laminal cells that are about twice as deep as wide (Pursell 1984, 1994). In the Guianas it is represented by *F. elegans, F. pallidinervis, F. guianensis, F. intramarginatus, F. neglectus, F. ramicola* and *F. weirii*.
Several of the synonyms listed as *F. guianensis* in Musci I in Fl. Suriname 6: 47. 1964, proved to be *F. elegans*.

58

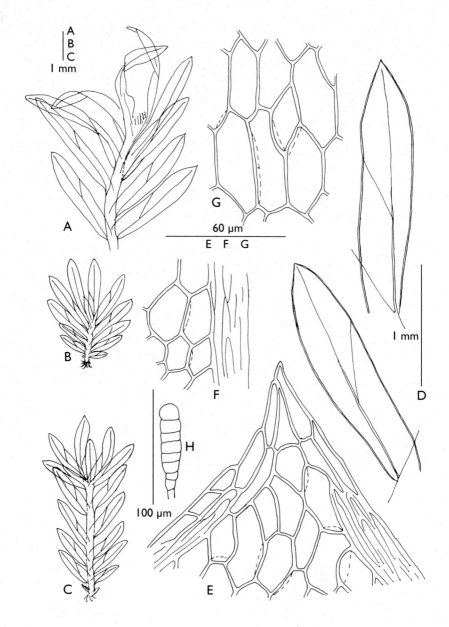

Fig. 19. *Fissidens flaccidus* Mitt.: A, upper part female plant; B, frondiform female plant; C, pinnate female plant; D, leaves; E, leaf apex; F, margin mid dorsal lamina; G, mid dorsal laminal cells; H, gemma (A, C-G, Florschütz-de Waard 4651; B, Newton & Florschütz-de Waard 3596; H, Florschütz 261).

8. **Fissidens flaccidus** Mitt., Trans. Linn. Soc., Bot. London 23: 56. 1860. Type: (Trop. W.) Africa, Niger, Vogel s.n. (holotype NY, isotype BM). – Fig. 19

Fissidens mollis Mitt., J. Linn. Soc., Bot. 12: 600. 1869. Type: Jamaica, Wilson 152 (holotype NY).

Plants dark green, growing scattered or in mats, small plants frondiform, larger ones pinnate, 4-15 mm long, 3-5 mm wide, often branched. Leaves imbricate to distant, to 15 pairs, crisped when dry, lanceolate, oblong-lanceolate and elliptical, acute, apiculate-acuminate, 2-4 mm long, 0.4-0.7 mm wide, 5-6 times as long as wide, limbate; limbidia confluent at leaf apex, ± reaching insertion of dorsal lamina, 3-4-stratose; vaginal laminae 3/5 the leaf length, equal; dorsal lamina tapering towards insertion; costa ending 2-8 cells below apex; laminal cells large, hexagonal, flat to lowly convex, smooth, 30-48 μm long and 13.5-16.5 μm wide; basal juxtacostal vaginal laminal cells oblong, to 93 μm long. Axillary, stalked, multicellular, elongated gemmae frequent. Dioicous. Perichaetia terminal, perichaetial leaves to 4.2 mm long. Perigonia and sporophyte not seen.

D i s t r i b u t i o n : Pantropical, widespread in the Neotropics, including the Guianas, also known from Africa, Australia and New Guinea.

E c o l o g y : On rocks or soil, frequently in wet places, alt. 0-700 m; rare.

E x a m i n e d s p e c i m e n s : Guyana: Region Potaro-Siparuni, Paramakatoi, Karibon Cr. Trail, Newton & Florschütz-de Waard 3596 (L). Suriname: Sarakreek, above Drie Gebroeders, Florschütz 261 (L); Brownsberg, Ireneval, at base of weeping rocks, Florschütz 4651 (L). French Guiana: Eaux Claires, 5 km N of Saül, tropical lowland forest on laterite soil, Florschütz-de Waard 5851 (L).

N o t e s : *F. flaccidus* is characterized by limbate, 2-4 mm long leaves, 3-4-stratose limbidia that are confluent at leaf apex and large laminal cells. Costae can be percurrent to ending as much as 18 cells below apex (Pursell 2007); costae of Guianan specimens invariably end 4-8 cells below apex.
F. flaccidus looks like a large *F. palmatus*. Both have large laminal cells, limbate leaves and short costae. *F. palmatus* has thinner (1-2-stratose) limbidia that become weak in the leaf apex and shorter leaves (1-2.2 mm long).

9. **Fissidens geijskesii** Florsch., Fl. Suriname 6: 30. 1964. Type: Suriname, Paloemeu R., between high and low watermark in Trombaka Falls, on rocks in shade, with capsules, Geijskes s.n. (holotype L, isotypes BM, BR). – Fig. 20

Plants black, young shoots green, growing in fascicles, pinnate, to 100 mm long with branches, 2.2 mm wide, heavily branched. Leaves closely imbricate, numerous pairs, stiff, hardly altered when dry, mostly worn off, lanceolate with acute apex, 2.1-2.5 mm long, 0.4-0.5 mm wide, 5 times as long as wide, 2-3-stratose (vaginant laminae 1-2-stratose), elimbate; margin entire, slightly thicker than inner lamina; vaginant laminae 3/5-2/3 the leaf length, subequal; dorsal lamina tapering towards insertion, often consisting of a single row of cells in basal part, reaching insertion; costa ill-defined, hardly protruding, ending 4-10 cells below apex; outer cortical cells of costa short and wide, not or hardly distinct from laminal cells; laminal cells roundish, smooth, 4.5-10.5 µm long and 6 µm wide; basal juxtacostal cells of vaginant laminae to 26.5 µm long. Axillary bundles of heavily branched multicellular, filamentous gemmae frequent. Gonioautoicous. Perigonia axillary, bud-shaped, 1-2 per axil; perichaetia terminal on branches and plants, perichaetial leaves to 2.7 mm long; calyptra wide, smooth, 0.5 mm long. Sporophyte 1 per perichaetium, seta thick, short, 0.4 mm long, theca 0.7-1 mm long and 0.7-0.8 mm wide, immersed, dehiscence zone not differentiated, peristome irregular, teeth 67.5 µm wide at the base, operculum rostrate, 0.5 mm long; spores 37.5 µm.

Distribution: Suriname, known only from the type locality.

Ecology: On frequently inundated rocks, elevation not indicated; rare.

Examined specimens: Suriname: Paloemeu R., between high and low watermark in Trombaka Falls, Geijskes s.n. (2 x) (L, BM, BR).

Notes: The rheophilous *F. geijskesii* is unlikely to be confused with any other species. It is characterized by blackish plants with pale green new shoots, stiff, elimbate, pluristratose leaves, costae covered by chlorophyllose cells, axillary bundles of branched, filamentous gemmae, short, thick setae and immersed thecae.
This is the only species of subgenus Pachyfissidens section *Pachyfissidens* in the area. Species of this section typically have elimbate leaves, taxifolius-type costae, taxifolius-type peristomes and theca with more than 40 files of exothecial cells. *F. geijskesii* has an anomalous costa and peristome.

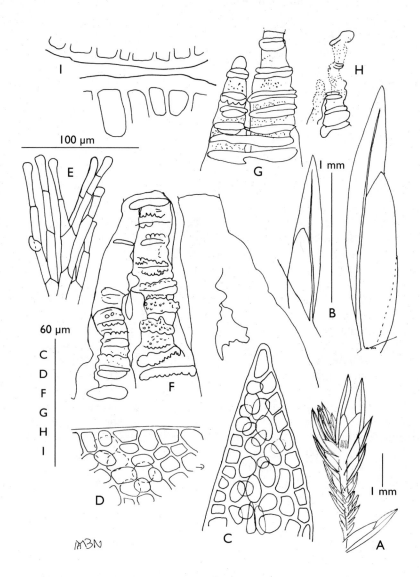

Fig. 20. *Fissidens geijskesii* Florsch.: A, female branch with short secondary branch (top left); B, leaves; C, leaf apex; D, margin mid dorsal lamina (arrow points to leaf apex); E, gemmae; F, two peristome teeth from the interior side, the right tooth with interior layer removed showing the perforated exterior layer; G, exterior side of peristome at bifurcation; H, exterior side peristome filament; I, detail of mouth of peristome and basal cell row of operculum (A-I, Geijskes s.n.).

10. **Fissidens guianensis** Mont., Ann. Sci. Nat., Bot. sér. 2, 14: 340. 1840. Type: French Guiana, Leprieur 315 (holotype PC, isotypes BM, BR).

KEY TO THE VARIETIES

1   Limbidia extending 1/5-1/2 the length of the vaginant laminae . . . .
. . . . *10a. F. guianensis* var. *guianensis*
Limbidium extending 9/10 or more the length of the vaginant laminae, often extending slightly onto the proximal part of the apical lamina . . . .
. . . . *10b. F. guianensis* var. *pacaas-novosensis*

10a. var. **guianensis**   – Fig. 21 A-E

Plants green, forming loose, often large mats, pinnate, 2.5-7 mm long, 1.4-1.6 mm wide, simple or branched. Leaves ± imbricate, to 16 pairs, when dry ± flat with somewhat inrolled tips that remain slightly inflexed when moistened, oblong, oblong-lanceolate or oblong-oblanceolate, at apex widely acute to obtuse, rounded, occasionally with short, wide apiculus, 0.4-1 mm long, 0.1-0.4 mm wide, 2.5-4 times as long as wide, unistratose or partly bistratose, margin serrulate-crenulate, limbate on 1/5-1/2 the vaginant laminae of most leaves; limbidia unistratose, in basal part marginal, more distally sometimes intramarginal; vaginant laminae 1/2 the leaf length, subequal; dorsal laminae slightly rounded to truncate at base, reaching insertion; costa ending 2-4 cells below apex, often papillose distally; laminal cells pluripapillose, 1-4 papillae per cell, 4.5-10.5 µm long and 4.5-6 µm wide; basal juxtacostal vaginant laminal cells to 15 µm long. Cladautoicous, gonioautoicous and rhizautoicous. Perigonia terminal on short plants or branches or bud-shaped in leaf axils; perichaetia terminal on main stems and branches, perichaetial leaves to 1.2 mm long; calyptra slightly scabrous. Sporophyte 1 per perichaetium, seta 2-3 mm long, theca 0.55 mm long and 0.25 mm wide, erect to inclined, peristome scariosus-type, teeth 39 µm wide at base, operculum rostrate; spores 12 µm, finely papillose.

Distribution: Neotropics, widespread; the Guianas.

Ecology: Corticolous on tree bases, trees, branches, lianas, roots, twigs, also on dead wood, trunks, logs, poles, rotten wood, rarely on stone or clay, alt. sea-level to 800 m; one of the most common *Fissidens*-taxa in the Guianas.

Selected specimens: Guyana: Mabura Hill, 180 km SSE of Georgetown, near Yaya Cr., Cornelissen & ter Steege 501 (L). Suriname:

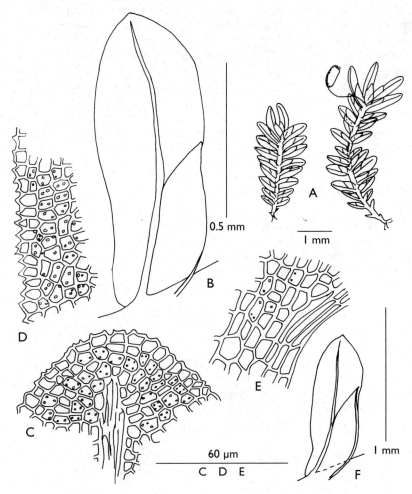

Fig. 21. A-E. *Fissidens guianensis* Mont. var. *guianensis*: A, 2 plants, one with sporophyte; B, leaf; C, leaf apex; D, margin mid dorsal lamina; E, insertion of vaginant lamina (A-E, Montfoort & Ek 586). F. *Fissidens guianensis* Mont. var. *pacaas-novosensis* Pursell & W.D. Reese: F, leaf (F, Allen 25079).

along railroad near km 121, on weathered rock, Florschütz 1790 (L); Nickerie Distr., area of Kabalebo Dam project, Florschütz-de Waard & Zielman 5657 (L). French Guiana: Bassin de Ba, Cremers & Gautier 12105 (L); Saül, 2 km SW of village, Montfoort & Ek 586 (L).

Notes: *F. guianensis* var. *guianensis* is recognized by oblong leaves with wide, obtuse, rounded, slightly inflexed tips, pluripapillose laminal cells, costae ending 2-4 cells below leaf apex and short limbidia on the proximal parts of the vaginant laminae of most leaves. It can be confused with *F. leptophyllus* which also has wide, oblong, partially limbate leaves and differs by smooth to lowly unipapillose laminal cells mixed with a few bipapillose ones. *F. guianensis* var. *guianensis* also resembles those forms of *F. pallidinervis* in which the costae end close to the apex. In *F. pallidinervis* all leaves except the perichaetial and subtending ones are elimbate.

F. guianensis sensu Musci I in Fl. Suriname 6: 47. 1964, includes 4 species (Pursell 1984, 1994): *F. guianensis, F. elegans* (leaves ending in clear apical cell), *F. pallidinervis* (limbidia restricted to the perichaetial and subtending pairs of leaves) and *F. intramarginatus* (vaginant laminae and often basal part of apical laminae limbate).

10b. var. **pacaas-novosensis** Pursell & W.D. Reese in Pursell, J. Hattori Bot. Lab 55: 245, fig. 38. 1984. Type: Brazil, Rondônia, 2-4 km above the first rapids of the Rio Pacaás Novos, mature forest with sandstone and granitic exposures, ridges of Serras dos Pacaás Novos along the river, to ca. 400 m, Reese 13437 (holotype INPA).
– Fig. 21 F

Distribution: Brazil and Suriname.

Ecology: As in typical variety, alt. 227 m; rare.

Examined specimens: Suriname, Sipaliwini, Kayserberg Airstrip Area, trail along creek E of airstrip, on stem of treelets by creek, Allen 25079 (MO, herb.Bruggeman-Nannenga); Suringar s.n. (L, slide).

Note: var. *pacaas-novosensis* can be confused with *F. intramarginatus* which also has completely limbate vaginant laminae and pluripapillose laminal cells. Leaves of *F. intramarginatus* are elliptic-lanceolate with acute-acuminate tips, those of *F. guianensis* are oblong with obtuse or widely acute apices that are distally rounded or end in short, wide apiculus. Musci I in Fl. Suriname 6: 49, fig. 12B, d-e, 1964, depict this variety.

11. **Fissidens hornschuchii** Mont., Ann. Sci. Nat., Bot. sér. 2, 14: 342. 1840. Lectotype (Pursell 1989): Brazil, Martius s.n. (M). – Fig. 22

Plants brownish, growing scattered or in mats, pinnate, 2-3 mm long, 0.7-1 mm wide, simple or branched. Leaves imbricate, to 15 pairs, inrolled from tip when dry, not completely flattening when moistened, oblong-lanceolate and oblong, acute, widely acute or obtuse, often ending in a

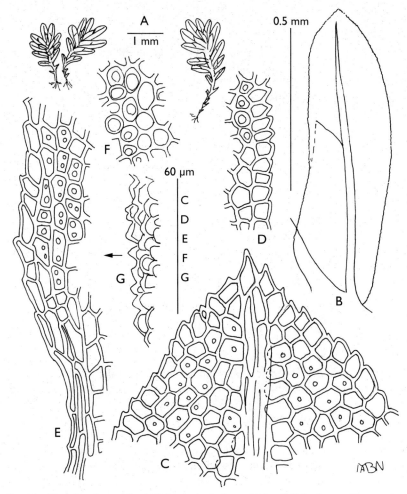

Fig. 22. *Fissidens hornschuchii* Mont.: A, 3 plants; B, leaf; C, leaf apex; D, margin mid dorsal lamina; E, insertion vaginant lamina; F, mid dorsal laminal cells; G, fold mid dorsal lamina showing one bipapillose and several unipapillose cells (arrow indicates exterior side) (A-G, Cornelissen & ter Steege C 147).

clear cell, 0.5-1 mm long, 0.2-0.3 mm wide, 3-3.5 times as long as wide, limbate on 3/4 the length of vaginant laminae of upper leaves of female stems, other leaves elimbate or with shorter limbidia; elimbate margins serrulate to denticulate; limbidium marginal or partly intramarginal, reaching insertions of vaginant laminae or not; vaginant laminae 3/5 the leaf length, unequal; dorsal lamina slightly to strongly rounded at insertion; costa ending 1-3 cells below apex to short excurrent, distally often papillose; laminal cells unipapillose, a few bipapillose, 4.5-12.5 μm long and 4.5-7.5 μm wide; basal juxtacostal vaginant laminal cells 9-18 μm long. Cladautoicous and dioicous. Perigonia terminal on long or short plants or branches; perichaetia terminal, perichaetial leaves longer than stem leaves; calyptra smooth to weakly papillose. Sporophyte 1 per perichaetium, seta 2-3.5 mm long, theca 0.35-0.4 mm long and 0.25 mm wide, peristome scariosus-type, teeth 37.5 μm wide at base, operculum rostrate, 0.3 mm long, spores 10.5-15 μm, papillose.

D i s t r i b u t i o n : Widespread in the Neotropics; the Guianas.

E c o l o g y : On ground, rock, clay, dead tree and on an old termitary, alt. 0-525 m; uncommon.

S e l e c t e d   s p e c i m e n s : Guyana: Mabura Hill, 180 km SSE of Georgetown, mixed lowland rain forest near Yaya Cr., Cornelissen & ter Steege 147, pp (L). Suriname: Rock savanna along trail S of Pakapaka, Saramacca R. to Ebbatop, van Asch van Wijck Mts., on clay near fallen tree, Florschütz 1584, pp (L); Area of Kabalebo Dam project, road to Amatopo, km 117, terrestrial on old termitary, Florschütz-de Waard & Zielman 5629 (NY); Tibiti Savanna, on ground, Lanjouw & Lindeman 1751, pp (L). French Guiana: Saül, vicinity of Eaux Claires, Buck 18799 (NY).

N o t e s : *F. hornschuchii* is recognized by unipapillose laminal cells (often mixed with a few bipapillose ones), oblong-lanceolate to oblong leaves and limbidia reaching to 3/4 the vaginant laminae of perichaetial and upper leaves of female stem, shorter or absent on other leaves. Sterile, elimbate forms of *F. hornschuchii* can be confused with *F. serratus* or *F. prionodes*. *F. prionodes* has narrower leaves with percurrent to long excurrent costae. *F. serratus* is distinct by frondiform plants, leaves without clear apical cell, slightly larger laminal cells and irregularly serrate margins of vaginant laminae.
The neotype of *F. hornschuchii* designated in Musci I in Fl. Suriname 6: 62. 1964, has been replaced by a lectotype (Pursell 1989); the neotype has smooth cells and is *F. pellucidus*.

12. **Fissidens inaequalis** Mitt., J. Linn. Soc., Bot. 12: 589. 1869. Type: Brazil, Amazonas, São Gabriel, Rio Negro, Spruce 538 (holotype NY). – Fig. 23, 24

Fissidens pauperculus M. Howe var. *surinamensis* Florsch., Fl. Suriname 6: 39. 1964. Type: Suriname, Peperpot, Florschütz 1707A (holotype L).

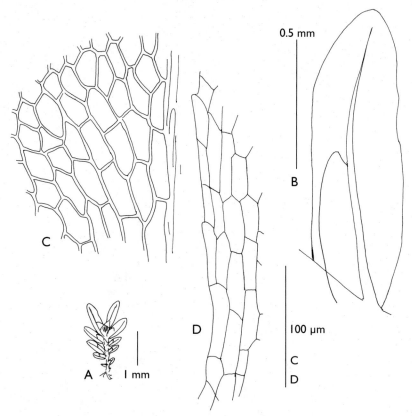

Fig. 23. *Fissidens inaequalis* Mitt.: A, plant with archegonia; B, leaf; C, mid dorsal laminal cells (costa on the right); D, margin basal vaginant lamina (A-D, Cornelissen & ter Steege 137).

Plants brown or pale green, growing scattered, rarely in small tufts, pinnate, 2.5-4 mm long, 1.5-1.7 mm wide, simple. Leaves mostly loosely imbricate, to 10 pairs, with reflexed margins when dry, elliptic-lanceolate, oblong and obovate, acute to obtuse, acuminate or rounded with apiculus, 0.85-1.4 mm long, 0.2-0.3 mm wide, 3-5 times as long as wide, often weakly limbate on basal 1/2 or less of vaginant laminae; elimbate margins entire, crenulate or serrate, infrequently with small bistratose parts; vaginant laminae less than 1/2 the leaf length, unequal, lesser lamina ending halfway between margin and costa; dorsal lamina slightly rounded at insertion; costa ending 1-4(-7) cells below apex, cortical cells in basal part often lax; laminal cells smooth, guttulate or not, often arranged in oblique rows making an angle of ± 30° with the costa, 18-45 μm long and 15-21 μm wide; 1-2 marginal rows often smaller, 7.5-

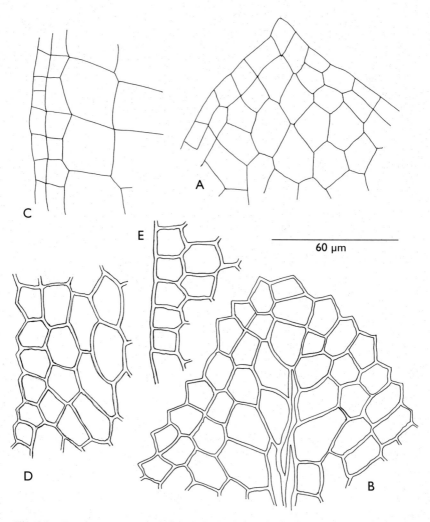

Fig. 24. *Fissidens inaequalis* Mitt.: A-B, variations in leaf apex; C-E, variation in margin of dorsal lamina (A, C, Maas 3366; B, D, E, Cornelissen & ter Steege 137).

15 μm long; juxtacostal cells of mid dorsal lamina 22.5-60 μm long; basal vaginant laminal cells to 30-40 μm long and 15 μm wide. Rhizautoicous. Perigonia terminal on dwarf plants at base of female stems; perichaetia terminal, perichaetial leaves longer than stem leaves with emarginate vaginant laminae; calyptra smooth, campanulate. Sporophyte 1 per perichaetium, seta 2-2.5 mm long, theca 0.4-0.5 mm long and 0.2-0.3 mm wide, peristome scariosus-type, teeth 32.5 μm wide at base, operculum unknown; spores 10.5-12.0 μm.

Distribution: S America and West Indies; the Guianas.

Ecology: Most often on clay, also collected on old termite mound and on weathered rock, alt. 0-1050 m; common.

Selected specimens: Guyana: savanna forest beyond Aratak Mission, loam, Florschütz-de Waard 6108 (L); Mabura hill, 180 km SSE of Georgetown, mixed riverine rain forest, Cornelissen & ter Steege 137 (L). Suriname: Area of Kabalebo Dam project, Marsh forest with Astrocaryum, on old termitary, Florschütz-de Waard & Zielman 5296, pp (L); Bakhuis Mts., track along Kabalebo and Zandkreek, 1-10 km upstream of airstrip, on clay between palm roots, Florschütz & Maas 2489 (L). French Guiana: Saül, primary forest near Boeuf Mort, in little cave, Aptroot 15315 (L).

Notes: The variable *F. inaequalis* is characterized by leaves elimbate or weakly limbate on basal part of the vaginant laminae, large laminal cells and costae ending 1-4(-7) cells below the leaf apex. *F. ornatus* also has large laminal cells and short limbidia on the vaginant laminae, but is distinct by excurrent costae. Expressions of *F. inaequalis* with small laminal cells can be hard to distinguish from expressions of *F. pellucidus* with large cells. However, juxtacostal dorsal laminal cells in *F. inaequalis* are usually much longer than inner dorsal laminal cells, whereas juxtacostal mid-dorsal laminal cells of *F. pellucidus* differ hardly from inner ones. A specimen with cells of intermediate size is Florschütz 2288. *F. pauperculus* var. *pauperculus* sensu Musci I in Fl. Suriname 6: 39. 1964 is *F. inaequalis. F. pauperculus* M. Howe is not.

13. **Fissidens intramarginatus** (Hampe) A. Jaeger, Enum. Fissident. 14. 1869. – *Conomitrium intramarginatum* Hampe, Linnaea 31: 531. 1862. – *C. intromarginatum* Hampe, Flora 45: 458. 1862. – *Fissidens intromarginatus* (Hampe) Mitt., J. Linn. Soc., Bot. 12: 594. 1869. Lectotype (Pursell 1994): Colombia, Cudinamarca, Bogotá, Tequendama, 7500 ft., Lindig 2149 (NY). – Fig. 25

Plants dirty green, growing scattered, pinnate, to 10 mm long, 2.4-2.7 mm wide, simple or branched. Leaves loosely imbricate, to 22 pairs, hardly altered to crisped when dry, lanceolate with acute, acuminate apex, 1.2-1.6 mm long, 0.3.5-0.45 mm wide, 3-4 times as long as wide, margin crenulate or denticulate, vaginant laminae limbate; limbidium unistratose fusing at apex of vaginant laminae, often extending slightly onto apical lamina, infrequently a few limbidial cells on the dorsal lamina; vaginant laminae 3/5-3/4 the leaf length, subequal to equal; dorsal lamina truncate to rounded below; costa percurrent to short-excurrent; laminal cells

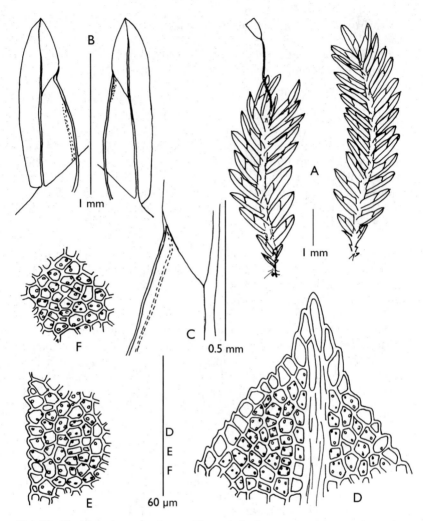

Fig. 25. *Fissidens intramarginatus* (Hampe) A. Jaeger: A, plants; B, leaves; C, upper part of vaginant laminae; D, leaf apex; E, margin mid dorsal lamina; F, mid dorsal laminal cells (A-F, Florschütz-de Waard & Zielman 5724).

pluripapillose, 2 or more papillae per cell, 4.5 μm long and 3-4.5 μm wide; basal juxtacostal vaginant laminal cells to 16.5 μm long. Rhizautoicous and gonioautoicous. Perigonia bud-shaped, axillary on female stems or at base of female stems; perichaetia terminal on main stem and branches, perichaetial leaves not differentiated; calyptra not seen. Sporophyte 1-2 per plant, seta 2.5-12 mm long, theca 0.6-0.8 mm long and 0.3-0.4 mm wide, peristome scariosus-type, operculum rostrate, 0.6 mm long; spores ± 9 μm.

Distribution: Africa, Mexico, C America, West Indies, S America; Suriname and French Guiana.

Ecology: On soil, sea-level; rare.

Examined specimens: Suriname: Area of Kabalebo Dam project, road km 212, rainforest, on vertical wall of creek, Florschütz-de Waard & Zielman 5724 (L). French Guiana: Cayenne, Sagot s.n. (NY).

Notes: This species is characterized by completely limbate vaginant laminae, percurrent to excurrent costae and pluripapillose laminal cells. It resembles *F. submarginatus* which is distinct by its unipapillose laminal cells.
F. intramarginatus belongs to the *F. elegans*-complex (see note under *F. elegans*).
There is a big difference in length of setae between the 2 Guianan collections: 2.5 mm in the Suriname specimen and 10-12 mm and more in Cayenne specimen.

14. **Fissidens lagenarius** Mitt., J. Linn. Soc., Bot. 10: 184. 1868. Type: Samoa, Tutuila, on Cyathea leucolepsis, Powell 22 (holotype NY, isotypes NY).

KEY TO THE VARIETIES

1   Laminal cells mammillose . . . . *14a. F. lagenarius* var. *lagenarius*
     Laminal cells with long, ± branched papillae. . . . .
     . . . . *14b. F. lagenarius* var. *muricatulus*

14a. var. **lagenarius**   – Fig. 26 C, E, G, I-J

Plants pale green, in loose mats or scattered, pinnate, to 6 mm long, 1-1.3 mm wide, branched or not; rhizoids brown or bright red. Leaves loosely imbricate, to 22 pairs, hardly altered when dry, elliptic-lanceolate to oblong with acute to obtuse, often acuminate apex, 0.7-0.9 mm long, 0.25-0.3 mm wide, ± 3 times as long as wide, margin crenulate to strongly serrate, limbate on proximal half or less of vaginant laminae of upper leaves of female stems; vaginant laminae 3/5 the leaf length, unequal; dorsal lamina somewhat rounded below, reaching insertion; costa often flexuose, ending 2-3(-5) cells below apex; laminal cells mammillose, 6-10.5 μm long and 4.5-7.5 μm wide; basal juxtacostal cells of vaginant laminae to 15 μm long. Cladautoicous and dioicous. Perigonia terminal on plants or on branches from upper part of stem; perichaetia terminal on stems or short branches from upper part of stem, outer perichaetial leaves

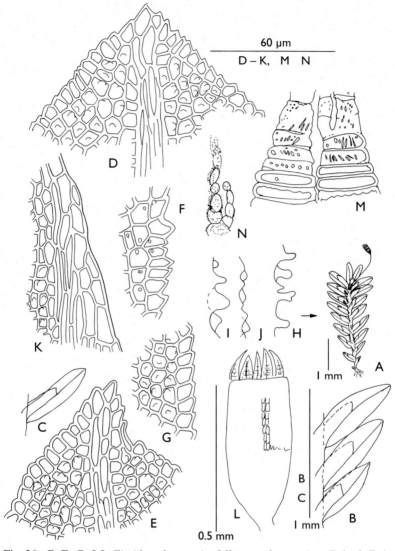

Fig. 26. C, E, G, I-J. *Fissidens lagenarius* Mitt. var. *lagenarius*: C, leaf; E, leaf apex; G, margin mid dorasal lamina; I-J, sideview papillae in mid dorsal lamina (C, E, G, J, Allen 25486; I, Buck 37786). A-B, D, F, H, K-N. *Fissidens lagenarius* Mitt. var. *muriculatus* (Spruce ex Mitt.) Pursell: A, plant with sporophyte and 2 axillary perigonia (upper left); B, leaves; D, leaf apex; F, margin in mid dorsal lamina; H, papillae in mid dorsal lamina (sideview, arrow indicates exterior side); K, limbate margin of vaginant lamina of perichaetial leaf; L, capsule with 2 files of exothecial cells partially drawn; M, 2 peristome teeth, left one undivided, right one divided; N, distal part peristome filament (A, B, D, K, L, Buck 25769; F, H, Newton *et al.* 3334; M-N, Richards 4).

not differentiated, inner ones often shorter and narrower than stem leaves; calyptra cucullate, papillose. Sporophyte 1-2 per perichaetium, setae 1.4 mm long, theca narrowly cylindrical, 0.6 mm long and 0.2 mm wide, peristome teeth ± straight, undivided to irregularly divided, vertically ridged to finely papillose below, densely papillose distally, 30 μm wide at the base, operculum rostrate; spores 15-16.5 μm, papillose.

Distribution: Pantropical; Suriname and French Guiana.

Ecology: On bark of living trees, alt. 200-750 m.

Selected specimens: Suriname: Sipaliwini, Eilerts de Haan Mts., along trail, mountain summit, on bark, Allen 25486 (MO). French Guiana: Saül, Mt. Galbao, Buck 25569 (NY); Mt. de l'Inini, sur arbre en sous-bois, Cremers *et al*. 9146 (BR, L); Kaw Mts., Trésor Reserve, Buck 37786 (NY).

Notes: This corticolous species is characterized by unipapillose laminal cells, narrow, cylindrical capsules, erect, undivided peristome teeth with characteristic ornamentation and limbidia restricted to basal parts of vaginant laminae of upper leaves of male and female stems.

14b. var. **muricatulus** (Spruce ex Mitt.) Pursell, Bryologist 102: 126. 1999. – *Fissidens muriculatus* Spruce ex Mitt., J. Linn. Soc., Bot. 12: 593. 1869. Lectotype (Grout 1943): Brazil, Amazonas, São Gabriel, ad ramulos, Spruce 473 (NY).
– Fig. 26 A-B, D, F, H, K-N

Fissidens diplodus Mitt., J. Linn. Soc., Bot. 12: 598. 1869. Type: Ecuador, Rio Bombonasa, in ramulos, 1,200 ped., Spruce 475 (holotype NY). *Moenkemeyera richardsii* R.S. Williams in P.W. Richards, Kew Bull. 8: 319. 1934. – *Fissidens muriculatus* Spruce ex Mitt. var. *richardsii* (R.S. Williams) Florsch., Fl. Suriname 6: 51. 1964. – *Fissidens diplodus* Mitt. var. *richardsii* (R.S. Williams) Pursell, Bryologist 83: 256. 1980. Type: Guyana, Moraballi Cr., swampy Mora forest, trunk of Miconia pluckenettii, Richards 118 (holotype NY, isotype BM).

Distribution: Neotropics; Guyana and French Guiana.

Ecology: On bark of living trees, alt. 200-750 m; uncommon in the Guianas.

Selected specimens: Guyana: without locality, Richards 41 (BM); Bartica, on bark, Linder 7 (NY); Kaieteur Falls National Park, W of Kaieteur Falls, path to old pump house, Newton *et al*. 3334 (L). French Guiana: Saül, Mt. Galbao, Buck 25769 (NY).

15. **Fissidens leptophyllus** Mont., Ann. Sci. Nat., Bot. sér. 2, 14: 344.
1840. Type: French Guiana, Leprieur 285 (holotype PC). – Fig. 27

Plants dark green, growing scattered or in loose mats, pinnate, to 6 mm
long, 2.5 mm wide, simple. Leaves imbricate, to16 pairs, little changed
when dry, oblong to oblanceolate with widely acute to obtuse, often
bluntly acuminate or rounded-apiculate apex, 1.0-1.4 mm long,
0.3-0.4 mm wide, 3-3.5 times as long as wide, margin denticulate,
limbate on basal 1/3-1/2 part of vaginant laminae of most leaves,
rarely elimbate on all leaves of a plant; limbidium marginal to weakly
intramarginal; vaginant laminae 1/2 or slightly more the leaf length,
subequal; dorsal lamina at insertion wide and rounded; costa ending
2-3 cells below apex to percurrent; laminal cells flat to slightly convex,
smooth or unipapillose mixed with a few bipapillose ones, 7.5-12 µm
long and 4.5-7.5 µm wide; basal cells of vaginant laminae to 16.5 µm
long. Gonioautoicous. Perigonia bud-shaped, axillary; perichaetia
terminal on stems or branches from upper part of stem, perichaetial
leaves 1.4-1.5 mm, often with short excurrent costa. Sporophyte 1 per
perichaetium, seta 3 mm long, theca 0.7-0.9 mm long and 0.4-0.6 mm
wide, peristome scariosus-type, teeth 45 µm wide at base, operculum
rostrate, 0.5 mm long; spores 16.5 µm, papillose.

Distribution: U.S.A. (Florida and Louisiana), C America, S America
(southernmost collection from Paraguay); Suriname and French Guiana,
not known from Guyana.

Ecology: On twigs, rotten wood and roots, lowland rainforest; rare.

Selected specimens: Suriname: Toso Cr. near Brokoboto Cr., on
twigs, Florschütz 482 (L); Kabalebo, Avanavero-Fall, N-bank, on roots,
Florschütz 2207 (L). French Guiana: Leprieur s.n. (BR).

Notes: This species is characterized by oblong to oblanceolate, obtuse-
rounded to broadly acute, bluntly acuminate or apiculate leaves, smooth
to unipapillose laminal cells and limbidia extending 1/3-1/2 the vaginant
laminae of all or most leaves. *F. leptophyllus* can be confused with *F.
obtusissimus* which is distinct by limbidia extending the whole length of
the vaginant laminae.

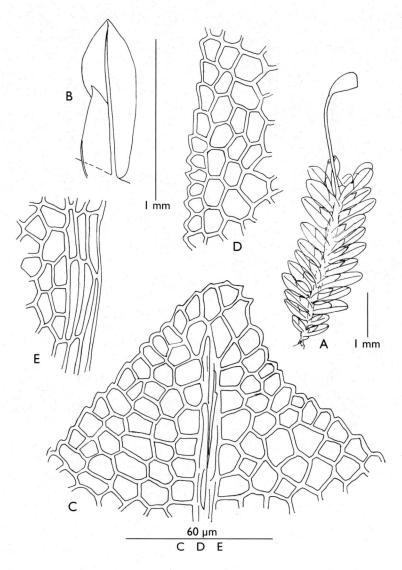

Fig. 27. *Fissidens leptophyllus* Mont.: A, plant with sporophyte; B, leaf; C, leaf apex; D, margin mid dorsal lamina; E, proximal margin of vaginant lamina with limbidium (A-E, Florschütz 482).

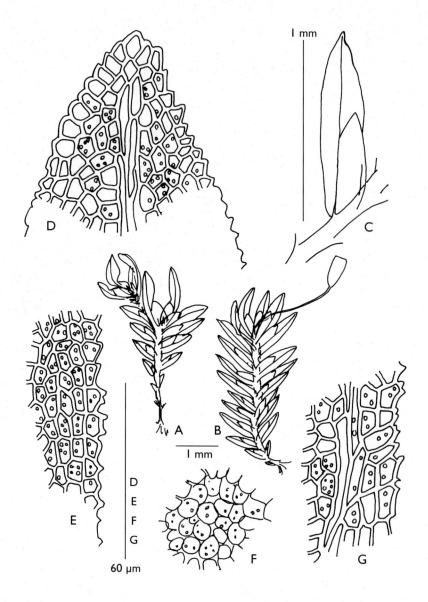

Fig. 28. *Fissidens neglectus* H.A. Crum: A, female plant with archegonial branch; B, plant with sporophyte; C, leaf; D, leaf apex; E, margin mid dorsal lamina; F, mid dorsal laminal cells; G, margin vaginant lamina with intralaminal limbidium (A-G, Cremers 6806).

16. **Fissidens neglectus** H.A. Crum, Bryologist 63: 95. 1960. Type: Cuba, Wright 23 (holotype FH).   – Fig. 28

Plants dull-green, growing in dense mats, pinnate, to 7 mm long, 1-2 mm wide, simple or branched. Leaves imbricate, to 20 pairs, somewhat crisped when dry, lanceolate to elliptical, acute often acuminate with rounded tip, 1.0-1.3 mm long, 0.25-0.3 mm wide, 4-5 times as long as wide, margin denticulate, limbate on lower 1/3-2/3 vaginant laminae of perichaetial and most upper leaves, lower leaves limbate or elimbate; limbidia intramarginal, bordered by 1-2 rows of papillose, chlorophyllose cells, on perichaetial leaves marginal; vaginant laminae 1/2 or slightly more the leaf length, subequal; dorsal lamina truncate at insertion; costa ending 2-4 cells below the apex; laminal cells obscure, pluripapillose, 4-6 papillae per cell, hexagonal, 6-7.5 µm long and 4.5-6 µm wide; basal juxtacostal vaginant laminal cells to 10.5 µm long. Dioicous. Perigonia and perichaetia terminal, perichaetial leaves slightly longer than stem leaves. Sporophyte 1 per perichaetium, seta 2-3.5 mm, theca 0.7-0.8 mm long and 0.3 mm wide, peristome scariosus-type, teeth 45 µm wide at base, operculum rostrate, 0.5 mm long; spores 12-13.5 µm, papillose.

Distribution: Mexico, C America, West Indies, S America.

Ecology: Saxicolous and lignicolous, alt. 0-700 m; in the Guianas rare, not collected in Guyana.

Examined specimens: Suriname: area of Kabalebo Dam project, road km 212, rainforest, on vertical face of granite rock in cascade, Florschütz-de Waard & Zielman 5760 (L). French Guiana: 45 km SE of Saül, Sommet Tabulaire, Cremers 6806 (L); Mt. de l'Inini, zone centrale, versant E, bois tombé à terre, forêt sur pente au vent, Cremers 9061 (BR) (Cremers 9061 in L is *F. guianensis*).

Notes: This species is recognized by pluripapillose laminal cells, apex acute often ending in acumen with rounded tip, and intramarginal limbidia on the lower 1/2-2/3 of the vaginant laminae of some or all leaves. Limbidia on perichaetial leaves are often marginal. *F. neglectus* belongs to the *F. elegans*-complex (see note under *F. elegans*). *F. elegans* differs from *F. neglectus* by acute leaf apices ending in a clear apical cell.

17. **Fissidens oblongifolius** Hook.f. & Wilson, London J. Bot. 3: 547.
1844. Lectotype (Bruggeman *et al.* 1994): New Zealand, North
Island, Bay of Islands, Hooker 321B (BM).   – Fig. 29

Fissidens similiretis Sull., Proc. Amer. Acad. Arts 5: 274. 1861 [1862].
Type: Cuba, Sullivant Musci Cubensis 18 (holotype FH, isotype L).

Bright green plants, growing scattered or in loose mats, pinnate, to
10 mm long, 2.7 mm wide, branched. Leaves loosely imbricate, to 16
pairs, flat with incurved tips when dry, tips not completely flattening
when moistened, narrowly oblong with acute apex ending in acumen
with blunt, rounded tip, 1-2.5 mm long, 0.3-0.5 mm wide, 4-5 times
as long as wide, elimbate, margin crenulate to denticulate; vaginant
laminae 1/2 the leaf length, subequal; dorsal lamina at insertion wide
and rounded; costa oblongifolius-type, ending 2-4 cells below apex;
laminal cells smooth, strongly convex, 7.5-10.5 µm long and 6-9 µm
wide; basal cells of vaginant laminae to 18 µm long. Cladautoicous and
synoicous. Perigonia, synoicia and perichaetia terminal on plants or on
short branches from upper part of stem, perichaetial leaves 1.5-2.2 mm
long. Sporophyte 1 per perichaetium, seta 3-4.5 mm long, theca inclined
to curved, 0.6-0.8 mm long and 0.4-0.5 mm wide, peristome similiretis-
type, teeth 48-57 µm wide at base; operculum long rostrate, 0.8-0.9 mm
long; spores 13.5-15 µm, papillose.

Distribution: Pantropical; the Guianas.

Ecology: On trees, twigs and rocks, alt. 0-650 m; common.

Selected specimens: Guyana: Maruba region, Holder Falls, Ek *et
al.* 964 (L). Suriname: Bakhuis Mts., near camp 3, Florschütz & Maas
3068 (L); Nassau Mts., along creek with rapids, Lanjouw & Lindeman
2891 (L). French Guiana: near trail from St-Laurent to Apatou, 65 km
from St-Laurent, in rainforest, Cornelissen & ter Steege 276 (L).

Notes: This species is characterized by narrowly oblong, elimbate
leaves, acute leaf tips ending in a blunt, distally rounded acumen, smooth,
strongly convex laminal cells, costa ending 2-4 cells below apex and
inclined to curved thecae. It can be confused with *F. radicans* and *F.
dendrophilus* which also have elimbate leaves, smooth, convex laminal
cells and oblongifolius-type costae. Both have shorter, wider leaves (less
than 1.5 mm long, 3-4 times as long as wide).
F. oblongifolius belongs to subgenus <u>Pachyfissidens</u> section *Amblyothallia*
(see note under *F. radicans*).
Florschütz (1964) confused *F. oblongifolius* and *F. asplenioides* Hedw.

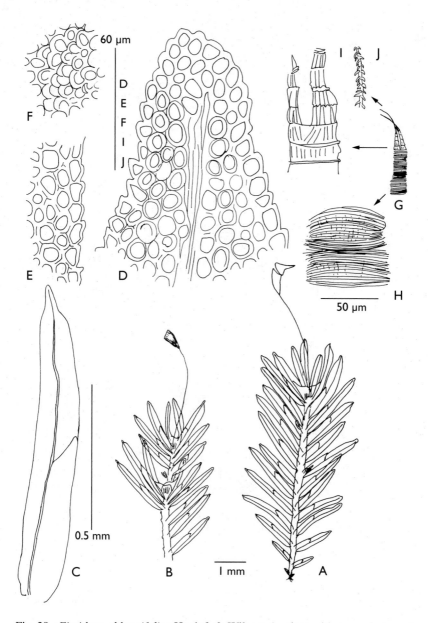

Fig. 29. *Fissidens oblongifolius* Hook.f. & Wilson: A, plant with sporophyte and axillary perigonial buds; B, upper part of female plant (some leaves removed) with sporophytic branch and axillary perigonial buds (part of the leaves removed); C, leaf; D, leaf apex; E, margin mid dorsal lamina; F, mid dorsal laminal cells; G, peristome tooth; H-J, details peristome tooth: H, basal undivided part exterior side; I, bifurcation exterior side; J, detail of filament (A-J, Lanjouw & Lindeman 2891).

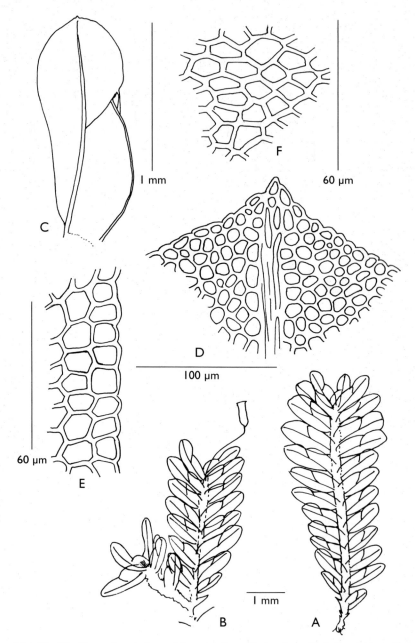

Fig. 30. *Fissidens obtusissimus* (Florsch.) Pursell: A, plant; B, 2 female branches (several leaves removed), right branch with sporophyte; C, leaf; D, leaf apex; E, margin mid dorsal lamina; F, mid dorsal laminal cells (A-B, D-F, Cayenne, sine collector; C, Florschütz 417).

The latter is characterized by strongly unequal vaginant laminae with the smaller ones rounded above and ending at or near the costa. Minor vaginant laminae of *F. oblongifolius* are acute distally and end near the leaf margin. From the synonyms listed by Florschütz (1964) *F. guadelupensis* Schimp. ex Besch. is indeed *F. asplenioides* Hedw. and *F. similiretis* is *F. oblongifolius* (Bruggeman-Nannenga & Berendsen 1987; Bruggeman-Nannenga *et al.* 1994).

18. **Fissidens obtusissimus** (Florsch.) Pursell, Bryologist 92: 524. 1989. – *Fissidens leptophyllus* Mont. fo. *obtusissimus* Florsch., Fl. Suriname 6: 52. 1964. Type: Suriname, Pasjensi, Tosokreek, on bark of small trees and roots, Florschütz 417, pp (holotype L).

– Fig. 30

Plants brownish green, growing scattered, pinnate, to 10 mm (with branches to 15 mm) long, 3 mm wide, fertile plants strongly branched. Leaves imbricate, to 21 pairs, ± flat when dry, lingulate with obtuse, rounded, frequently apiculate apex, 1.4-1.75 mm long, 0.45-0.55 mm wide, ± 3 times as long as wide, margin weakly crenulate, limbate on vaginant laminae of all leaves, limbidium often extending slightly onto apical lamina; vaginant laminae 2/3 the leaf length, subequal; dorsal lamina wide and rounded below, reaching the insertion; costa ending 1-2 cells below apex; laminal cells smooth, guttulate or not, 9-12 μm long and 4.5-9 μm wide; basal juxtacostal cells of vaginant laminae to 19.5 μm long. Polyoicous: cladautoicous, synoicous and dioicous. Solitary bud-shaped perigonia and clusters of perigonial, perichaetial and synoicous branches in axils of perichaetial and subperichaetial leaves, perichaetia and synoicia also terminal on stems, perichaetial leaves 1.4-2 mm long; calyptra unknown. Sporophyte 1-2 per perichaetium, setae 1.7-2 mm long, theca erect, 0.5-1 mm long and 0.15-0.4 mm wide, peristome scariosus-type, teeth 37.5-45 μm wide at base, operculum unknown; spores 16.5-18 μm in diam.

Distribution: Endemic in Suriname and French Guiana (Musci I in Fl. Suriname 6: 52. 1964 cites a specimen from Guyana: Richards 599, pp (NY, K) (not located)).

Ecology: On bark of small trees and roots, alt. 0-200 m; rare.

Selected specimens: Suriname: surroundings Pasjensi, Tosokreek, on bark of small tree and roots, Florschütz 417, pp (L). French Guiana: Cayenne, without collector (BM); Saül, ca. 10 km N of Eaux Claires, on moist tropical flood plain forest along small river, Buck 33088 (NY), 33124 (NY, with sporophytes).

Notes: This species is characterized by oblong-lingulate leaves with rounded-apiculate apex, limbidia extending the whole length of the vaginant laminae and smooth laminal cells. It resembles *F. leptophyllus* which has shorter limbidia extending at most half the length of the vaginant laminae. In Musci I in Fl. Suriname 6: 52. 1964 it was described as a form of *F. leptophyllus*.

19. **Fissidens ornatus** Herzog, Arch. Bot. São Paulo 1: 57. 1925. Type: Brazil, Hoehne 410 (holotype JE).   – Fig. 31

Plants brown, growing scattered, 2-4 mm long, 1-2 mm wide, basal part distantly foliate, upper part frondiform, unbranched, with or without persistent protonemata. Leaves distant on basal part of stem, imbricate in upper part, to 10 pairs, hardly altered when dry, elliptic-lanceolate to elliptical, upper leaves sometimes oblanceolate with acute mucronate, symmetrical or asymmetrical apex, 1.1-1.5 mm long, 0.3 mm wide, 3.5-5 times as long as wide; margin denticulate, limbate on basal 1/2 of vaginant laminae of most leaves; limbidium ill-defined to distinct, marginal or intramarginal; vaginant laminae 1/2 or less the leaf length, unequal; dorsal lamina narrow, mostly slightly rounded at insertion; costa brown, long excurrent; laminal cells firm-walled, guttulate, hexagonal to oblong, smooth, 15-37.5 μm long and 12-18 μm wide; 1-3 rows of marginal cells smaller, 7.5-15 μm long, forming a distinct border; basal juxtacostal cells of vaginant laminae 27-52.5 μm long. Dioicous. Perichaetia terminal, perichaetial leaves 1.7 mm long with emarginate vaginant laminae. Sporophyte and perigonia not seen.

Distribution: Neotropics; the Guianas.

Ecology: Scattered on clay, loam and weathered rock, alt. 200-1050 m; uncommon.

Selected specimens: Guyana: Region Potaro-Siparuni, Paramakatoi, ± 3 km SW of village, Newton *et al.* 3550 (L). Suriname: Brownsberg plateau, on clay, Florschütz 4799 (L); Brownsberg, on plateau, sparsely in mat of Calypogeia miquelii, Bekker s.n. (L); Florschütz 2491 (L, only prepared slide seen); Sipaliwini, Tafelberg National Park, Allen 20894 (MO). French Guiana: Cayenne, Mt. de Mahury, Cornelissen & ter Steege 233 (L).

Note: This species is characterized by large laminal cells, differentiated leaf margins of 1-3 rows of smaller cells and excurrent costae. *F. inaequalis* also has large laminal cells, but is distinct by costae ending below leaf apex.

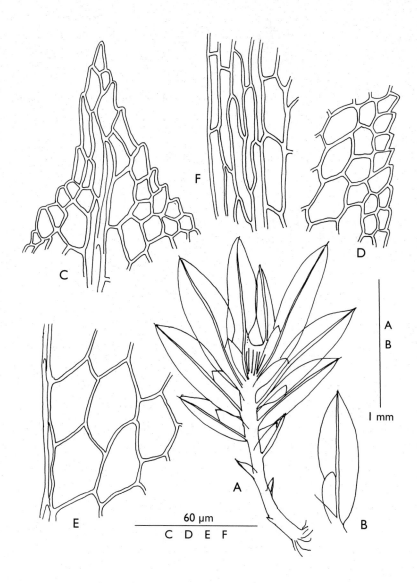

Fig. 31. *Fissidens ornatus* Herzog: A, archegonial plant; B, leaf; C, leaf apex; D, margin mid dorsal lamina; E, mid dorsal laminal cells (costa to the left); F, limbidium mid vaginant lamina (A-F, Allen 20894).

20.　**Fissidens pallidinervis** Mitt., J. Linn. Soc., Bot. 12: 592. 1869. Type: Peru, Andes, Tarapoto ad rivulum Marona-yacu, 2000 ped., Spruce 536 (holotype NY, isotypes BM, H, NY).　　　　 – Fig. 32

Plants growing scattered, pinnate, to 6 mm long with branches, 1.5 mm wide, unbranched or branched. Leaves distant to close, to 14 pairs, not to slightly inflexed when dry, oblong to oblong-lanceolate with acute to broadly acute often distally rounded apex, 0.7-1.2 mm long, 0.2-0.4 mm wide, 3-5 times as long as wide, margin crenulate to serrulate, limbate on basal 1/2-3/4 vaginant laminae of perichaetial and subtending leaves, infrequently weakly limbate on mid leaves of female stems and upper leaves of sterile stems; limbidium marginal or intramarginal; vaginant laminae 3/5 the leaf length, subequal; dorsal lamina slightly rounded to rounded-truncate below, reaching insertion; costa ending 2-11 cells below apex; laminal cells 1-4 papillose, 3-9 μm long and 3-7.5 μm wide; basal juxtacostal cells of vaginant laminae to 16.5 μm, mostly less than 10.5 μm long. Dioicous, cladautoicous and rhizautoicous. Perigonia terminal on plants, branches or dwarf plants at base of female plants; perichaetia terminal, perichaetial leaves to 1.6 mm long; calyptra cucullate, smooth. Sporophyte 1 per perichaetium, seta 2.5-3 mm long, theca 0.5 mm long and 0.25 mm wide, peristome scariosus-type, teeth 30 μm wide at the base, operculum rostrate; spores 10.5-12 μm, finely papillose.

D i s t r i b u t i o n : Pantropical; Suriname and French Guiana.

E c o l o g y : Mostly on rock, also soil and wood, alt. 200-500 m; uncommon.

S e l e c t e d　s p e c i m e n s : Suriname: Brownsberg Massif, 15 km SW of Brownsberg, Pursell 11859 (MO). French Guiana: Mt. Galbao, stream valley just below fork between NW and SE peak, rocky ravine with small waterfall over granitic rocks, Buck 25719 (NY); Canton de Approuague-Kaw, Pic Matécho, narrow peninsula on S side of peak, Buck 38084 (NY).

N o t e s : This species is characterized by pluripapillose laminal cells, costae ending 2-11 cells below the leaf apex and limbidia restricted to lower 3/4 or less of perichaetial and subtending leaves. Most Guianan specimens have relatively long costae ending 2-4 cells below the apex. Such specimens can be confused with *F. guianensis* which, however, has limbidia on the lower part of the vaginant laminae of all leaves and grows on wood. *F. pallidinervis* may occur on wood, but is also found on other substrates. Expressions of *F. pallidinervis* with short costae, ending 8-11 cells below leaf apex, are unlikely to be confused with any other species known from the Guianas.
This species belongs to the *F. elegans*-complex.

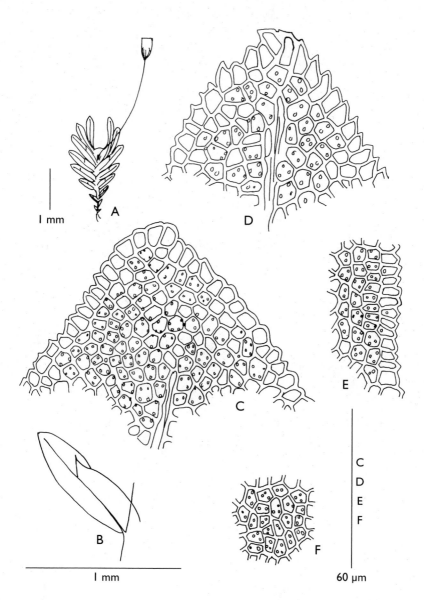

Fig. 32. *Fissidens pallidinervis* Mitt.: A, plant with sporophyte; B, leaf; C, leaf apex with short costa; D, leaf apex with long costa; E, margin mid dorsal lamina; F, mid dorsal laminal cells (A-C, E-F, Pursell 11859; D, Buck 38084).

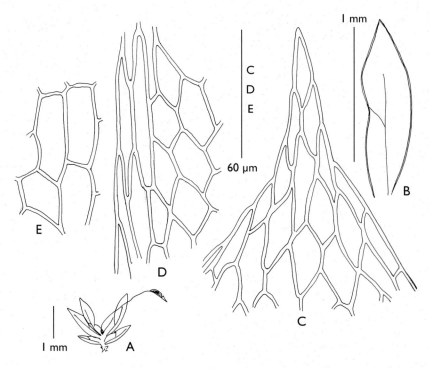

Fig. 33. *Fissidens palmatus* Hedw.: A, plant with sporophyte; B, leaf; C, leaf apex; D, margin mid dorsal lamina; E, mid dorsal laminal cells (A-E, Florschütz-de Waard & Zielman 5064).

21. **Fissidens palmatus** Hedw., Sp. Musc. Frond. 154. 1801. Type: Jamaica, loco terrestri, Swartz s.n. (holotype G, isotypes H, S).
– Fig. 33

Conomitrium reticulosum Müll. Hal., Syn. Musc. Frond. 2: 525. 1851. – *Fissidens reticulosus* (Müll. Hal.) Mitt., J. Linn. Soc., Bot. 12: 603. 1869. Type: Mexico, Liebmann s.n. (NY).

Plants light green, growing scattered, often frondiform, to 3 mm long, 2.5 mm wide, simple. Leaves imbricate, 3-4 pairs, crispate when dry, oblanceolate, acuminate ending in slender, sharp terminal cell, 1-2.2 mm long, 0.4-0.6 mm wide, 3.5-5 times as long as wide, limbate; limbidium 1-2-stratose, reaching insertion of dorsal lamina, often confluent at apex of vaginant laminae, becoming lax near leaf apex and insertions of vaginant laminae, infrequently confluent at leaf apex; vaginant laminae 1/2 the leaf length, equal to unequal; dorsal lamina gradually narrowed towards insertion; costa ending 8-16 cells below apex; laminal cells smooth, hexagonal, flat to weakly convex, thin-walled, 22.5-49.5 long and 10.5-37.5 μm wide, 1.5-4 times as long as wide; limbidium cells in mid dorsal lamina 67.5-97.5 μm long and 4.5-6 μm wide; basal juxtacostal vaginant laminal cells and basal dorsal laminal cells 60-105 μm long and 15 μm wide. Dioicous and rhizautoicous. Perigonia terminal on 1 mm tall plants or on bud-shaped plants at base of female plants; perichaetia terminal, perichaetial leaves to 2.3 mm; calyptra mitrate, smooth. Sporophyte 1 per perichaetium, seta 5 mm long, theca inclined to curved, 0.5 mm long and 0.4 mm wide, peristome scariosus-type, teeth 37.5 μm wide at base, operculum rostrate, 0.4 mm long; spores 25 μm.

Distribution: Neotropics; Suriname and French Guiana.

Ecology: Terrestrial, on clay and loam, alt. 0-200 m; uncommon.

Selected specimens: Suriname: area of Kabalebo Dam Project, near camp road 23, rainforest, riverside, on loamy soil, Florschütz-de Waard & Zielman 5064 (L); Bakhuis Mts., 20 km downstream of Kabalebo airstrip on east-bank, track, on clay, Florschütz & Maas 2540 (L). French Guiana: Saül, Buck 18517 (NY).

Notes: This species is characterized by frondiform plants, limbate leaves and large laminal cells. The length of the costa varies from ending as many as 25 cells below apex to percurrent (Pursell 2007). Costae of Guianan specimens end 6-16 cells below the apex.
F. palmatus can be confused with other species with large cells and limbate leaves. *F. amazonicus* has pinnate plants, elimbate to weakly limbate leaves and longer, narrower laminal cells. *F. flaccidus* has thicker, 3-4-stratose, limbidia.

22. **Fissidens pellucidus** Hornsch., Linnaea 15: 146. 1846. Neotype (Pursell 1994): Brazil, Sta. Catarina, Itajahy, Erdboden im Walde, Ule 67 (H).

KEY TO THE GUIANAN VARIETIES

1   Laminal cells smooth, flat or slightly convex. . . . .
. . . . 22a. *F. pellucidus* var. *pellucidus*
Laminal cells prorate . . . . 22b. *F. pellucidus* var. *papilliferus*

22a. var. **pellucidus**   – Fig. 34 A-F

Fissidens flexinervis Mitt., J. Linn. Soc., Bot. 12: 588. 1869. – *Fissidens prionodes* Mont. fo. *flexinervis* (Mitt.) Florsch., Fl. Suriname 6: 61. 1964. Type: Brazil, Spruce 493 (holotype NY).

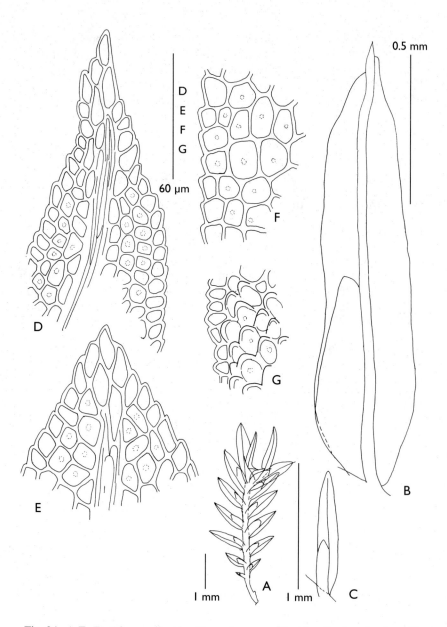

Fig. 34. A-F. *Fissidens pellucidus* Hornsch. var. *pellucidus*: A, female plant; B, C, leaves; D, E, leaf apices; F, margin mid dorsal lamina (A, B, Allen 20811; C, E, Florschütz 4886; D, Richards 323; F, Florschütz-de Waard & Zielman 5086). G. *Fissidens pellucidus* Hornsch. var. *papilliferus* (Broth.) Pursell: G, margin mid dorsal lamina (G, Lanjouw & Lindeman 803).

Fissidens saülensis Pursell & W.R. Buck, Brittonia 48: 27. 1996. Type: French Guiana: St-Laurent du Maroni, Canton de Maripasoula, Commune de Saül, ca. 10 km ENE of Eaux Claires, along upper tributary of Arataye R., ca. 400 m, non flooded moist forest on hill above stream, on bare soil, growing with Fissidens prionodes Mont., Buck 25534 (holotype NY, isotypes PAC, CAY), syn. nov.

Plants pale green to reddish, growing in dense mats or scattered, pinnate, to 6 mm long, 1.2-2.2 mm wide, usually simple, infrequently with persistent protonemata; rhizoids smooth or rarely papillose. Leaves distant to imbricate, to 14 pairs, little changed when dry, lanceolate to elliptical, acute, 0.8-1.3 mm long, 0.15-0.4 mm wide, 3-5.5 times as long as wide, margin crenulate to serrulate, elimbate or weakly limbate on proximal parts of vaginant laminae of some or all leaves; limbidia marginal to intramarginal; vaginant laminae 1/2 the leaf length, unequal, minor lamina ending on leaf or costa; dorsal lamina straight to slightly rounded below, mostly reaching insertion; costa often reddish, percurrent to long excurrent or ending just below leaf apex; laminal cells smooth, hexagonal, firm-walled, flat to low convex, guttulate or not, 9-16.5(-27) µm long and 4-15 µm wide, marginal cells often smaller, 6-9 µm long and 4-6 µm wide; basal juxtacostal cells frequently larger, to 30 µm long. Dioicous. Perigonial plants very small to long, perichaetia and perigonia terminal, perichaetial leaves longer than stem leaves, with emarginate vaginant laminae; calyptra smooth to slightly prorate. Sporophyte 1-2 per perichaetium, seta 2-3.5 mm long, theca erect to slightly inclined, 0.4-0.6 mm long and 0.2-0.25 mm wide, peristome scariosus-type, teeth 25-37 µm wide at base, operculum rostrate, 0.25-0.5 mm long; spores 7.5-10.5 µm long, smooth to finely papillose.

Distribution: Pantropical; the Guianas.

Ecology: On clay, sand, loam, old termit nest, rock, bark, roots and tree bases, infrequently in periodically flooded places, alt. 0-705 m; one of the most common *Fissidens* species in the Guianas.

Selected specimens: Guyana: Mabura Hill, Mora forest, along creek, terrestrial, Florschütz-de Waard 6014 (L); Upper Mazuruni Distr., at confluence of Arobaru and Kako Rs. on steep riverbank, periodically flooded, Gradstein 4967 (L). Suriname: Snake Cr. (Marataka), on bark of living tree, Maas *et al.* F3277 (L); Brownsberg, on clay, Florschütz 4486 (L); area of Kabalebo Dam project, rainforest, Florschütz-de Waard & Zielman 5086 (L). French Guiana: Saül, 2 km SW of village, "Sentier Limonade", lowland moist forest on lateritic soil, epiphytic, Montfoort & Ek 512 (L); Mts. de Kaw, Kaw caves, on rock, Florschütz-de Waard 6173A (L).

Notes: *F. pellucidus* is a common, variable species recognized by clear, 9-16.5(-27) μm long, firm walled, often guttulate laminal cells, lanceolate to elliptical leaves that are elimbate or have short, weak limbidia on the vaginant laminae of some or all leaves, terminal perigonia and perichaetia and perichaetial leaves with emarginate vaginant laminae. Costae can end below leaf apex, be percurrent or long excurrent. Leaf apices with long excurrent costae are often asymmetrical.

F. saülensis is here subsumed under *F. pellucidus* var. *pellucidus*. The two are not sharply distinct and there is no strong correlation between the features that are considered diagnostic for *F. saülensis* (asymmetric leaf apices, excurrent costae, limbidia on proximal parts of vaginant laminae of many leaves and persistent protonemata).

F. prionodes Mont. fo. *hornschuchii* sensu Musci I in Fl. Suriname 6: 62. 1964 is *F. pellucidus*. See note under *F. hornschuchii*.

22b. var. **papilliferus** (Broth.) Pursell, Bryologist 97: 262. 1994. – *Fissidens papilliferus* Broth., Hedwigia 45: 266. 1906. Lectotype (Pursell 1994C): Brazil, Amazonas, Rio Madeira, an Baumstämmen am Marmellos, Ule 2337 (H). – Fig. 34 G

Similar to var. *pellucidus* but laminal cells prorate.

Distribution: Bolivia, northwestern Brazil, Venezuela and Suriname.

Ecology: On soil, elevation not indicated; rare.

Examined specimen: Suriname: Grote Zwiebelzwamp, E of km 12.6, on soil, Lanjouw & Lindeman 803 (L).

Note: This is the only taxon with prorate laminal cells known from the Guianas. It can be confused with *F. prionodes* expressions with eccentric papillae. *F. prionodes* has narrower, linear-lanceolate, leaves that are often irregularly bistratose.

In Musci I in Fl. Suriname 6: 61. 1964 *F. papilliferus* was listed as synonym of *F. prionodes* fo. *flexinervis* (= *F. pellucidus* var. *pellucidus*).

23. **Fissidens perfalcatus** Broth., Bih. Kongl. Svenska Vetensk.-Akad. Handl. 26 Afd. 3(7): 13. 1900. Type: Brazil, Matto Grosso, ad ligna et truncos marginis silvatici amnis "Sangrador" prope Cuyabá, Lindman 404 (holotype H, isotype S). – Fig. 35

Plants bright green, growing in loose mats, pinnate, to 4 mm long, 1.2 mm wide, mostly simple. Leaves distant to imbricate, to 9 pairs, crispate when dry, falcate-ovate, acute to obtuse, ending in a sharp cell,

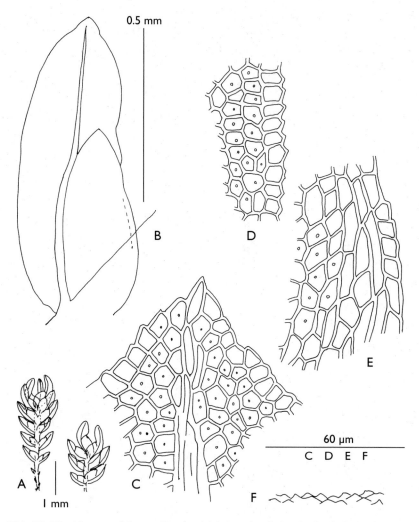

Fig. 35. *Fissidens perfalcatus* Broth.: A, 2 female plants; B, leaf; C, leaf apex; D, margin mid dorsal lamina; E, vaginant lamina of upper leaf of female stem with weak intralaminal limbidium; F, side view of dorsal laminal cells (A-F, Allen 25034).

0.6-0.8 mm long, 0.25-0.3 mm wide, 2.5 times as long as wide, margin crenulate to denticulate, limbate on 2/3 or less the vaginant laminae of perichaetial and subtending leaf pairs, lower leaves elimbate or with a few limbidial cells; limbidia ending well above insertion of vaginant laminae; vaginant laminae 2/3 the leaf length, subequal; dorsal lamina tapering below, reaching insertion; costa ending 2-4 cells below apex; laminal cells unipapillose, a few bipapillose, 4.5-7 µm long and 4.5-6 µm

92

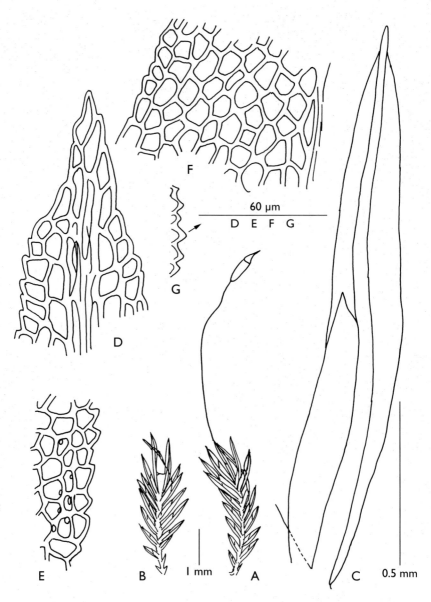

Fig. 36. *Fissidens prionodes* Mont.: A, plant with sporophyte; B, male plant; C, leaf; D, leaf apex; E, margin mid dorsal lamina; F, detail mid dorsal lamina (costa on the right, margin on the left); G, laminal cell in side view (A, B, Leprieur 5; C-F, Cornelissen *et al.* C 094).

wide; basal juxtacostal cells often larger, to 16.5 μm long. Dioicous. Perigonial plants short, perichaetia and perigonia terminal, perichaetial leaves to 1.25 mm long. Calyptra and sporophyte not seen.

Distribution: NW Brazil, Venezuela and Suriname.

Ecology: On sandstone, alt. 227 m; rare.

Examined specimen: Suriname: Sipaliwini, Kayserberg Airstrip area, on sandstone boulders, Allen 25034 (MO, herb Bruggeman-Nannenga).

Notes: Typical *F. perfalcatus* is characterized by sharply unipapillose laminal cells, wide, recurved, falcate-ovate leaves, wide, ± 3/4 limbate (completely limbate in perichaetial leaves) vaginant laminae (Pursell 2007). The Guiana-specimen has lower, less sharp papillae and weaker limbidia restricted to the upper and middle leaves of female plants. This expression is unlikely to be confused with any other species in the area.

24. **Fissidens prionodes** Mont., Ann. Sci. Nat., Bot. sér. 2, 3: 200. 1835.
  – *Fissidens prionodes* Mont. fo. *prionodes* (Mont.) Florsch., Fl. Suriname 6: 59. 1964. Type: French Guiana, Leprieur 5 (holotype PC, isotypes BM, L). – Fig. 36

Fissidens marmellensis Broth., Hedwigia 45: 267. 1906. Lectotype (Florschütz 1964): Brazil, Amazonas, im Walde an den Wasserfällen des Marmellos, Rio Madeira, III 1902, Ule 2334 (H, BM).

Plants reddish brown, young ones pale green, growing in mats, pinnate, 3-5.5 mm long, 1.4 mm wide, simple. Leaves imbricate, to 16 pairs, hardly altered when dry, often irregularly bistratose, linear-lanceolate with narrowly acute, infrequently asymmetrical apex, 1.1-1.4 mm long, 0.2-0.25 mm wide, 5-7.5 times as long as wide; margin denticulate to crenulate, rarely weakly limbate on proximal parts of vaginant laminae of female plants; vaginant laminae 1/2 the leaf length, unequal; dorsal lamina tapering, straight or rounded at insertion, ending at or just above insertion; costa reddish, in mid leaf to 35 μm wide, percurrent to long excurrent; laminal cells roundish-hexagonal, strongly unipapillose, papillae often eccentric, 6-15 μm long and 6-10.5 μm wide; basal cells of vaginant laminae smooth, oblong, to 30 μm long. Rhizautoicous. Perigonia terminal on small bud-shaped or longer plants at base of female stems; perichaetia terminal, perichaetial leaves to 1.7 mm long; calyptra smooth. Sporophyte 1-2 per perichaetium, seta 5-6 mm long, theca 0.55 mm long and 0.35 mm wide, peristome scariosus-type, teeth 30 μm wide at the base, operculum not seen; spores 7.5-10.5 μm.

Distribution: Colombia, northwestern Brazil and the Guianas.

Ecology: On clay, rotten tree, soil, tree base, alt. 0-600 m; common.

Selected specimens: Guyana: Moraballi Cr. near Bartica, on clayey soil and on very rotten tree stump, Richards 258 (BM); Mabura hill, 180 km SSE of Georgetown, Cornelissen *et al*. 94 (L). Suriname: Brownsberg, Plateau-road at km 17, on very dark clay-bank, Florschütz 4663 (L); Lely Mts., Florschütz 4839 (L). French Guiana: Montsinery, ca. 20 km W of Cayenne, on clay heap in swamp forest, Gradstein 5783 (L).

Notes: *F. prionodes* is characterized by red-brown plants, elimbate, linear-lanceolate, often in part bistratose leaves with narrow, acute tips, percurrent to long excurrent costae and unipapillose laminal cells. Leaves are typically elimbate, but short, weak limbidia may be present on the proximal parts of the vaginant laminae of some or all leaves of female stems. This species resembles *F. allionii*. For differences see under the latter species.

25. **Fissidens radicans** Mont., Ann. Sci. Nat., Bot. sér. 2, 14: 345. 1840.
     Type: French Guyana, Leprieur 310 (PC).   – Fig. 37

Plants yellowish green, growing scattered, pinnate, to 10 mm long, 2 mm wide, mostly branched. Leaves imbricate, to 25 pairs, ± flat with inrolled leaf tips when dry, caducous, oblong with broadly acute to rounded-obtuse apex, 1.2 mm long, 0.3-0.4 mm wide, 3-4 times as long as wide, margin crenulate, elimbate; vaginant laminae ± 1/2 the leaf length, subequal; dorsal lamina narrow and straight to wide and truncate at base; costa oblongifolius-type, ending 6-9 cells below leaf apex; laminal cells smooth, strongly convex, 7.5-12 μm long and 6-9 μm wide; basal juxtacostal cells of vaginant laminae to 16.5 μm long. Cladautoicous. Perichaetia, synoicia and perigonia terminal on plants or short or long branches, perichaetial leaves 1.5 mm; calyptra mitrate. Sporophyte 1 per perichaetium, seta 3 mm long, theca cylindrical, 0.7-0.9 mm long and 0.3 mm wide, peristome ± similiretis-type, teeth undivided or irregularly divided, 35 μm wide at base, operculum 0.3-0.5 mm long; spores 16-19.5 μm, finely papillose.

Distribution: Neotropics, widespread in the Guianas.

Ecology: On living trees, dead poles, rotting wood, rarely on concrete, alt. 0-800 m; common.

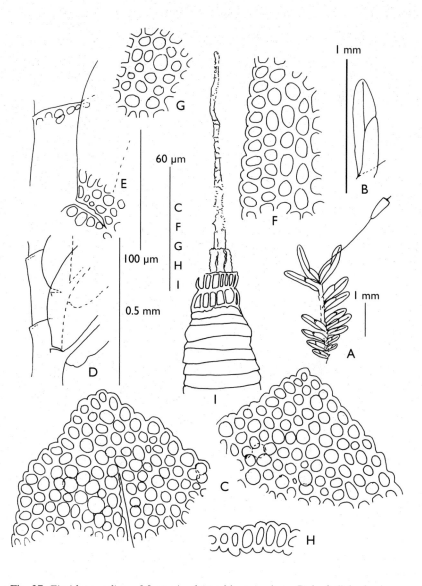

Fig. 37. *Fissidens radicans* Mont.: A, plant with sporophyte; B, leaf; C, leaf apices; D, part of stem with 2 leaf scars (left) and 2 leaves partly detached (right); E, detail of basal part of D; F, margin mid dorsal lamina; G, mid dorsal laminal cells; H, cross-section of dorsal lamina; I, peristome tooth (A-G, Aptroot 12546; H, I, Lanjouw & Lindeman 2288).

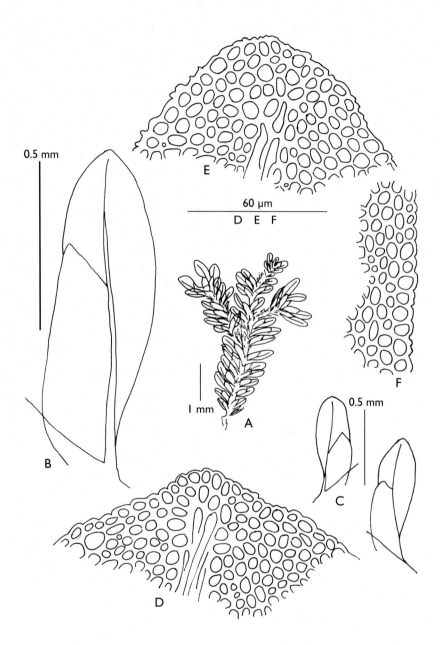

Fig. 38. *Fissidens ramicola* Broth.: A, plant with archegonial branches; B, C, leaves; D, E, leaf apices; F, margin mid dorsal lamina (A-F, Florschütz 2073).

Selected specimens: Guyana: N Rupununi Savanna, dry savanna forest, on bark of living tree, Florschütz-de Waard 6042 (L); Kaieteur Falls National Park, Korume Cr., Newton & Florschütz-de Waard 3367 (L). Suriname: area of Kabalebo Dam project, near camp road km 212, epiphytic, Florschütz-de Waard & Zielman 5664 (L); Nassau Mts., in forest on slope, epiphytic, Lanjouw & Lindeman 2288 (L). French Guiana: Saül, primary forest near Boeuf Mort, on tree, Aptroot 15246 (L); Mts. de Kaw, 40 km SE of Cayenne, on dead pole, Gradstein 5913 (L).

Notes: This species is recognized by caducous, oblong, elimbate leaves, smooth, strongly convex laminal cells, oblongifolius-type costae ending 6-9 cells below leaf apex and similiretis-type peristomes with undivided or imperfectly divided teeth.
It resembles *F. dendrophilus* which differs by non-caducous leaves and divided peristome teeth. *F. guianensis* superficially resembles *F. radicans* but has pluripapillose laminal cells, short limbidia on all or most leaves, bryoides- type costae that end 2-4 cells below apex and scariosus-type peristomes.
Subgenus Pachyfissidens section *Amblyothallia* is represented by 3 species in the Guianas, *F. radicans, F. dendrophilus* and *F. oblongifolius*. This section is characterized by elimbate leaves, oblongifolius-type costae, smooth, convex or flat laminal cells, peristomes similiretis-type (infrequently taxifolius-type, not in the Guianas) and thecae with 40 or more files of exothecial cells.

26. **Fissidens ramicola** Broth., Hedwigia 45: 268. 1906. Lectotype (Pursell 1986): Brazil, Amazonas, Rio Juruá, Juruá Miry, an Zweigen, Ule 2263 (H).   – Fig. 38

Plants brownish green, growing in loose mats, pinnate, 3.5-5 mm long, 1.7 mm wide, simple or with short, fertile branches near the apex. Leaves imbricate, to 18 pairs, slightly crisped when dry, oblong to lingulate with rounded-obtuse, slightly inflexed apex, 0.6-1.1 mm long, 0.2-0.4 mm wide, 3 times as long as wide; margin subentire to slightly crenulate, limbate on basal 1/2 of vaginant laminae of perichaetial and subperichaetial leaves; vaginant laminae 2/3 the leaf length, subequal; dorsal laminae ending far above insertion or narrow (1-3 cells wide) in the basal part and reaching insertion; costa ending 2-4 cells below apex; laminal cells, smooth or finely pluripapillose, irregularly hexagonal to quadrate, in mid 6-7.5 µm long and 4.5-7.5 µm wide; basal juxtacostal cells of vaginant laminae to 13.5 µm long. Gonioautoicous and cladautoicous. Perigonia terminal or bud shaped in axils of upper leaves of female plants; perichaetia terminal on plants or short branches in upper leaf axils, perichaetial branches numerous, perichaetial leaves longer than stem leaves, to 1.2 mm long, with rounded apex; calyptra smooth. Sporophyte 1 per perichaetium, seta 1.5 mm long, theca 0.6 mm long and 0.4 mm wide, peristome scariosus-type, teeth 32.5 µm wide at base, operculum rostrate, 0.35 mm long; spores 18-21 µm, papillose.

Distribution: Mexico, Brazil, Suriname and Guyana.

Ecology: On liana, elevation not indicated; rare.

Specimens examined: Guyana: S Rupununi Savanna near Dadanawa, moist savanna bush on bank of dry creek near Rupununi R., Florschütz-de Waard 6020 (L). Suriname: Lumberman's camp Bartika at Kabalebo R., on liana, Florschütz 2073 (BR, L).

Notes: This species is characterized by oblong to lingulate leaves with rounded-obtuse tips, dorsal laminae often ending well above insertion, laminal cells finely pluripapillose or smooth and limbidia restricted to basal half of the vaginant laminae of upper leaves of female stems. It can be confused with other species with lingulate leaves and dorsal laminae that end well above the insertion. *F. brevipes* Besch. is distinct by costae that are distally obscured by chlorophyllose cells, whereas *F. subramicola* Broth. has smooth to weakly unipapillose laminal cells and ca. 50 (instead of ca. 32) files of exothecial cells on the theca.
F. brevipes Besch. in Musci I in Fl. Suriname 6: 55. 1964 is a mixture of *F. ramicola* (Florschütz 2073) and *F. subramicola* (Florschütz 1105A).

27. **Fissidens scariosus** Mitt., J. Linn. Soc., Bot. 12: 599. 1869. Type: Peru, Spruce 537 (holotype NY). – Fig. 39

Plants bright green or brownish, growing scattered, pinnate, to 5 mm long, 2.2 mm wide, simple. Leaves imbricate, to 9 pairs, little changed when dry, elliptical to elliptic-lanceolate with acute-acuminate apex, 1.7-1.9 mm long, 0.3-0.4 mm wide, 5-6 times as long as wide, weakly limbate; limbidia 1-2 cells wide on the apical and dorsal laminae, to 6 cells wide on the vaginant laminae; vaginant laminae 1/2-3/5 the leaf length, subequal; dorsal lamina narrow at insertion, weakly rounded; costa ending ± 7 cells below apex, branched distally or not; laminal cells smooth, prosenchymatous, 79.5-165 µm long and 7.5-12 µm wide; marginal cells 1.5-7.5 µm wide in mid dorsal lamina; basal juxtacostal cells of vaginant laminae to 72 µm long and 18 µm wide. Rhizautoicous. Perigonia terminal on small plants at base of female plants; perichaetia

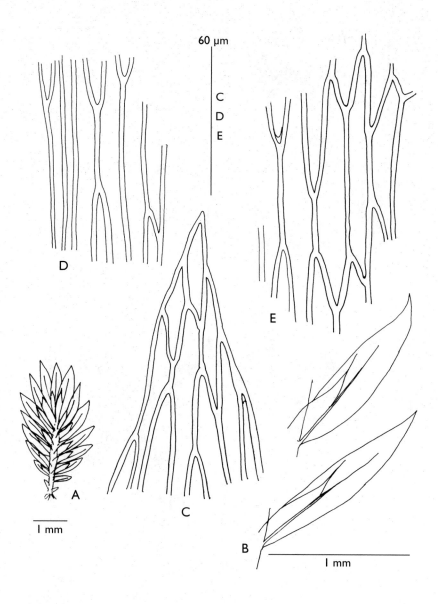

60 µm

C
D
E

D

A

1 mm

C

E

B

1 mm

Fig. 39. *Fissidens scariosus* Mitt.. A, plant; B, leaves; C, leaf apex; D, margin mid dorsal lamina (limbidium on the left); E, mid dorsal laminal cells (A-E, Onraedt 8952).

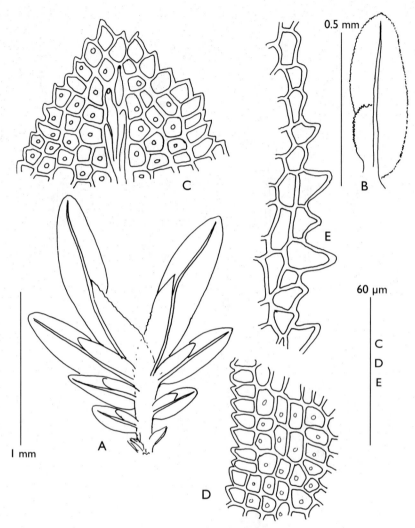

Fig. 40. *Fissidens serratus* Müll. Hal.: A, plant; B, leaf; C, leaf apex; D, margin mid dorsal lamina; E, margin vaginant lamina (A-E, Gradstein 4938).

terminal, perichaetial leaves to 2 mm long; calyptra not seen. Sporophyte 1 per perichaetium, seta 3.5 mm long, theca 0.7 mm long and 0.3 mm wide, peristome scariosus-type, teeth 33-39 μm wide at base, operculum rostrate, 0.4 mm long; spores 9 μm.

Distribution: Neotropics; the Guianas.

Ecology: On clay, soil and rocks, alt. 500-810 m; rare.

Selected specimens: Guyana: Paramakatoi, Karibon Cr., trail NW of village, on soil, Newton & Florschütz-de Waard 3571 (L). Suriname: Lely Mts., path from Plateau I to Plateau II, on vertical clay, Florschütz 4843 (L); Brownsberg, on plateau, rainforest, on vertical clay wall at base of fallen tree, Florschütz-de Waard & Zielman 5026 (L). French Guyana: Mt. de l'Inini, extrémité NW, Onraedt 8952 (L; BR as Cremers 8952).

Note: This species is characterized by prosenchymatous laminal cells that are 7-20 times as long as wide, weak limbidia on all laminae and subequal vaginant laminae (in lower leaves often equal). *F. scariosus* resembles *F. amazonicus* which is distinct by wider laminal cells, 3.5-6 times as long as wide, equal vaginant laminae and shorter costae.

28. **Fissidens serratus** Müll. Hal., Bot. Zeitung (Berlin) 5: 804. 1847. Neotype (Bruggeman-Nannenga & Pursell 1995): Indonesia, Java, Tjibodjas, Fleischer s.n. (FH). – Fig. 40

Conomitrium puiggari Geh. & Hampe in Hampe, Vidensk. Meddel. Naturhist. Foren. Kjøbenhavn, ser. 4, 1: 161. 1879. – *Fissidens puiggari* (Geh. & Hampe) Paris, Index Bryol. 482. 1896. – *Fissidens prionodes* Mont. fo. *puiggari* (Geh. & Hampe) Florsch., Fl. Suriname 6: 63. 1964. Lectotype (Pursell 1994): Brazil, São Paulo, Apiahy, Puiggari 407 (BM, NY).

Plants small, green, growing in tufts, typically frondiform, 1.5-2 mm tall, 1.5 mm wide, unbranched. Leaves imbricate, 3-4 pairs, weakly curled in from tip when dry, oblong to elliptic-oblong with broadly acute apex, middle leaves 0.6 mm long, 0.2 mm wide, 3 times as long as wide, elimbate, margin serrate, on vaginant laminae often coarsely and irregularly serrate; vaginant laminae 1/2-3/5 the leaf length, strongly unequal, minor one ending half way between margin and costa or on costa; dorsal lamina somewhat rounded or straight at insertion; costa ending 1-3 cells below leaf apex, distally smooth or papillose; laminal cells unipapillose, 6-12 μm long and 6-9 μm wide; basal juxtacostal cells of vaginant laminae to 21 μm long. Dioicous (rhizautoicous?). Perigonia terminal on short plants; perichaetia terminal, perichaetial leaves to 1.4 mm long; calyptra scabrose, campanulate. Sporophyte 1 per perichaetium, seta 2.5-3 mm, theca 0.4 mm long and 0.25 mm wide, peristome scariosus-type, operculum not seen; spores ± 12 μm, papillose.

Distribution: Pantropical; Guyana and Suriname.

Ecology: On rotting stem and tree trunks, alt. 0-650 m; rare.

Examined specimens: Guyana: Upper Mazaruni Distr., Jawalla, at confluence of Kukui and Mazaruni Rs., on trunk base, Gradstein 4928 (L). Suriname: Lely Mts., Plateau V, savanna forest, on rotting stem, Florschütz 4850 (L).

Notes: *F. serratus* is recognized by its small, mostly frondiform plants, elimbate leaves, unipapillose laminal cells and leaf margins that are evenly serrate on dorsal and apical lamina, and irregularly serrate on the vaginant laminae.

29. **Fissidens submarginatus** Bruch in C. Krauss, Flora 29: 133. 1846. Lectotype (Pursell 1994): S Africa, Natal, in faucibus sylvarum nataliensium prope litus, Krauss s.n. (BM, PC). – Fig. 41

Fissidens intermedius Müll. Hal., Linnaea 21: 181. 1848. Lectotype (Florschütz 1964): Suriname, sine loco, Kegel 501 (GOET).

Plants usually dark-green, growing in dense mats, pinnate, 4 mm long, 1.7 mm wide, simple. Leaves imbricate, to 10 pairs, slightly crispate when dry, lanceolate with acute-acuminate apex, 1.0-1.2 mm long, 0.3-0.4 mm wide, 2.5-3 times as long as wide, limbate on vaginant laminae; limbidum extending the complete length of the vaginant laminae and often slightly onto the apical lamina, infrequently a few limbidial cells on mid dorsal lamina; vaginant laminae 3/5 the leaf length, equal; dorsal lamina wide, slightly rounded to rounded-truncate at insertion; costa percurrent to long excurrent; laminal cells sharply unipapillose, 6-7 µm long and 4.5-6 µm wide; basal juxtacostal cells of vaginant laminae oblong, to 28.5 µm long. Gonioautoicous. Perigonia bud-shaped, axillary; perichaetia terminal, perichaetial leaves 1.2 mm long; calyptra smooth. Sporophyte 1 per perichaetium, seta 2.5-3 mm long, theca 0.5 mm long and 0.2 mm wide, peristome scariosus-type, teeth 30-36 µm wide at base, operculum rostrate, 0.3-0.4 mm long; spores 10.5-12 µm long.

Distribution: Pantropical; the Guianas.

Ecology: On clayey soil, once collected on a termite mound and once on sandy soil, no altitudes indicated; common in Suriname, rare in Guyana and French Guiana.

Selected specimens: Guyana: S Rupununi savanna, Mt. Tawatawun, granitic outcrop 5 km N of Dadanawa Ranch, Florschütz 6029 (L). Suriname: Bakhuis Mts., track along Kabalebo and Zandkreek, on clay, Florschütz & Maas 2490 (L); bank of Kabalebo R. near Doksen Savanna, Florschütz 2255 (L). French Guiana: Iles de Salut, Ile Royale, sur sol argileux en sous-bois, Cremers 8641 (BR, L).

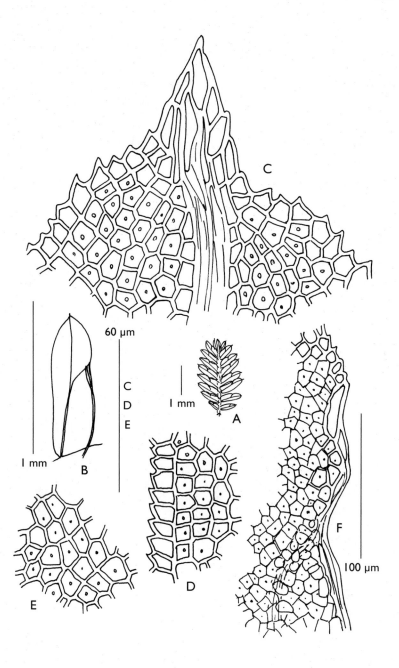

Fig. 41. *Fissidens submarginatus* Bruch: A, plant; B, leaf; C, leaf apex; D, margin mid dorsal lamina; E, mid dorsal laminal cells; F, upper part vaginant laminae (A-F, Florschütz & Maas 2490).

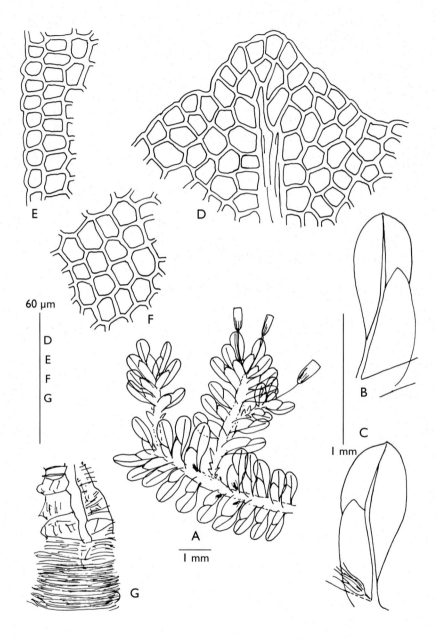

Fig. 42. *Fissidens subramicola* Broth.: A, plant with sporophytes and several axillary perigonia; B, leaf; C, leaf with axillary perigonium; D, leaf apex; E, margin mid dorsal lamina; F, mid dorsal laminal cells; G, peristome tooth at bifurcation (A-F, Florschütz de Waard 6020; G, Cremers 5364).

Notes: This species is easily recognized by its completely limbate vaginant laminae, per- to excurrent costae and sharply unipapillose laminal cells. The limbidia frequently extend slightly onto the apical laminae.

30. **Fissidens subramicola** Broth., Hedwigia 45: 268. 1906. Type: Brazil: Estado de Amazonas, Rio Juruá, Juruá Miry, auf Stacheln einer Bactris, Ule 2275 (holotype H). – Fig. 42

Fissidens austro-americanus Pursell & W.D. Reese, Brittonia 37: 335. 1985. Type: Suriname, Saramacca-oever, tussen Kwakoegron en Doorsnee, op liaan in zwaar oerbos, Florschütz 1105A (holotype L).

Plants brownish green, pinnate, to 20 mm long, 2.7 mm wide, in upper part often with many short branches. Leaves imbricate, to 30 pairs, hardly changed when dry, broadly lingulate, apex obtuse often with low wide apiculus, 1.5-1.7 mm long, 0.5-0.7 mm wide, 2.5-3 times as long as wide, margin crenulate, indistinctly limbate on basal half of the vaginant laminae of perichaetial leaves; vaginant laminae 2/3 the leaf length, unequal; dorsal lamina tapering towards insertion, reaching insertion or not; costa ending 2-4 cells below leaf apex; laminal cells smooth or lowly unipapillose, irregularly hexagonal, 7.5-13.5 μm long and 7.5-9 μm wide; basal juxtacostal cells of vaginant laminae to 30 μm long. Gonioautoicous. Perigonia bud-shaped, axillary; perichaetia terminal on plants and branches, perichaetial leaves not differentiated; calyptra smooth, cucullate. Sporophyte 1-2 per perichaetium, seta 1-1.5 mm long, theca 0.7 mm long and 0.25 mm wide, peristome type not clear (see note below), teeth 39-45 μm wide at base, operculum 0.3-0.4 mm long; spores 16-24 μm, minutely papillose.

Distribution: Northern Brazil, French Guiana and Suriname.

Ecology: Corticolous, elevation not indicated; rare.

Examined specimens: Suriname: Saramacca-oever, tussen Kwakoegron en Doorsnee, op liaan in zwaar oerbos, Florschütz 1105A (holotype *F. austro-americanus*, L). French Guiana: SE Cayenne, Cacao, bord de la rivière, sur écorce d'arbre, Cremers 5364 (BR, L).

Notes: This species is recognized by lingulate leaves, smooth or lowly unipapillose laminal cells and limbidia restricted to the basal half of the vaginant laminae of perichaetial leaves.
It can be mistaken for other species with lingulate leaves, viz. *F. brevipes* and *F. ramicola*. The first has costae with obscured distal ends, whereas *F. ramicola* has pluripapillose to smooth laminal cells. *F. guianensis* too

has wide leaves with broad tips but has pluripapillose laminal cells and partially limbate vaginant laminae.

F. subramicola has characters of both subgenus Aloma and subgenus Fissidens. Vertical walls at the bifurcation of the peristome lacking (Fig. 42 G) and the 52 files of exothecial cells on the theca are typical for subgenus Fissidens. On the other hand, lamellar ridges of basal parts of peristome as high as trabeculae (Fig. 42 G) are typical of subgenus Aloma. Moreover, unipapillose laminal cells and elimbate leaves are regularly found in subgenus Aloma but rarely in subgenus Fissidens. Therefore, I consider this species best placed in subgenus Aloma.

F. brevipes in Musci I in Fl. Suriname 6: 55. 1964 is in part (Florschütz 1105A) this species.

31. **Fissidens weirii** Mitt., J. Linn. Soc., Bot. 12: 602. 1869. Type: Brazil, Weir 24 (holotype NY).   – Fig. 43

> *Fissidens weirii* Mitt. var. *bistratosus* W.R. Buck, Mem. New York Bot. Gard. 76(3): 18, fig. 21. 2003. Type: French Guiana: Saül, base of Mt. Galbao, along upper part of Mana R., Buck 33207 (holotype NY).

Plants yellowish green, growing scattered, pinnate, to 6 mm long (including branches to 20 mm long), 2.7 mm wide, simple or branched. Leaves imbricate, to 17 pairs, slightly crispate when dry, unistratose or partly to completely bistratose, oblong-lanceolate with acute apex, 1.2-1.7 mm long, 0.4-0.5 mm wide, 3-4 times as long as wide, limbate; limbidium pluristratose, ending below leaf apex and above insertion of dorsal lamina, at apex of vaginant laminae confluent or not; vaginant laminae 1/2 the leaf length, mostly subequal; dorsal lamina wide and slightly rounded at insertion; costa ending 2-3 cells below the apex; laminal cells obscure, pluripapillose, 2-4 papillae per cell, 4.5-6 µm long and 3-4.5 µm wide; basal juxtacostal cells to 9 µm long. Dioicous and synoicous. Perigonia in axils of apical leaves, bud-shaped; perichaetia and synoicia terminal, perichaetial leaves 1.4-1.9 mm long. Sporophyte 1-2 per perichaetium, seta 3 mm long, theca 0.5 mm long and 0.2 mm wide, peristome scariosus-type, teeth 33-39 µm wide at base; spores 9-10.5 µm.

Distribution: Neotropics and tropical Africa; the Guianas.

Ecology: On roots and bases of trees, alt. 180-500 m; uncommon.

Selected specimens: Guyana: Richards 45 (slide, L). Suriname: railway Paramaribo-Kabel, km 121, Florschütz 1832 (L). French Guiana: Saül, 2 km SW of village, "Sentier Limonade", epiphytic on tree base, Montfoort & Ek 596 (L).

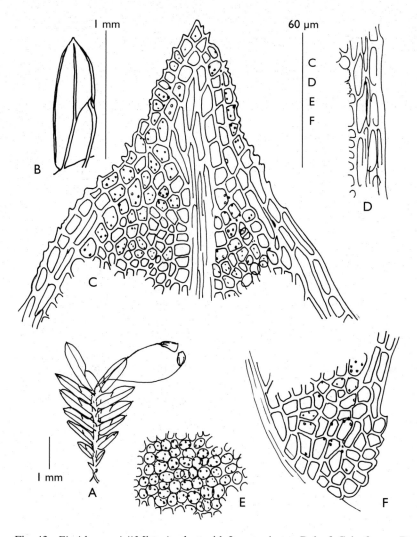

Fig. 43. *Fissidens weirii* Mitt.: A, plant with 2 sporophytes; B, leaf; C, leaf apex; D, margin mid dorsal lamina; E, mid dorsal laminal cells; F, base dorsal lamina (A-F, Montfoort & Ek 596).

Note: *F. weirii* is the only pluripapillose species with limbidia on all three laminae in the Guianas.

Buck (2003) described a bistratose form of *F. weirii* as var. *bistratosus* W.R. Buck. Such forms are not sharply distinct from *F. weirii* var. *weirii* and therefore not recognized here.

32.   **Fissidens zollingeri** Mont., Ann. Sci. Nat., Bot. sér. 3, 4: 114. 1845.
Type: Indonesia, Java, Zollinger 1604 (holotype PC).   – Fig. 44

Fissidens kegelianus Müll. Hal., Syn. Musc. Frond. 1: 49. 1849. Type:
Suriname, Kegel 1226 (holotype GOET).

Plants yellowish green, growing in tufts or scattered, typically frondiform,
1.5.-2.5 mm long, 1.5-1.7 mm wide, usually simple, with small axillary
nodules. Leaves imbricate, 5-7 pairs, hardly changed when dry, linear-
lanceolate to elliptical with acute-acuminate to long-mucronate apex,
1.2-1.5 mm long, 0.2-0.3 mm wide, 5-6.5 times as long as wide, upper
leaves longest, limbate; limbidia confluent at leaf apex or ending just
below, reaching or almost reaching insertion of the dorsal lamina, often
confluent at apex of vaginant laminae; vaginant laminae 1/2-2/3 the leaf
length, equal; dorsal lamina in basal part narrow and straight, reaching
insertion to ending far above; costa percurrent to long excurrent; laminal
cells hexagonal, smooth, flat, 9-13.5 μm long and 4.5-6 μm wide; basal
juxtacostal cells much larger, to 54 μm long. Rhizautoicous. Perigonia
terminal on small plants at base of female plants; perichaetia terminal,
perichaetial leaves 1.6 mm long; calyptra smooth. Sporophyte 1 per
perichaetium, seta 4-4.5 mm, theca cylindrical, erect, 0.6 mm long and
0.3 mm wide, peristome scariosus-type, teeth 27-36 μm wide at base,
operculum rostrate, 0.6 mm long; spores ± 12 μm.

D i s t r i b u t i o n : Pantropical; Suriname and French Guiana.

E c o l o g y : Typically on soil, clay or loam, infrequently on rotten wood
or stone, alt. 0-500 m; common.

S e l e c t e d   s p e c i m e n s : Suriname: Kabalebo R., trail from Avanavero
Falls to Red-Hill Falls, on stone, Florschütz 2130 (L); Brokopondo,
Phedra along Suriname R., on bare soil, Pursell 11800 (PAC). French
Guiana: Eaux Claires, 5 km N of Saül, on loamy soil, Florschütz-de
Waard 5956 (L); Saül, on Mt. Galbao, rainforest, on stone in creek,
Bekker 2305-2, pp (L).

N o t e : This species is characterized by frondiform plants, limbate leaves,
smooth laminal cells, vaginant laminae with enlarged, clear cells, small
but distinct axillary nodules and scariosus-type peristomes. It can be
confused with other limbate species. *F. angustifolius* also has frondiform
plants, axillary nodules and scariosus-type peristomes. It differs by
unipapillose laminal cells. *F. anguste-limbatus* has limbate leaves and
smooth laminal cells and is distinct by longer, pinnate plants without
axillary nodules and bryoides-type peristomes.

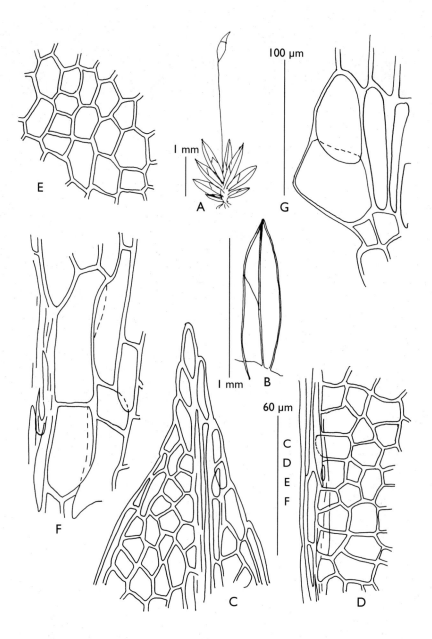

Fig. 44. *Fissidens zollingeri* Mont.: A, plant with sporophyte; B, leaf; C, leaf apex;
D, margin mid dorsal lamina (limbidium on the left); E, mid dorsal laminal cells;
F, basal juxtacostal cells costa on the left; G, axillary nodule (A-G, Pursell 11800).

Order **Dicranales**

Plants acrocarpous. Stems erect, simple or sparsely branched. Leaves narrow, often subulate, costa single, strong; laminal cells short or elongate, at basal angles usually differentiated. Sporophytes terminal, capsules erect and symmetric or inclined and asymmetric, peristome single, teeth 16, forked or deeply divided.

KEY TO THE FAMILIES

1 Plants whitish green. Leaves mainly composed of a multilayered costa, with the reduced lamina restricted to the base 2
Plants not whitish green. Lamina extending to the apex 3

2 Costa without bundle of stereids. . . . .*6. LEUCOBRYACEAE*
Costa in cross-section with a longitudinal bundle of stereids. . . . .
. . . . *7. LEUCOPHANACEAE*

3 Alar cells differentiated in shape or colour. Peristome teeth divided in 2 flat, pitted-striate divisions . . . . *4. DICRANACEAE*
Alar cells not differentiated. Peristome teeth deeply cleft in 2 terete divisions . . . .*5. DITRICHACEAE*

4. DICRANACEAE

Description see Musci I in Fl. Suriname 6: 69. 1964.

D i s t r i b u t i o n : Worldwide; in the Guianas 6 genera.

N o t e : *Atractylocarpus* Mitt. and *Dicranodontium* Bruch, reported from Guyana (Musci I in Fl. Suriname 6: 83. 1964) were collected at the Venezuelan side of Mt. Roraima. These genera are not known from the Guianas.

LITERATURE

Allen, B.H. 1990. A preliminary treatment of the Holomitrium complex in Central America. Trop. Bryol. 3: 59-71.
Frahm, J.P. 1980. Synopsis of the genus Campylopus in North America north of Mexico. Bryologist 83: 570-588.
Frahm, J.P. 1982. A reinterpretaion of Bryohumbertia P. de la Varde & Thér., Cryptog. Bryol. Lichénol. 3: 365-369.

Frahm, J.P. 1991. Dicranaceae: Campylopodioideae. Flora Neotropica Monograph 54: 37-196.
Frahm, J.P. & S.R. Gradstein. 1987. The genera Bryohumbertia and Campylopus in the Guianas. Cryptog. Bryol. Lichénol. 8: 311-319.
Gradstein, S.R. & H.J.M. Sipman. 1987. Taxonomy and world distribution of Campylopus introflexus and C. pilifer (= C. polytrichoides). Bryologist 81: 114-121.

KEY TO THE GENERA

1   Costa broad, 1/3 width of leaf base or more 2
     Costa narrower, less than 1/3 width of leaf base 3

2   Stems interruptedly foliate. Leaves to 15 mm long, with narrow-acuminate apex and long-excurrent costa. Seta straight or slightly twisted . . . .
     . . . . 1. *Bryohumbertia*
     Stems equally foliate or comose at apex (if interruptedly foliate: leaves shorter than 10 mm). Seta cygneous . . . . 2. *Campylopus*

3   Leaves bordered with elongate, hyaline cells, at least in lower half . . . .
     . . . . 6. *Leucoloma*
     Leaves without hyaline border 4

4   Leaves erect or secund when dry; cells in basal lamina thin-walled, alar cells not differentiated. Perichaetia inconspicuous . . . . 3. *Dicranella*
     Leaves strongly crispate or spirally twisted when dry; cells in basal lamina incrassate and pitted, alar cells inflated and coloured. Perichaetia with long, convolute leaves 5

5   Leaves acuminate with long-tubulose or hyaline apex; upper laminal cells elongate, incrassate and pitted . . . . 4. *Eucamptodontopsis*
     Leaves lanceolate with serrate, acute apex; upper laminal cells rounded quadrate . . . . 5. *Holomitrium*

1.   **BRYOHUMBERTIA** P. de la Varde & Thér., Bull. Soc. Bot. France 86: 422. 1939.
     Type: B. metzlerelloides P. de la Varde & Thér. [= Campylopus flavicoma Müll. Hal.]

Slender plants, interruptedly foliate. Leaves linear-lanceolate, slenderly acuminate; costa broad, in cross-section with dorsal and ventral bands of stereids; laminal cells rectangular, alar cells coloured and inflated. Dioicous. Seta long, straight or slightly twisted, operculum as long as the capsule, annulus present, peristome well developed.

Note: *Bryohumbertia* was described based on the African species *B. metzlerelloides*. This species, later synonymized with *Campylopus flavicoma*, proved to be closely related to the neotropical species *Campylopus filifolius*. Frahm (1982) concluded that both species are considerably distinct from other *Campylopus* species. He restored the genus *Bryohumbertia*, mainly based on the following characters: the straight seta, the very long operculum, the presence of an annulus and the smooth inner surface of the peristome teeth (papillose in *Campylopus*). Except for the interruptedly foliate stems (also occurring in *Campylopus*) the only gametophytic character to distinguish the genus is the rather uniform areolation of the lamina, not clearly differentiated in the upper and lower laminae.

Distribution: Tropical Africa, SE Asia, Neotropics; in the Guianas 1 species.

1. **Bryohumbertia filifolia** (Hornsch.) Frahm, Cryptog. Bryol. Lichénol. 3: 366. 1982. – *Dicranum filifolium* Hornsch. in Mart., Fl. Bras. 1(2): 12. 1840. – *Campylopus filifolius* (Hornsch.) Mitt., J. Linn. Soc. Bot. 12: 76. 1869. Lectotype (Frahm 1991): Brazil, prope Novo-Friburgum, Beyrich s.n. (JE).

Campylopus nanofilifolius (Müll. Hal.) Paris, Index Bryol. Suppl. 94. 1900. – *Dicranum nanofilifolium* Müll. Hal., Nuovo Giorn. Bot. Ital. 4: 35. 1897. Lectotype (Frahm 1991): Brazil, Sta. Catarina, Ule 439 (H).

Description see Musci I in Fl. Suriname 6: 81. 1964, as *Campylopus nanofilifolius*.

Distribution: Mexico, West Indies, C America, tropical and subtropical S America, tropical Africa.

Ecology: In the Guianas only collected at higher altitudes (700-1200 m), in cloud forests on roots and stems of trees, also terrestrial on moist rock slabs; not collected in French Guiana.

Note: This species is distinguished from *Campylopus trachyblepharon* specimens with interruptedly foliate stems by the more slender growth form, the leaf tufts are more distantly arranged along the slender, not-tomentose stems, the leaves with filiform, flexuose acumen are longer and the laminal cells more elongate.

2. **CAMPYLOPUS** Brid., Muscol. Recent. Suppl. 4: 71. 1819.
    Type: C. flexuosus (Hedw.) Brid. (Dicranum flexuosum Hedw.)

Description see Musci I in Fl. Suriname 6: 70. 1964.

Distribution: Worldwide; in the Guianas 13 species.

KEY TO THE SPECIES

1   Basal laminal cells above the auricles with strongly incrassate and pitted
    walls 2
    Basal laminal cells thin-walled or slightly thickened but not pitted. 7

2   Leaves with a conspicuous hyaline hairpoint. . . . . *9. C. richardii*
    Leaves without hyaline hairpoint or with only a few hyaline cells at end
    of costa. 3

3   Costa ribbed, with conspicuous lamellae at back .4
    Costa without lamellae at back 5

4   Stems equally foliate. Leaves to 15 mm long. . . . . *6. C. lamellinervis*
    Stems often interruptedly foliate. Leaves to 7 mm long. . . . .
    . . . . *13. C. trachyblepharon*

5   Robust plants with leaves 8-13 mm long, costa in cross- section with dorsal
    stereids only . . . . *11. C. subcuspidatus*
    Leaves not over 9 mm long, costa with dorsal and ventral stereids . . . . . . 6.

6   Upper laminal cells quadrate to rhomboid; costa short-excurrent . . . .
    . . . . *2. C. arctocarpus*
    Upper laminal cells more elongate, oval-oblong; costa long-excurrent,
    coarsely serrate . . . . *4. C. cubensis*

7   Basal laminal cells quadrate to short-rectangular, along the margin smaller,
    subquadrate. 8
    Basal laminal cells more elongate, towards margin narrower 9

8   Leaves 3-5 mm long, costa narrower (1/3 of leaf base), ending in or just
    beyond apex . . . . *5. C. dichrostis*
    Leaves 3-9 mm long, costa strong and excurrent, serrate and sometimes
    hyaline at apex . . . . *10. C. savannarum*

9   Leaves to 12 mm long with a broad basal lamina, abruptly contracted to a
    long, narrow subula, longer than the basal lamina . . . . *7. C. luteus*
    Leaves to 7 mm long, lamina gradually narrowing to the apex 10

10   Costa slender, ca. 1/3 of leaf width at base........................ 11
     Costa broad, more than 1/2 of leaf width at base 13

11   Plants usually interruptedly foliate. Costa at back with 2-3 cell high lamellae,
     in cross-section with stereids at ventral and dorsal side...............
     .... *13. C. trachyblepharon*
     Plants equally foliate or with only a comal tuft. Costa at back not lamellose,
     at most ridged, in cross-section with hyalocysts at ventral and stereids at
     dorsal side.. 12

12   Plants equally foliate, leaves wide-spreading. Upper laminal cells more
     elongate, oblong, incrassate .... *1. C. angustiretis*
     Fullgrown plants with appressed stem leaves and a comal tuft of spreading
     leaves at the end. Upper laminal cells short-rectangular or rhomboidal, little
     incrassate .... *12. C. surinamensis*

13   Costa short-excurrent, only dentate at the tip, sometimes hyaline; lamina
     gradually narrowed in the tubulose apex .... *3. C. bryotropii*
     Costa long-excurrent in a serrate, hyaline hairpoint; lamina abruptly narrowed
     and involute at apex .... *8. C. pilifer*

1.   **Campylopus angustiretis** (Austin) Lesq. & James, Man. Mosses
     N. America 80. 1884. – *Dicranum angustirete* Austin, Bot. Gaz. 4:
     150. 1879. – *Campylopus surinamensis* Müll. Hal. var. *angustiretis*
     (Austin) Frahm, Bryologist 83: 582. 1980. Type: U.S.A., Florida,
     Jacksonville, Austin s.n. (NY).   – Fig. 45

Slender plants growing in dense cushions. Stems to 4 cm high, little
tomentose, unbranched, equally foliate. Leaves wide-spreading,
slenderly lanceolate, 3-6 mm long, apex keeled; costa 1/3 of leaf width
at base, short-excurrent, toothed at apex, in cross-section with small
hyalocysts at ventral side and groups of pseudo-stereids at dorsal side;
alar cells coloured and inflated in conspicuous auricles, inner basal
laminal cells rectangular, thinwalled or slightly incrassate, narrower and
hyaline towards margin, quickly becoming more incrassate and irregular
distally, upper laminal cells regularly oblong, incrassate, to 45 µm long
and 8 µm wide along the costa, shorter and rhomboidal along the margin.
Sporophyte unknown.

Distribution: Florida, Guyana, French Guiana, SE Brazil (the
Guianan collections fill a gap between the Brazilian and the Carribean
distribution areas (Frahm 1991)).

Ecology: Terrestrial on exposed rocks.

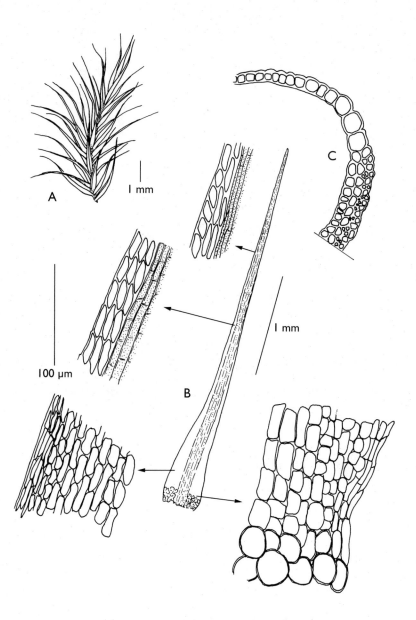

Fig. 45. *Campylopus angustiretis* (Austin) Lesq. & James: A, upper part of plant; B, leaf; C, cross-section of leaf (A-C, Newton *et al*. 3456).

Selected specimens: Guyana: Kaieteur Falls, along path from guesthouse to Johnson's View, alt. 420 m, Newton *et al.* 3456 (L, US). French Guiana: Savanne Roche de Virginie, Bassin de l'Approuague, alt. 100 m, Florschütz-de Waard 6150 (L).

Note : This species, previously considered as a variety of *C. surinamensis* (Frahm 1980), was later reïnstated as a distinct species (Frahm 1991). It is distinguished from *C. surinamensis* by the equally foliate stems with wide-spreading leaves; moreover the leaves are more slenderly acuminate with a narrow, keeled apex; the alar cells are more inflated and coloured forming distinct auricles, whereas in *C. surinamensis* the auricles are often indistinct.

2. **Campylopus arctocarpus** (Hornsch.) Mitt., J. Linn. Soc. Bot. 12: 87. 1869. – *Dicranum arctocarpus* Hornsch. in Mart., Fl. Bras. 1(2): 12. 1840. Type: Uruguay, Montevideo, Sellow s.n. (BM).

Description and synonymy see Musci I in Fl. Suriname 6: 73. 1964.

Distribution: Widespread in the Neotropics: Mexico, West Indies, C and (sub)tropical S America.

Ecology: Terrestrial or epiphytic on tree trunks; in the Guianas restricted to higher altitudes.

Note : This species is closely related to *C. cubensis*. The 2 species are not easy to separate because the differentiating characters are not consistent (see under *C. cubensis*).

3. **Campylopus bryotropii** Frahm, Nova Hedwigia 39: 152. 1984. Type: Peru, Ancash, Laguna Llanganuco, Frahm 825118 (B, L, NY, PC).

– Fig. 46

Plants green in upper part, brownish below, in loose tufts. Stems to 4 cm long, tomentose below. Leaves to 7 mm long, appressed when dry, erect-spreading when moist, lanceolate, tubulose in upper half; costa filling 1/2-2/3 of the leaf base, in upper part with lamellae at back of 2-3 cells high, excurrent, dentate and sometimes hyaline at the tip, in cross-section hyalocysts at ventral and stereids at dorsal side; alar cells coloured and inflated, basal laminal cells rectangular, thin-walled, narrower towards margins forming an indistinct border, extending halfway up the leaf length, upper laminal cells incrassate, oval-oblong, more or less sigmoid. Sporophyte unknown.

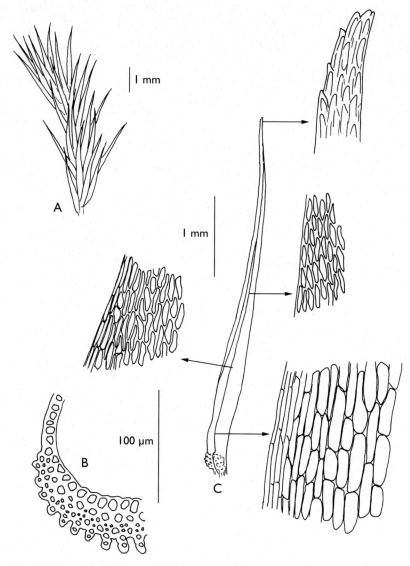

Fig. 46. *Campylopus bryotropii* Frahm: A, upper part of plant; B, cross-section of leaf; C, leaf (A-C, Hoffman 2860).

Distribution: Venezuela, Colombia, Peru, Guyana.

Ecology: Terrestrial on wet rocks and sand, at high altitudes; in Guyana collected once, in crevices of cliff face along river.

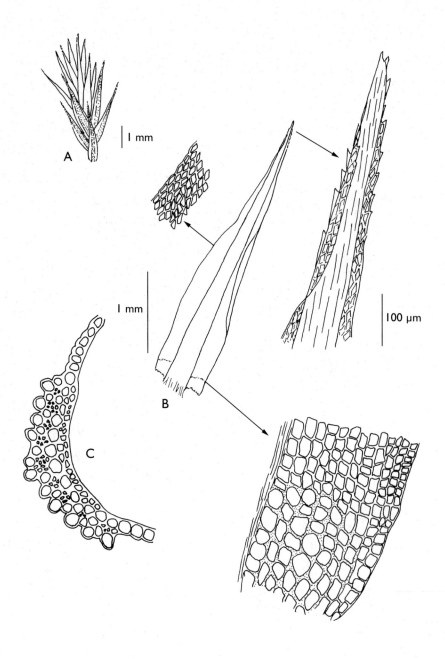

Fig. 47. *Campylopus dichrostis* (Müll. Hal.) Paris: A, upper part of plant; B, leaf; C, cross-section of leaf (A-C, de Granville *et al*. 9425).

Specimen examined: Guyana: Pakaraima Mts., upper Mazaruni R., alt. 525-575 m, Hoffman 2860 (L).

Note: This species resembles *C. pilifer* in the involute upper lamina, smooth or only slightly serrate. It is different in the distinctly coloured and inflated alar cells and the more elongate and incrassate upper laminal cells.

4. **Campylopus cubensis** Sull., Proc. Amer. Acad. Arts 5: 278. 1861. Type: Cuba, Wright 39 (NY).

Campylopus harrisii (Müll. Hal.) Paris, Index Bryol. Suppl. 92. 1900. – *Dicranum harrisii* Müll. Hal., Bull. Herb. Boissier 5: 553. 1897. Type: Jamaica, Harris 11008A (NY).

Description see Musci I in Fl. Suriname 6: 74. 1964, as *C. harrisii*.

Distribution: West Indies, Panama, Venezuela, the Guianas.

Ecology: Terrestrial on wet rocks and humic soil in low savanna vegetation; in the Guianas at higher altitudes.

Note: Typical *C. cubensis* is well distinguished from *C. arctocarpus* by the more elongate upper laminal cells and the long-excurrent, dentate costa. In the collections from the Guianas both characters show a broad variation, which makes the separation difficult. Collections from the Pakaraima Mts. in Guyana include specimens with elongate upper laminal cells and a rather short apex as well as specimens with short upper laminal cells and a long-excurrent, sharply toothed costa. The distribution of *C. cubensis* is mainly Caribbean and *C. arctocarpus* is widespread in the Neotropics (Frahm & Gradstein 1987). Frahm (1991) suggested that *C. cubensis* developed as a subspecies of *C. arctocarpus* by isolation on the Caribbean Islands.

5. **Campylopus dichrostis** (Müll. Hal.) Paris in Engler & Prantl, Nat. Pflanzenfam. 1(3): 333. 1901. – *Dicranum dichroste* Müll. Hal., Hedwigia 39: 255. 1900. Lectotype (Frahm 1991): Brazil, Ule 1359 (H, BR).   – Fig. 47

Slender plants growing in dense tufts. Stems to 4 cm high, erect or ascending from a prostrate base, little branched, densely tomentose in lower part. Leaves appressed when dry, erect-spreading when moist, 3-5 mm long, ovate-lanceolate with acute apex; costa ca. 1/3 of the leaf base, percurrent or short-excurrent, with serrate ridges at back, in cross-section with a dorsal band of stereids and at ventral side a row of small cells or substereids; laminal cells just above the auricles firm-walled, quadrate to short-rectangular, near the costa 20-50 µm long and 20-25 µm wide, towards margin much smaller, quadrate; upper laminal cells irregular-quadrate or rhomboid. Sporophyte not known.

Distribution: Brazil, the Guianas.

Ecology: Terrestrial, on rock and charcoal; in the Guianas not common, confined to white sand savannas and open rock vegetations.

Selected specimens: Guyana: Santa Mission, white sand savanna, Florschütz-de Waard 6110 (L); Kaieteur Falls, open rock area, alt. 400 m, Newton *et al.* 3356 (L, US). Suriname: Granite outcrop, 1 km NW of Voltzberg, Schulz & van Donselaar 10545 (L). French Guiana: Roche Koutou, Bassin du Haut-Marouini, alt. 480 m, Hoff 5262; idem, de Granville *et al.* 9425 (L, CAY).

Note: This species is distinguished from *C. savannarum* by the shorter and broader leaves with the lamina extending to the apex; the costa may be percurrent or short-excurrent, varying along one stem. In this respect the difference with *C. savannarum* in the modification "*bartlettii*", distinguished by the same characters, is not sharply marked. Other distinguishing characters could not be observed. Perhaps this species could as well be interpreted as an extreme modification of *C. savannarum* (Frahm 1991).

6. **Campylopus lamellinervis** (Müll. Hal.) Mitt., J. Linn. Soc. Bot. 12: 82. 1869. – *Dicranum lamellinerve* Müll. Hal., Syn. Musc. Frond. 1: 390. 1849. Type: Jamaica, Wilson s.n. (destroyed at B, isotypes not located). – Fig. 48

Robust plants growing in loose tufts. Stems to 8 cm long, seldom divided, densely tomentose in lower part. Leaves squarrose-spreading, flexuose when dry, slenderly lanceolate, to 15 mm long, long-acuminate, margin sharply serrate in upper leaf half; costa filling ca. 1/3 of leaf base, with conspicuous lamellae at back, excurrent and toothed at apex, in cross-section with groups of stereids at ventral and dorsal side, lamellae 2-3 cells high; alar cells inflated, brownish, forming distinct auricles, basal laminal cells elongate-rectangular, strongly incrassate and porose, towards margins narrower and not porose, forming a short hyaline border, laminal cells at midleaf incrassate, along the costa elongate-rectangular, towards margins and in the narrow upper lamina rectangular to quadrate with oval lumen. Sporophyte not seen.

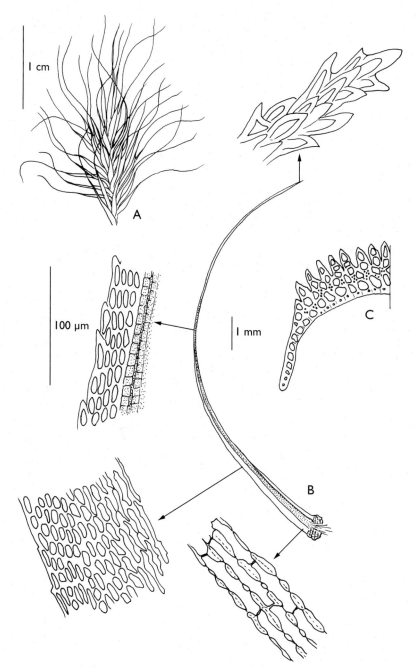

Fig. 48. *Campylopus lamellinervis* (Müll. Hal.) Mitt.: A, upper part of plant; B, leaf; C, cross-section of leaf (A-C, Boom *et al.* 7620).

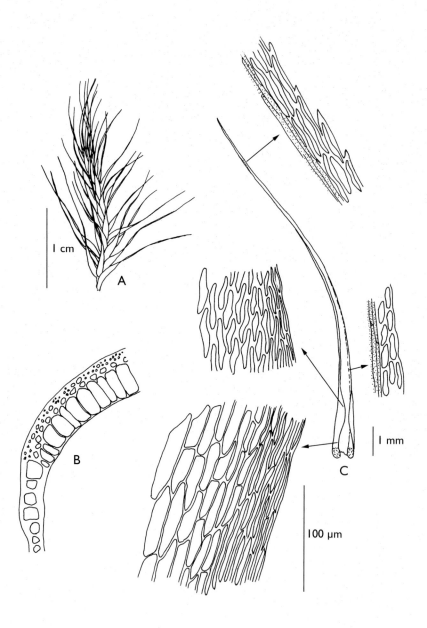

Fig. 49. *Campylopus luteus* (Müll. Hal.) Paris: A, upper part of plant; B, cross-section at leaf base; C, leaf (A-C, Gradstein 5336).

Distribution: West Indies, Venezuela, Colombia, Ecuador, Bolivia, Peru, Brazil, Guyana, French Guiana.

Ecology: Terrestrial and epiphytic, in moist savanna forest, at higher altitudes.

Specimen examined: Guyana: Upper Mazaruni R., Karowtipu Mt., alt. 1000 m, Boom *et al*. 7620 (NY).

Note: Buck (2003) reported *C. lamellinervis* var. *exaltatus* (Müll. Hal.) Frahm for French Guiana. This variety is distinguished from the species by straight, lanceolate leaves, 6-10 mm long, serrulate at extreme apex.

7. **Campylopus luteus** (Müll. Hal.) Paris, Index Bryol. 254. 1894. – *Thysanomitrium luteum* Müll. Hal., Linnaea 42: 470. 1879. Lectotype (Frahm 1991): Venezuela, Tovar, Fendler 39 (H, BR). – Fig. 49

Robust plants growing in loose tufts. Stems 10 cm or more long, tomentose below. Leaves spreading, to 12 mm long and 1.2 mm wide, slenderly lanceolate with an ovate basal part ca. 1/4 of the leaf length, abruptly narrowed to a long, narrow subula; costa filling more than half the leaf base, excurrent and toothed at apex, in cross-section with wide hyalocysts ventrally and dense groups of stereids dorsally; alar cells inflated and coloured forming large auricles, inner basal laminal cells just above the auricles thin-walled or slightly incrassate, rectangular, towards margin linear, forming a hyaline border along the basal lamina, cells in upper part of the basal lamina incrassate, sigmoid-oblong; upper lamina very narrow, 2-3 cells wide, extending nearly to apex, sharply serrate in upper part. Sporophyte (description after Frahm 1991): seta 8-10 mm long, capsule curved, 2 mm long, operculum obliquely rostrate; calyptra fringed at base.

Distribution: Colombia, Venezuela, Guyana, Bolivia.

Ecology: Terrestrial, on wet rocks; collected in the Guianas only from a single locality in Guyana, at high altitude in humid mossy forest.

Specimens examined: Guyana: N slope of Mt. Roraima, alt. 1200-1600 m, Aptroot 17112, Gradstein 5336, 5379 (L).

Note: This large moss with a predominant Andean distribution occurs in Guyana at the same elevation as the Neotropical *C. subcuspidatus*. It is similar in size but it is different in the more slender leaves with flexible subula and in the thin-walled basal laminal cells; the cells just above the auricles are thin-walled, but quickly become incrassate in distal direction.

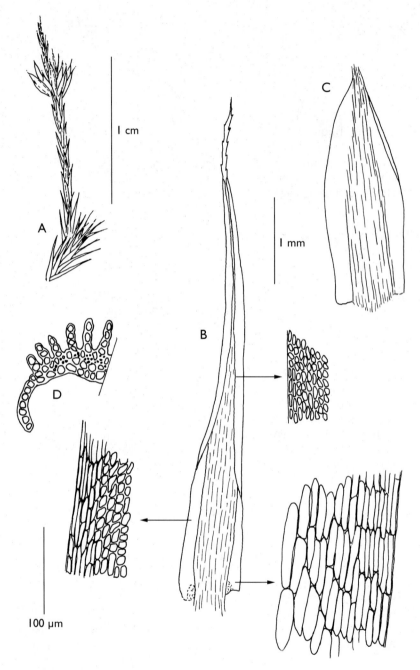

Fig. 50. *Campylopus pilifer* Brid.: A, part of plant; B, stem leaf; C, comal leaf; D, cross-section of stem leaf (A-D, Gradstein 5293).

8. **Campylopus pilifer** Brid., Muscol. Recent. Suppl. 4: 72: 1819. Lectotype (Gradstein & Sipman 1978): Italy, Ischia, Bridel s.n. (B).
– Fig. 50

Slender plants growing in dense cushions. Stems to 5 cm high, equally foliate but often forming comal tufts with perichaetia or young innovations. Stem leaves appressed when dry, erect when moist, lanceolate, 3-6 mm long, apex broad-acute, often appearing acuminate by the involute upper margins, costa filling 1/2-2/3 of the leaf base, excurrent in a hyaline, serrate hairpoint (variable in length and often absent in older leaves), in cross-section with ventral hyalocysts and dorsal stereids, at back with prominent, 2-4 cell high lamellae; comal leaves broader, oblong, cuspidate or piliferous at apex; alar cells little differentiated, sometimes partly coloured and inflated; basal laminal cells hyaline, thin-walled, long-rectangular, towards margins linear, in distal direction rather abruptly changing in the small, oval or irregularly rhombic upper laminal cells. Sporophyte (description after Frahm 1991): seta 3-5 mm long, capsules 1.5 mm long, ovoid; calyptra fringed at base.

Distribution: Tropical and warm-temperate regions in N and S America, Africa, India and Europe; in the Neotropics only at higher altitudes (to 4800 m).

Ecology: Commonly on exposed rocks or on dry soil of road banks; in the Guianas collected only once, on rotten log.

Specimen examined: Guyana, N slope of Mt. Roraima, alt. 1200-1600 m, Gradstein 5293 (L).

Note: Variable in appearance; very typical if the spreading comal leaves are present at the end of the appressed-foliate stem, but not easy to recognize without comae; also the hyaline hairpoint is not always distinct. For differences with *C. bryotropii* see under that species.

9. **Campylopus richardii** Brid., Muscol. Recent. Suppl. 4: 73: 1819. Type: Guadeloupe, Parker s.n. (B). – Fig. 51

Campylopus atratus Broth., Trans. Linn. Soc. London, Bot. ser. 2, 6: 89. 1901. Type: Venezuela, Mt. Roraima, McConnell & Quelch 527 (BM).

Plants blackish with light-green tips, growing in dense tufts. Stems to 6 cm high, usually shorter, tomentose below. Stem leaves appressed, in fertile plants more or less comose at the end of the stem, lanceolate,

126

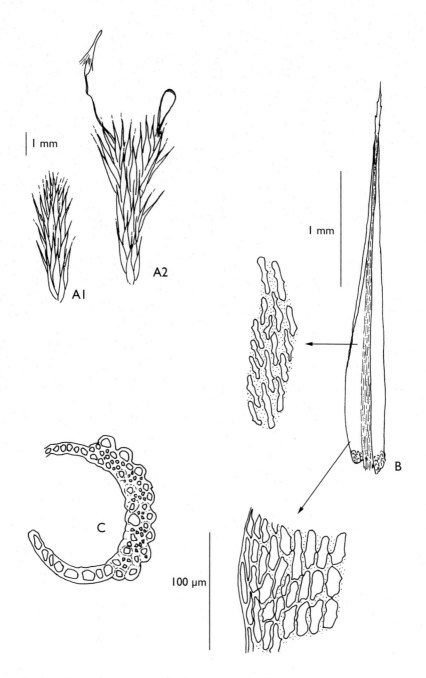

Fig. 51. *Campylopus richardii* Brid.: A, upper part of plant; B, leaf; C, cross-section of leaf (A1, B-C, Gradstein 5358; A2, Frahm, Campylopodes Brasiliae exs. 13).

3-8 mm long, tubulose at apex with a conspicuous hyaline, serrate hairpoint; costa filling 1/3-1/2 of the leaf base, in upper part ridged at back, in cross-section with stereids at dorsal and substereids at ventral side; alar cells coloured and inflated, laminal cells all strongly incrassate and pitted, basal cells rectangular, narrower towards margin, sometimes forming a short hyaline border of 1-2 cells wide, upper laminal cells oval-oblong or rhomboidal in oblique rows. Sporophyte (description after Frahm 1991): seta sinuose or curved, 7-8 mm long, capsule 1.5 mm long, scabrous at base, operculum short-rostrate, peristome teeth narrow, hyaline, filiform, split nearly to base; calyptra ciliate at base.

Distribution: Neotropics, at higher altitudes (900-3700 m).

Ecology: Terrestrial on rock and exposed gravelly ridges; in the Guianas only known from Mt. Roraima.

Collection examined: Guyana: N slope of Mt. Roraima, alt. 2000-2300 m, Gradstein 5358 (L).

Note: This species, reported for Guyana as *C. atratus* in Musci I in Fl. Suriname 6: 73. 1964, was only known from the Venezuelan side of Mt. Roraima. The hyaline hairpoint, in combination with the rather homogeneous incrassate cell pattern, distinguish this species from other Guianan species with a hairpoint.

10. **Campylopus savannarum** (Müll. Hal.) Mitt., J. Linn. Soc. Bot. 12: 85. 1869. – *Dicranum savannarum* Müll. Hal., Syn. Musc. Frond. 2: 596. 1851. Type: Suriname, Kegel s.n. (L, BM).

Campylopus bartlettii E.B. Bartram, J. Washington Acad. Sci. 22: 477. 1932. – *Campylopus savannarum* (Müll. Hal.) Mitt. subsp. *bartlettii* (E.B. Bartram) Florsch., Fl. Suriname 6: 79. 1964. Type: Honduras, Bartlett 12973 (FH, NY).

Description see Musci I in Fl. Suriname 6: 78. 1964.

Distribution: Pantropics.

Ecology: Terrestrial and on rocks; in the Guianas rather common on white sand savannas in xeromorphic scrub and light savanna forest.

Note: Subsp. *bartlettii* was described for specimens with short and rigidly appressed leaves with excurrent (sometimes hyaline) costa (Florschütz 1964). Frahm considered this as a modification in drier habitats, since all possible intergradations occur (Frahm & Gradstein 1987).

11. **Campylopus subcuspidatus** (Hampe) A. Jaeger, Ber. St. Gallischen
Naturwiss. Ges. 2 :441. 1872. – *Dicranum subcuspidatum* Hampe,
Vidensk. Meddel. Naturhist. Foren. Kjøbenhavn ser. 3, 2: 273. 1870.
Type: Brazil, Rio de Janeiro, Tijuca, Glaziou 7096 (BM).

Campylopus praealtus (Müll. Hal.) Paris, Index Bryol. Suppl. 96. 1900.
– *Dicranum praealtum* Müll. Hal., Hedwigia 37: 227. 1898. Type: Puerto
Rico, Sierra de Luquillo, Sintensis s.n. (NY).

Description see Musci I in Fl. Suriname 6: 70. 1964, as *C. praealtus*.

Distribution: West Indies, C and tropical S America.

Ecology: Terrestrial in xeromorphic scrub vegetation on white sand
savannas; collected in Guyana at higher altitudes (900-2000 m).

Note: The broad, rigid leaves, gradually narrowing to the apex and the
distinct border of thin-walled, hyaline cells, extending more than half
the leaf length, distinguish this species from *C. luteus* (see also under
that species).

12. **Campylopus surinamensis** Müll. Hal., Linnaea 21:186. 1848. Type:
Suriname, near Paramaribo, Kegel 516 (GOET, PC).

Campylopus gracilicaulis Mitt., J. Linn. Soc., Bot. 12: 83: 1869. Type:
Brazil, Rio Negro, Spruce 60 (NY).

Description see Musci I in Fl. Suriname 6: 75. 1964.

Distribution: SE U.S.A., West Indies, C America, tropical S America.

Ecology: Terrestrial on sand, clay or organic debris; in the Guianas
rather common on granitic rockplateaus and in low savanna vegetations.

Notes: This species is easy to recognize in its full-grown form by
the comal tuft of spreading leaves at the end of an elongate stem with
appressed leaves. Young plants, consisting of a low rosette, occasionally
with only the distantly foliate stem, are sometimes difficult to identify. For
differences with the closely related *C. angustiretis* see under that species.
In Musci I *Campylopus arenicola* (Müll. Hal.) Mitt. was listed as a
synonym, but in view of the lamellae at the back of the costa it should be
considered a young form of *C. trachyblepharon* (Frahm 1991).

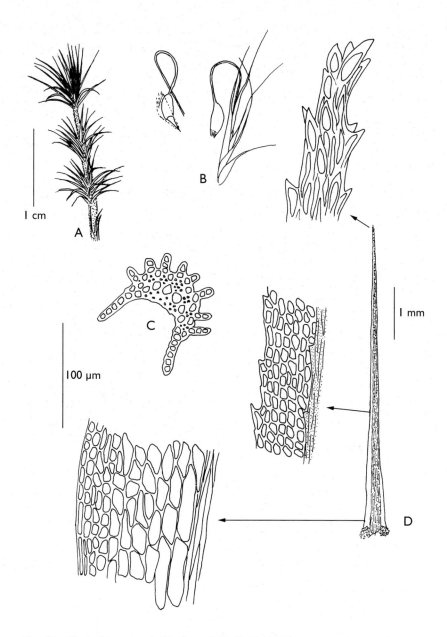

Fig. 52. *Campylopus trachyblepharon* (Müll. Hal.) Mitt.: A, upper part of plant; B, perichaetium and capsules; C, cross-section of leaf; D, leaf of comal tuft (A-D, Florschütz-de Waard 6058).

13. **Campylopus trachyblepharon** (Müll. Hal.) Mitt., J. Linn. Soc., Bot. 12: 80. 1869. – *Dicranum trachyblepharon* Müll. Hal., Syn. Musc. Frond. 1: 389. 1848. Type: Brazil, Beyrich s.n. (NY).   – Fig. 52

Campylopus arenicola (Müll. Hal.) Mitt., J. Linn. Soc., Bot. 12: 77. 1869. – *Dicranum arenicola* Müll. Hal., Bot. Zeitung (Berlin) 13: 762. 1855. Type: Brazil, Serra do Itatiaia, Pabst s.n. (NY).

Rather large plants growing in loose tufts. Stems to 7 cm high, usually interruptedly foliate, densely tomentose, between the tufts scarcely foliate. Leaves in the comal and lower tufts spreading, slenderly lanceolate, to 7 mm long, between the tufts shorter, more or less appressed; apex acute, serrate in upper half, dentate at the tip, costa ca. 1/3 of leaf base, percurrent to short excurrent, in cross-section with stereids at ventral and dorsal side, at back with conspicuous, serrate, 2-3 cell high lamellae; alar cells coloured and inflated, in the smaller stem leaves less conspicuous; inner basal laminal cells rectangular, slightly incrassate, towards margin rounded quadrate; upper laminal cells subquadrate or rhomboid, incrassate. Perichaetial leaves ovate, abruptly narrowed to a long and narrow subula. Seta to 1 cm long, capsule curved, operculum obliquely rostrate. Calyptra cucullate, fringed at base.

Distribution: E Africa, Bermuda, Guyana, coastal plains of SE Brazil.

Ecology: Terrestrial in low savanna vegetation on white sand; not known from Suriname and French Guiana.

Selected specimens: Guyana: Timehri, Dakara Cr., Gradstein 3734 (L); Santa Mission, trail W of village, Florschütz-de Waard 6059 (L).

Note: In the Guyanan collections the verticillate foliation is not always distinct, in the Gradstein collections even absent; the prominent lamellae at the back of the costa and the strongly serrate upper lamina with subquadrate laminal cells distinguish this species from all other Guyanan species. The basal laminal cells are variably incrassate and pitted; the cells just above the auricles may be thin-walled but quickly become more incrassate in distal direction.

3. **DICRANELLA** (Müll. Hal.) Schimp., Coroll. Bryol. Eur. 13. 1856. – *Ångstroemia* sect. *Dicranella* Müll. Hal., Syn. Musc. Frond. 1:430. 1848.
Type: D. grevilliana (Brid.) Schimp. (Dicranum schreberi Sw. var. grevillianum Brid.)

Description see Musci I in Fl. Suriname 6: 94. 1964.

Distribution: Worldwide; in the Guianas 1 species.

1. **Dicranella hilariana** (Mont.) Mitt., J. Linn. Soc., Bot. 12: 31. 1869.
   – *Dicranum hilarianum* Mont., Ann. Sci. Nat., Bot. sér. 2, 12: 52.
   1839. Type: Brazil, Saint Hilaire s.n. (PC).

Description and synonymy see Musci I in Fl. Suriname 6: 95. 1964.

Distribution: S U.S.A., West Indies, Mexico, C and tropical S
America.

Ecology: Terrestrial in open areas on clay and loamy sand; common on
road sides and creek banks.

Note: Male plants with elongate stems and short, appressed leaves
usually have conspicuous lateral perigonia and are very different in
aspect from the female plants in which the leaves are longer and twisted
to falcate.

4. **EUCAMPTODONTOPSIS** Broth. in Engler, Nat. Pflanzenfam. ed.
   2, 10: 202. 1924.
   Type: E. pilifera (Mitt.) Broth. (Eucamptodon piliferus Mitt.)

Medium-sized to large plants, sparingly branched. Leaves ovate to
lanceolate, contorted to spirally twisted when dry; costa narrow, percurrent
or excurrent, margins entire to crenulate; laminal cells elongate, incrassate
and pitted, alar cells strongly differentiated. Perichaetial leaves sheathing,
long-acuminate. Seta long, straight; capsule erect, cylindical, peristome
teeth undivided.

Distribution: West Indies, C and tropical S America; in the Guianas
2 species.

Note: In Musci I in Fl. Suriname 6: 90. 1964 *Eucamptodontopsis* was
treated as a monospecific genus, based on *E. pilifera*. Later *E. tortuosa*
was described from Venezuela (Robinson 1965) and recorded from
Guyana by Buck (1990). Allen (1990) transferred *Dicranoloma brittonae*
E.B. Bartram to *Eucamptodontopsis*; the genus now includes 3 species.

KEY TO THE SPECIES

1 Leaves to 5 mm long, apex acuminate, costa excurrent, hyaline . . . .
. . . . *1. E. pilifera*
Leaves to 10 mm long, linear-lanceolate, apex acute, tubulose, costa
percurrent . . . . *2. E. tortuosa*

1. **Eucamptodontopsis pilifera** (Mitt.) Broth. in Engler, Nat.
Pflanzenfam. ed. 2, 10: 202. 1924. – *Eucamptodon pilifera* Mitt., J.
Linn. Soc., Bot. 12: 69. 1869. Type: Trinidad, Crüger s.n. (NY).

Description see Musci I in Fl. Suriname 6: 90. 1964.

Distribution: West Indies, Venezuela, Guyana, Suriname.

Ecology: Epiphytic in open, humid forests at higher altitudes; only
collected on N slope of Mt. Roraima, Tafelberg and Lely Mts.

2. **Eucamptodontopsis tortuosa** H. Rob., Acta Bot. Venez. 1: 74.
1965. Type: Venezuela, Cerro Venamo, Steyermark & Dunsterville
92260 (US). – Fig. 53

Yellowish green plants growing in loose tufts. Stems erect, to 4 cm long.
Leaves wide-spreading, erect at base, when dry strongly spirally contorted
in upper part, lanceolate, 8-10 mm long, sheathing at base, auriculate,
tubulose in upper 3/4 of the leaf, margins incurved, slightly crenulate
above, costa percurrent, indistinct in upper leaf; laminal cells elongate-
rectangular throughout, 80-120 μm long and ca. 12 μm wide, narrower
along the margins, strongly incrassate and pitted, alar cells shorter and
wider, coloured, incrassate and pitted. Perichaetial leaves long-sheathing,
inner ones to 15 mm long. Seta smooth, ca. 20 mm long, capsule narrow-
cylindrical, 2.5 mm long, operculum slenderly rostrate, to 3 mm long.

Distribution: Venezuela, Guyana.

Ecology: Epiphytic on tree trunks and logs at higher altitudes.

Specimens examined: Guyana: N slope of Mt. Roraima, alt. 700 m,
Gradstein 5095 (L); ibid. alt. 1200-1600 m, Gradstein 5219 (L); Mt.
Wokomung, alt. 1650 m, Boom et al. 9146 (NY); Mt. Ayanganna, alt.
1500 m, Tillett & Boyan 45099 (NY).

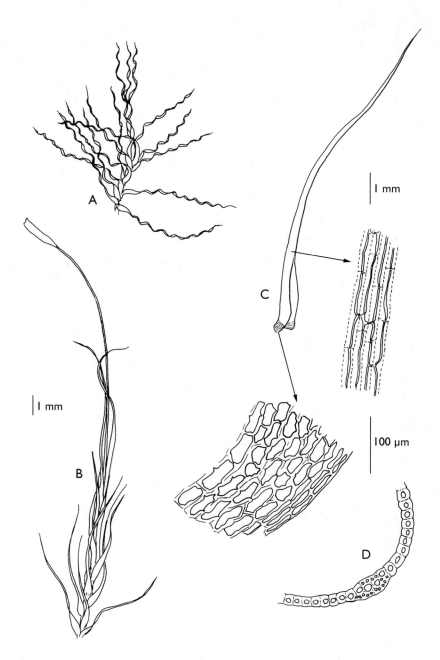

Fig. 53. *Eucamptodontopsis tortuosa* H. Rob.: A, upper part of plant (dry); B, upper part of plant with perichaetium and capsule; C, leaf; D, cross-section of leaf (A-D, Gradstein 5219).

N o t e : Contrary to the description of Robinson the costa is not excurrent in the collections from Guyana.

5. **HOLOMITRIUM** Brid., Bryol. Univ. 1: 226. 1826.
   Type: H. perichaetiale (Hook.) Brid. (Trichostomum perichaetiale Hook.)

Description see Musci I in Fl. Suriname 6: 94. 1964.

D i s t r i b u t i o n : Worldwide; in the Guianas 1 species.

1. **Holomitrium arboreum** Mitt., J. Linn. Soc., Bot. 12: 58.1869.
   Lectotype (Allen 1990): Peru, Andes Peruvianae, in monte Campana, Spruce 22 (NY).

Description see Musci I in Fl. Suriname 6: 94. 1964.

D i s t r i b u t i o n : West Indies, C America and northern S America.

E c o l o g y : Epiphytic, often in the canopy of lowland rainforest, also at higher altitudes.

N o t e : In Musci I two species of *Holomitrium* are recorded for the Guianas: *H. williamsii* E.B. Bartram was distinguished from *H. arboreum* by shorter leaves and a gradually tapering lamina, but these differences prove to be included in the broad variability of the species. Allen (1990) distinguished *H. williamsii* and *H. arboreum* by the changeover of the quadrate upper laminal cells to the elongate lower laminal cells: in *H. williamsii* in a straight line, whereas in *H. arboreum* the quadrate cells extend down the margin of the lower lamina. Examination of the types confirm this conclusion. As in all collections from the Guianas the quadrate cells extend downwards along the margins of the lower lamina we may conclude that *H. williamsii* does not occur in the Guianas.
The collection from French Guiana reported as *H. olfersianum* Hornsch. (Buck 2003) is not essentially different from *H. arboreum*. Only the short upper lamina resembles *H. olfersianum*, but in that species the lower lamina is much wider and abruptly narrowing to the very short upper lamina (slide "Olfers s.n. type?" (BM)).

6. **LEUCOLOMA** Brid., Bryol. Univ. 2: 218. 1827.
   Type: L. bifidum (Brid.) Brid. (Hypnum bifidum Brid.)

Description see Musci I in Fl. Suriname 6: 83. 1964.

Distribution: Pantropical; in the Guianas 4 species.

KEY TO THE SPECIES

1 Leaves quickly narrowed from an oblong or ovate base to a linear upper lamina with a bistratose, often rounded apex; hyaline border not extending beyond lower lamina .... *4. L. tortellum*
   Leaves gradually tapering to a subulate, unistratose apex; hyaline border extending nearly to apex. 2

2 Hyaline border in basal part distinctly separated from the chlorophyllose inner cells; margins serrate in upper part .... *3. L. serrulatum*
   Hyaline border in basal part gradually merging with inner laminal cells, not distinctly separated; margins smooth in upper part, only apex dentate . . . 3

3 Basal laminal cells more or less incrassate but not pitted; chlorophyllose cells in upper lamina quadrate or oblong, less than 20 μm long ....
   .... *1. L. cruegerianum*
   Basal laminal cells strongly incrassate and pitted; chlorophyllose cells in upper lamina elongate, 20 μm or more long .... *2. L. mariei*

1. **Leucoloma cruegerianum** (Müll. Hal.) A. Jaeger, Ber. St. Gallischen Naturwiss. Ges. 1870-71: 412. 1872. – *Dicranum cruegerianum* Müll. Hal., Syn. Musc. Frond. 2: 588. 1851. Type: Trinidad, Crüger s.n. (NY, PC).

Description see Musci I in Fl. Suriname 6: 88. 1964.

Distribution: Mexico, West Indies, C and tropical S America.

Ecology: Epiphytic on tree trunks and lower branches of the canopy, occasionally on wet rocks; in forest at higher elevations not uncommon.

2. **Leucoloma mariei** Besch., J. Bot. (Morot) 5: 145. 1891. Type: Guadeloupe, le Gommier, Marie s.n. (BM, PC).   – Fig. 54

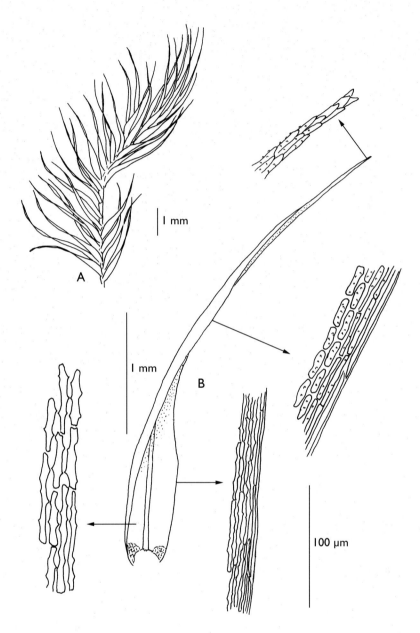

Fig. 54. *Leucoloma mariei* Besch.: A, upper part of plant; B, leaf (A-B, Newton *et al.* 3462).

Medium-sized plants, growing in loose mats. Stems irregularly branched, to 2 cm long; leaves falcate, patent-spreading in lower part, secund at the end of the stem, slenderly lanceolate, 4-6 mm long, gradually narrowed to the subtubulose apex, margins smooth, costa slightly excurrent, dentate at the tip; basal laminal cells linear, strongly incrassate, smooth, 40-80 μm long and ca. 8 μm wide, towards margin gradually narrower, forming an inconspicuous border of linear, hyaline cells, alar cells inflated, brownish; upper laminal cells pluripapillose, oblong-linear, 20-40 μm long, marginal cells smooth, linear and hyaline. Sporophyte unknown.

Distribution: West Indies, C America, the Guianas.

Ecology: Epiphytic on tree trunk in rainforest at higher altitudes.

Specimens examined: Guyana: forest near Kaieteur Falls, alt. 450 m, Newton et al. 3462 (L, US). Suriname: Tafelberg, near Lisa Cr., alt. 635 m, Allen 20661 (L, MO).

Note: This species with falcate-secund leaves at the end of the stem resembles a large specimen of *L. cruegerianum*. Also the smooth margins and the indistinctly separated hyaline border are similar; different are the strongly incrassate cells of the basal lamina and the elongate, faintly papillose upper laminal cells. The hyaline border is difficult to see and in fact restricted to 1 or 2 very narrow linear cells.

3. **Leucoloma serrulatum** Brid., Bryol. Univ. 2: 752. 1827. Type: Hispaniola, Desvaux s.n. (B).

Description see Musci I in Fl. Suriname 6: 85. 1964.

Distribution: Mexico, West Indies, C and tropical S America.

Ecology: Epiphytic in tropical rainforest at higher altitudes (400-900 m); regularly collected in French Guiana and Suriname, only once in Guyana.

4. **Leucoloma tortellum** (Mitt.) A. Jaeger, Ber. St. Gallischen Naturwiss. Ges. 1870-71: 413. 1872. – *Poecilophyllum tortellum* Mitt., J. Linn. Soc., Bot. 12: 94. 1869. Lectotype (Florschütz 1964): Brazil, Spruce 67 (NY).

Description see Musci I in Fl. Suriname 6: 84. 1964.

Distribution: West Indies, C and tropical S America.

Ecology: On rocks, occasionally on decaying wood in moist open areas of lowland rainforest; not common.

5. DITRICHACEAE

Small plants with erect stems, simple or forked. Leaves lanceolate, costa single, percurrent to long-excurrent; laminal cells smooth, quadrate to linear, alar cells not differentiated. Capsules immersed to long-exserted, erect or inclined, often asymmetric; peristome single, teeth 16, split into terete divisions or perforated nearly to base.

Distribution: Worldwide; in the Guianas 1 genus.

LITERATURE

Margadant, W.D. & A.H. Norkett. 1973. The nomenclatural tangle of Garckea phascoides. J. Bryol. 7: 439-441.

1. **GARCKEA** Müll. Hal., Bot. Zeitung (Berlin) 3: 865. 1845.
   Type: G. phascoides Müll. Hal.

Small plants growing gregarious or in loose tufts. Stems simple, lower leaves small, appressed to the stem, upper leaves larger, crowded in a comal tuft, lanceolate, acuminate, costa percurrent to slightly excurrent; laminal cells elongate throughout the leaf. Dioicous. Perichaetial leaves not differentiated, setae very short, capsules immersed, ovoid-cylindric, operculum conic-apiculate; peristome inserted below the mouth, teeth papillose, irregularly forked. Calyptra conic-mitrate.

Distribution: Tropical and subtropical Asia; in the Neotropics and in the Guianas 1 species.

1. **Garckea flexuosa** (Griff.) Margad. & Nork., J. Bryol. 7: 440. 1973.
   – *Grimmia flexuosa* Griff., Calcutta J. Nat. Hist. 2: 492. 1842. Type:
   East India, Griffith s.n., hb. Ward p.p. (BM).   – Fig. 55

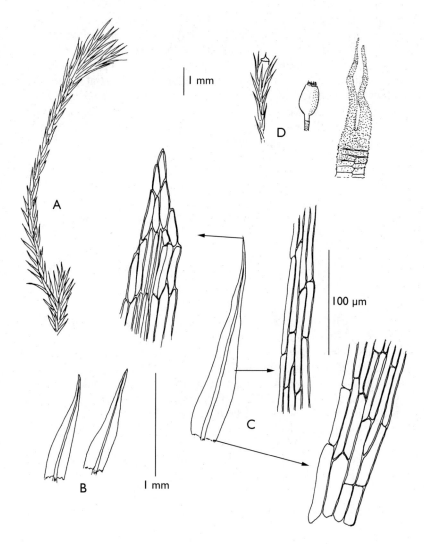

Fig. 55. *Garckea flexuosa* (Griff.) Margad. & Nork.: A, stem with comal tuft; B, stem leaves; C, comal leaf; D, capsule and peristome tooth (A-D, Allen 19327).

Plants gregarious, stems to ca. 1 cm high, erect. Lower leaves distant, appressed to the stem, triangular-lanceolate, acute, 0.8-1.2 mm long, comal leaves crowded, to 2 mm long, lanceolate, apex acuminate, margin smooth or bluntly serrulate in upper part, costa ending a few cells below apex; laminal cells thin-walled or slightly incrassate, linear, 100-170 μm

140

long and 6-10 μm wide, shorter and wider in apex and in a few basal rows. Perichaetia not differentiated, archegonia scattered at base of comal leaves. Capsules not seen from the Neotropics, description from Carr 12229 (New Guinea): seta to 0.5 mm long, capsule immersed, ovoid-cylindric, ca. 1 mm long, operculum conic-rostrate; peristome teeth ca. 150 μm high, split more than halfway down, densely papillose, trabeculate in lower part. Calyptra short, just covering the operculum, coarsely mammillose.

Distribution: In the Guianas rare, only collected in Suriname.

Ecology: On bare soil bank.

Specimen examined: Suriname: Phedra, along Suriname R., Allen 19324, 19327 (MO).

6. LEUCOBRYACEAE

Whitish-green or glaucous plants, sometimes reddish tinged, often growing in dense cushions. Leaves consisting of a thick, expanded costa and a short, thin-walled lamina restricted to the basal part of the leaf; costa in cross-section with several layers of large; hyaline, porose cells (leucocysts) and a median layer of small, chlorophyllose cells (chlorocysts). Dioicous. Seta straight, capsule immersed or exserted, erect or inclined, sometimes strumose, operculum short- to long-rostrate, peristome single with 8 or 16 lanceolate teeth or absent. Calyptra cucullate or mitrate, entire or fringed at base.

Distribution: Worldwide; in the Guianas 4 genera.

LITERATURE

Allen, B. 1992. A revision of Ochrobryum. Contr. Univ. Michigan Herb. 18: 113-130.
Florschütz, P.A. 1955. A short survey of the American species of the genus Octoblepharum. Mitt. Thüring. Bot. Ges. 1: 51-58.
Newton, A.E. & H. Robinson. 1994. The structure of the leaf and peristome of Holomitriopsis laevifolia (Broth.) H. Robins. illustrated with scanning electron microscopy. Trop. Bryol. 9: 111-116.
Peterson, W. 1994. Leucobryaceae. In A.J. Sharp, H. Crum & P.M. Eckel (eds.), The Moss Flora of Mexico. Mem. New York Bot. Gard. 69: 169-186.

Robinson, H. 1965. Venezuelan bryophytes collected by Julian A. Steyermark. Acta Bot. Venez. 1: 73-83.
Salazar Allen, N. 1991. A preliminary treatment of the Central American species of Octoblepharum. Trop. Bryol. 4: 85-97.

KEY TO THE GENERA

1   Leaves ligulate with an elliptic or obovate basal lamina, costa in cross-section at midleaf with a median row of triangular chlorocysts..........
.... *4. Octoblepharum*
Leaves lanceolate-linear with a narrow lamina of several rows of elongate, hyaline cells along the margin, costa in cross-section with quadrangular leucocysts... 2

2   Very small plants, leaves slender, 2-4(-6) mm long, linear; chlorocysts in leaf apex short and rounded forming green chains in surface view. Capsules immersed, hemisphaeric, peristome absent; budlike propagules on short stalks often present.............................*3. Ochrobryum*
Plants different. Capsules cylindric, exserted on an elongate seta, peristome present; asexual reproduction, if present, by reduced, deciduous leaves.. 3

3   Capsules erect, smooth, peristome teeth papillose, not vertically striate. Leaves with a blunt, cucullate apex; hyaline lamina conspicuous to the apex ....*1. Holomitriopsis*
Capsules arcuate, furrowed when dry, peristome teeth vertically striate. Leaves with a long, acuminate apex (more or less cucullate in *L. antillarum*); hyaline lamina indistinct or absent in upper half of leaf...........*2. Leucobryum*

1. **HOLOMITRIOPSIS** H. Rob., Acta Bot. Venez. 1: 73. 1965.
Type: H. laevifolia (Broth.) H. Rob. (Leucobryum laevifolium Broth.)

Distribution: Guayana Highlands; 1 species.

Note: In Musci I in Fl. Suriname 6: 101. 1964 *Leucobryum laevifolium* was reported for Guyana. Sporophytes were unknown at that time so the genus could not be determined with certainty. Robinson (1965) was the first to describe the capsule as erect and smooth with papillose, variably divided peristome teeth and mitrate calyptra. This is considerably different from the capsules of *Leucobryum* which are arcuate and furrowed when dry with vertically striate, bifid peristome teeth and cucullate calyptra.
The genus *Holomitriopsis* is intermediate between *Leucobryum* and *Schistomitrium* and closely related to the latter. The name is derived from *Holomitrium* referring to the similarity of the elongate perichaetial leaves.

1. **Holomitriopsis laevifolia** (Broth.) H. Rob., Acta Bot. Venez. 1: 75. 1965. – *Leucobryum laevifolium* Broth., Trans. Linn. Soc. London, Bot. ser.2, 6: 90. 1901. Syntypes: Venezuela, McConnell & Quelch 505, 536 (BM). – Fig. 56

Medium sized, dull-green to brownish plants growing in dense, often ball-shaped cushions. Stems prostrate, frequently branched, to 5 cm long; branches short, erect. Leaves 2-6 mm long, erect-spreading, gradually tapering from the flat basal part to the subtubulose, cucullate apex, blunt or apiculate at the tip; costa in cross-section with one layer of leucocysts at either side of the median layer of chlorocysts, chlorocysts nearest the dorsal side throughout the leaf; lamina consisting of elongate, hyaline cells with porose walls forming a border of 3-6 cells wide, extending to apex, broadest at midleaf. Perichaetial leaves convolute, to 5 mm long, acute. Seta to 12 mm long, capsule erect, cylindric, ca. 2 mm long; peristome single (description after Robinson 1965): teeth narrowly lanceolate, 250 μm long, papillose, variably bifid. Calyptra 3 mm long, cucullate, not fringed.

Distribution: Endemic in the Guyana Highlands of Venezuela and Guyana.

Ecology: Epiphytic, often as large balls on branches or slender tree trunks, also terrestrial on detritus; at high altitudes; locally common.

Specimens examined: Guyana: N slope of Mt. Roraima, alt. 700-2300 m, Gradstein 5097, 5153, 5259, 5346, 5450 (L).

Note: This species is easily recognized by the stiff, short leaves with cucullate apex. Noticeable characters are the conspicuous hyaline lamina extending to the apex and the rough dorsal surface of the leaf, caused by the collapsed dorsal leucocysts (visible in cross-section) as a result of resorption of the outer cell walls (Newton & Robinson 1994).

2. **LEUCOBRYUM** Hampe, Linnaea 13: 42. 1839.
   Type: L. vulgare Hampe, nom. illeg. [= L. glaucum (Hedw.) Ångstr. (Dicranum glaucum Hedw.)]

Description see Musci I in Fl. Suriname 6: 99. 1964.

Distribution: Worldwide; in the Guianas 4 species.

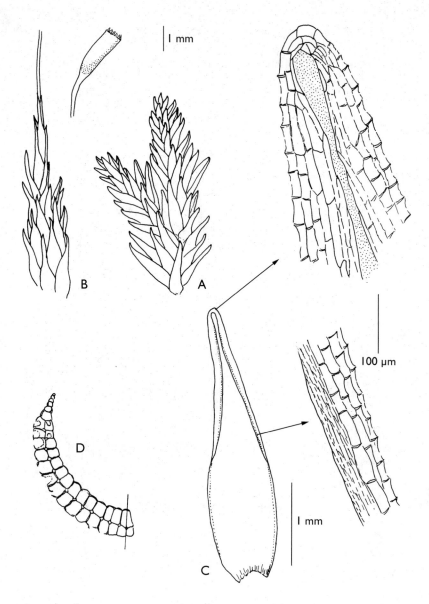

1 mm

100 µm

1 mm

Fig. 56. *Holomitriopsis laevifolia* (Broth.) H. Rob.: A, upper part of plant; B, branch with capsule; C, leaf; D, cross-section of costa with collapsed outer walls of dorsal leucocysts (A, C-D, Gradstein 5097; B, Gradstein 5346).

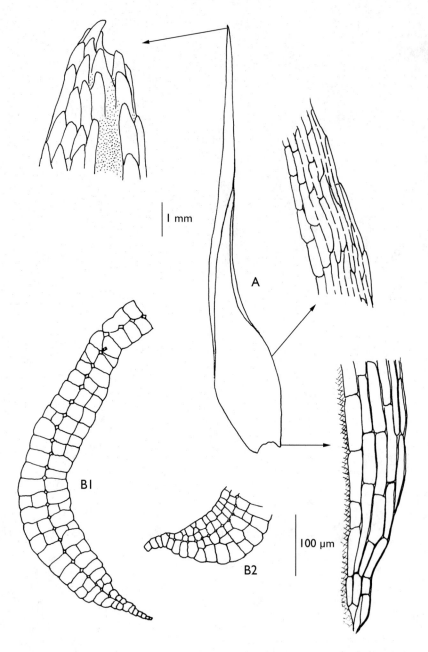

Fig. 57. *Leucobryum albicans* (Schwägr.) Lindb.: A, leaf; B, cross-sections: B1, in basal part of leaf; B2, in alar region at extreme leaf base (A, Gradstein 5225; B, Gradstein 5261).

Note: Usualy the number of leucocyst layers at leaf base is the most important character in the identification of *Leucobryum* species. In species with only one layer of leucocysts at either side of the chlorocyst layer, such as *L. martianum*, it is possible to see this in surface view. For all other species it is necessary to make a cross-section just above leaf base. As this is a very precise job, the results are variable. For that reason other characters are included in the key.

KEY TO THE SPECIES

1   Leaves to 15 mm long, basal part 1.5-2 mm wide . . . . *1. L. albicans*
     Leaves not over 10 mm long, basal part less than 1.5 mm wide 2

2   Leaf apex more or less cucullate, lamina extending nearly to apex 3
     Leaf apex acuminate, lamina extending not beyond midleaf 4

3   Leaves 5-8 mm long, abruptly narrowed to the subtubulose apex; some basal
     laminal cells subdivided, forming rows of quadrate cells . *2. L. antillarum*
     Leaves 2-6 mm long, gradually tapering to apex; all basal laminal cells
     elongate . . . . see *6-1. Holomitriopsis*

4   Leaves strongly flexuose or crispate, (4-)5-9 mm long, abruptly narrowed
     from an obovate or oblong base to a long, slender upper part; costa in cross-
     section with more than one layer of leucocysts in the alar region . . . .
     . . . . *3. L. crispum*
     Leaves falcate-secund or erect-spreading, 4-5(-7) mm long, gradually
     narrowed to the subtubulose upper part; costa in cross-section with one
     layer of leucocysts at either side of the chlorocysts throughout the leaf . . .
     . . . . *4. L. martianum*

1.   **Leucobryum albicans** (Schwägr.) Lindb., Öfvers. Kongl. Vetensk.-
     Akad. Förh. 20: 402. 1863. – *Dicranum albicans* Schwägr., Sp. Musc.
     Frond. Suppl. 2(2): 122. 1827. Type: Brazil, Beyrich s.n. (G).
                                                                            – Fig. 57

Large whitish green plants, growing in loose cushions. Stems undivided, to 7 cm high. Leaves to 15 mm long, leaf base more or less appressed to the stem, 1.5-2 mm wide, gradually narrowed to a flexuose, subtubulose upper part, apex acute, slightly dentate at the tip by protruding cell ends; lamina consisting of porose, elongate-rectangular and linear, hyaline cells, 5-8 rows in the basal part, gradually narrowing to 1 row, extending nearly to apex; costa in cross-section near base with 1-2 layers of leucocysts at either side of the chlorocysts, at extreme base in the alar region 2-3 layers. Sporophytes not seen.

Distribution: West Indies, Mexico, C and tropical S America.

Ecology: Epiphytic and terrestrial in moist habitats; in the Guianas only at higher altitudes.

Specimens examined: Guyana: Upper Essequibo R., Mt. Makarapan, alt. 800 m, Maas *et al.* 7477 (L); N slope of Mt. Roraima, alt. 1200-1600 m, Gradstein 5225, 5261 (L). French Guiana: Sommet Tabulaire, ca. 50 km SE of Saül, alt. 600-700 m, Cremers 6794 (CAY, L).

Note: *Leucobryum albicans* and *L. giganteum* Müll. Hal. are closely related species. Comparison of the types do not provide any gametophytic differences except the size. The specimens from the Guianas are intermediate and show a great variation in size: in collections from high altitudes leaves of 15 mm length occur together with leaves less than 10 mm on the same plant. According to Allen (1994) the leaf length in the 2 species overlaps but *L. albicans* can be distinguished by a narrower leaf base (1.5-2 mm in the Guianas). The main difference between *L. albicans* and *L. giganteum* seems to be in the annulus of the capsule, but no sporophytes could be observed in the collections from the Guianas.
In Musci I *L. giganteum* is reported from Mt. Roraima (McConnell & Quelch 351); this collection from the Venezuelan side of the mountain belongs probably also to *L. albicans*.

2. **Leucobryum antillarum** Schimp. ex Besch., Ann. Sci. Nat., Bot. sér. 6, 3: 190. 1876. Syntypes: Guadeloupe, Herminier 3, Husnot 119 (BM, PC).   – Fig. 58

Plants glaucous to whitish green, growing in dense cushions. Stems to 5 cm long, sparingly branched. Leaves (4-)5-8 mm long, erect-spreading to slightly flexuose, rather quickly narrowed to the subtubulose upper part, apex cucullate, acute, margin entire or slightly denticulate at apex; lamina at leaf base consisting of 5-8 rows of hyaline cells, porose and linear towards margins, gradually narrowing to 1 or 2 rows just below apex, in the basal lamina the rectangular cells are intermixed with short rows of quadrate cells; costa in cross-section near extreme base in the central part with 1 layer of leucocysts and in the alar region with 2-4 layers of leucocysts at either side of the chlorocyst layer. Sporophytes not seen (description after Peterson 1994): seta to 2 cm long, reddish brown, capsule strumose, ca. 2 mm long, operculum long-rostrate, peristome teeth bifid, vertically striate below, papillose above. Calyptra cucullate.

Distribution: SE U.S.A., West Indies, Mexico, C and northern S America.

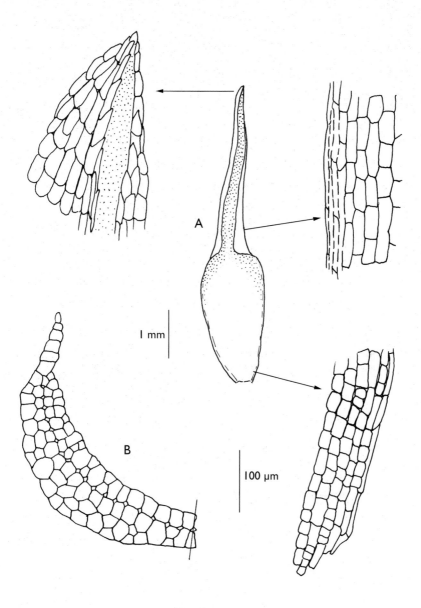

Fig. 58. *Leucobryum antillarum* Schimp. ex Besch.: A, leaf; B, cross-section at leaf base (A-B, Florschütz-de Waard & Zielman 5055).

Ecology: On decaying wood and tree bases, occasionally on rocks; in the Guianas only known from somewhat higher altitudes.

Selected specimens: Guyana: Mt. Latipu, alt. 1000 m, Gradstein 5643 (L). Suriname: Brownsberg, alt. 500 m, Florschütz 4655, 4677 (L). French Guiana: Mt. Galbao, alt. 650 m, Cremers 4134 (CAY, L).

Note: A distinct character to identify this species is the presence of quadrate cells in the basal lamina (in surface view), resulting from subdivision of the elongate laminal cells (Allen 1994).

3. **Leucobryum crispum** Müll. Hal., Syn. Musc. Frond. 1: 78. 1849. Type: Venezuela, Moritz 131 (L).

Description and synonymy see Musci I in Fl. Suriname 6: 103. 1964.

Distribution: West Indies, Mexico, C and tropical S America.

Ecology: On tree bases and decaying wood, also terrestrial and on rocks; in the Guianas usually at higher altitudes.

4. **Leucobryum martianum** (Hornsch.) Hampe, Linnaea 17: 317.1843. – *Dicranum martianum* Hornsch. in Mart., Fl. Bras. 1(2): 11. 1840. Type: Brazil, Martius s.n. (M).

Description and synonymy see Musci I in Fl. Suriname 6: 102. 1964.

Distribution: West Indies, Mexico, C and tropical S America.

Ecology: On tree bases and decaying wood, terrestrial and on rocks; very common in lowland rainforest, also at higher altitudes.

Note: When sporophytes are present this species can be recognized by the sharp-strumose capsules. Sometimes the leaves of *L. martianum* can be slightly crispate, which makes the identification doubtful. The size and shape of the leaf, gradually narrowing to the apex, are good characters to distinguish this species from *L crispum*. Very tiny specimens with leaves not over 3 mm long can be dubious. These "reduced" specimens may be mistaken for *Ochrobryum subulatum*; for differences see under that species.

3. **OCHROBRYUM** Mitt., J. Linn. Soc., Bot. 12: 107. 1869.
Type: O. gardneri (Müll. Hal.) Mitt. (Leucophanes gardneri Müll. Hal.)

Plants small, pale-green to whitish. Stems short, erect. Leaves narrowly oblong to lanceolate or linear, apex acute or obtuse, lamina consisting of a narrow border of elongate, hyaline cells; costa in cross-section with one layer of leucocysts at either side of the median layer of chlorocysts, in surface view chlorocysts in apex short and rounded, forming conspicuous green chains. Asexual reproduction by globose propagules on dorsal leaf surface, in leaf axils or on short terminal stalks. Perigonia terminal on short branches, perichaetia terminal on stem; seta short, capsule immersed to short-exserted, hemispheric to obconic, wide-mouthed when dry, annulus non-revoluble, operculum long-rostrate, peristome absent. Calyptra long-rostrate, mitrate, laciniate or fringed at base.

Distribution: Pantropical; in the Guianas 2 species.

Notes: This genus is mainly determined by the distinctive sporophytic characters. The presence of globose, budlike propagules on the leaves or on terminal stalks can be a good character to recognize *Ochrobryum* without sporophytes.
In *O. subobtusifolium* sporophytes are unknown which makes the generic position uncertain. Asexual reproduction by apical clusters of brood leaves rather than globose propagules made Allen (1992) decide to exclude this species from *Ochrobryum* and transfer it to *Leucobryum*. However, gametophytically this species has much in common with *O. subulatum* including the chains of short, rounded chlorocysts in the apex and the enlarged ventral leucocysts which give the apex a swollen appearance. For the present *Ochrobryum* is maintained for this species until more is known about the sporophyte.

KEY TO THE SPECIES

1   Leaf apex blunt, cucullate; lamina with 1-3 rows of hyaline cells extending
    nearly to apex . . . . *1. O. subobtusifolium*
    Leaf apex slenderly acuminate; lamina with 4-6 rows of hyaline cells along
    basal leaf half . . . . *2. O. subulatum*

1. **Ochrobryum subobtusifolium** Broth., Bih. Kongl. Svenska Vetensk.-Akad. Handl. 26 Afd. 3(7): 10. 1900. – *Leucobryum subobtusifolium* (Broth.) B.H. Allen, Contr. Univ. Michigan Herb. 18: 113. 1992. Type: Brazil, Serra da Chapada, Lindman 402 (H).
    – Fig. 59

150

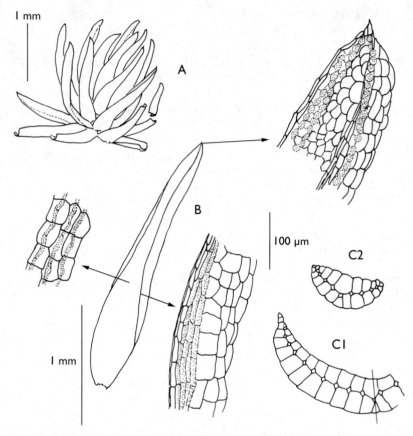

Fig. 59. *Ochrobryum subobtusifolium* Broth.: A, plant; B, leaf; C, cross-sections: C1, in basal part of leaf; C2, in leaf apex (A-C, Allen 20819).

Small plants growing in low mats. Stems to ca. 3 mm long. Leaves 2-3 mm long, lanceolate, tubulose in upper part, apex blunt, more or less cucullate, ending in 1-3 teeth at the tip (apex usually broken in older leaves), margin minutely serrulate at apex; lamina very narrow, consisting of 1-3 elongate-rectangular hyaline cells extending nearly to apex; costa in cross-section with the layer of chlorocysts nearer the ventral side except in apex where the ventral leucocysts are enlarged and filling the cucullate apex; chlorocysts in surface view rounded quadrate, arranged in chains. Sporophyte unknown.

Distribution: Brazil, Suriname.

Ecology: Epiphytic (type) and terrestrial on sandstone boulder (Suriname); apparently rare.

Specimen examined: Suriname: Tafelberg Nat. Park, Arrowhead Basin, alt. 695 m, Allen 20819 (MO).

2. **Ochrobryum subulatum** Hampe in Besch., J. Bot. (Morot) 11: 150. 1897. Type: Burma, Plumadoe, Kurz 2833 (BM). – Fig. 60

Slender plants in low tufts. Stems to 5 mm long. Leaves 2-4 mm long, linear-lanceolate, gradually tapering to a narrow, acuminate apex, margin entire; lamina consisting of elongate-rectangular to linear hyaline cells, 4-6 rows at base, gradually narrower and extending to midleaf; costa in cross-section with the layer of chlorocysts central between the two layers of leucocysts, towards apex nearer the dorsal surface due to the enlarged ventral leucocysts; in surface view the chlorocysts are elongate in lower leaf, rounded-quadrate in apex. Propagules globose, multicellular, produced in globose clusters on specialized stem extensions with rudimentary leaves. Sporophytes not seen (description after Allen 1992): seta 1 mm long, capsule cup-shaped to hemispherical, 0.5 mm long, operculum, 1.5 mm long. Calyptra 3-4 mm long, ciliate at base.

Distribution: Tropical Asia and Africa, Venezuela, Brazil, Bolivia, the Guianas.

Ecology: On rotting logs and tree bases in lowland rainforest; in the Guianas rare, mostly collected at higher altitudes.

Specimens examined: Guyana: Potaro-Siparuni region, near Paramakatoi, alt. 500 m, Newton et al. 3644 (L, US). Suriname: Tafelberg, Crystal Spring Valley, alt. 610 m, Allen 23219 (MO); Sipalawini, Kayserberg Airstrip, alt. 227 m, Allen 25102 (L, MO). French Guiana: Bassin du Haut Marouini, Monpé Soula, alt. 165 m, Hoff 5315 (CAY, L).

Note: When capsules or propagules are lacking this species is best recognized by the very small size and the narrow transparent leaves. From very small specimens of *Leucobryum martianum* it can be distinguished by the chains of rounded chlorocysts and the elongate rectangular leucocysts in the leaf apex (in surface view 3:1, in *L. martianum* hexagonal and 1.5-2:1).

152

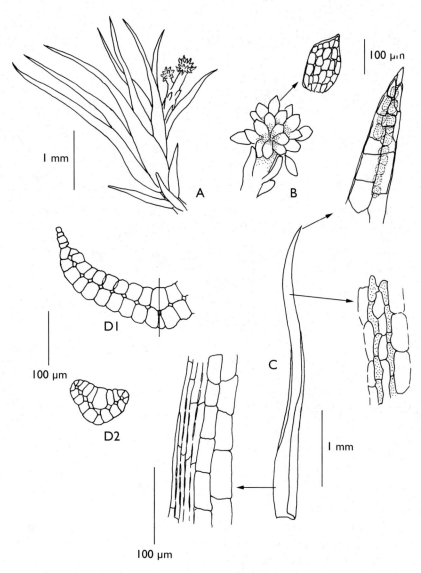

Fig. 60. *Ochrobryum subulatum* Hampe: A, plant with propagula; B, cluster of propagula; C, leaf; D, cross-sections: D1, in basal part of leaf; D2, in leaf apex (A-D, Newton *et al.* 3644).

4. **OCTOBLEPHARUM** Hedw., Sp. Musc. Frond. 50. 1801.
Type: O. albidum Hedw.

Description see Musci I in Fl. Suriname 6: 106. 1964.

Distribution: Pantropical; in the Guianas 8 species.

Note: In a preliminary treatment of the genus *Octoblepharum* in C
America (Salazar Allen 1991) some new characters are introduced to
distinguish the species, e.g. the "pseudocostal" area of elongate cells
visible as a central band in the leaves of *O. cocuiense*. Also the shape of
the lamina and the laminal cells, including the size of the pores, are good
additional characters.

KEY TO THE SPECIES

1   Costa in cross-section near midleaf rounded-triangular or semicircular . . . 2
     Costa in cross-section near midleaf biconvex, at base sometimes triangular . . 3

2   Leaves to 20 mm long, extremely fragile; pores in laminal cells small;
     superficial cells of upper costa quadrate-hexagonal (in surface view), not
     porose . . . .*5. O. erectifolium*
     Leaves usually much shorter, occasionally to 15 mm long, moderately
     fragile; pores in laminal cells large, about as wide as the cell walls;
     superficial cells of upper costa irregular-rectangular with large pores . . . .
     . . . .*2. O. ampullaceum*

3   Leaves when dry with a metallic luster, straw-coloured or light-green in
     upper part, dark red in basal part . . . . *7. O. stramineum*
     Leaves when dry whitish green, sometimes pink or purplish tinged at base 4

4   Leaves with a dark "pseudocostal" band of elongated chlorocysts; laminal
     cells incrassate and conspicuously porose, pores as wide as the cell walls
     . . . . *3. O. cocuiense*
     Leaves without dark "pseudocostal" band; laminal cells thin-walled or only
     slightly incrassate 5

5   Leaves in basal part keeled at back; upper laminal cells elongate-rectangular,
     linear along the margins, pores large (as wide as the lateral cell walls). . . .
     . . . .*8. O. tatei*
     Leaves not keeled at back; upper laminal cells different, pores smaller . . . 6

6   Leaves fragile, upper laminal cells irregular, small, 20-50 μm long. Peristome
     teeth 16 . . . . *6. O. pulvinatum*
     Leaves not fragile, upper laminal cells elongate-hexagonal or rectangular,
     50-80 μm long. Peristome teeth 8. 7

7   Lamina rounded in upper part, at least at one side, central laminal cells elongate, pores minute. Seta less than 10 mm long, capsule ovoid . . . . *1. O. albidum*
Lamina in upper part gradually tapering to the costa, central laminal cells regularly quadrate, pores distinct (about half the cell width). Seta at least 10 mm long, capsule cylindric . . . . *4. O. cylindricum*

1. **Octoblepharum albidum** Hedw., Sp. Musc. Frond. 50. 1801. Type: Jamaica, Swartz s.n. (G).

Description and synonymy see Musci I in Fl. Suriname 6: 112. 1964.

Distribution: Pantropical and subtropical.

Ecology: Epiphytic on stems and branches of solitary trees, in open vegetations or in the canopy of rainforest; common.

Note: This species is easily recognized by the ovoid capsules on a short seta with 8 triangular, non-trabeculate peristome teeth. Without capsules it could be confused with *O. cylindricum*, see key for the differences.

2. **Octoblepharum ampullaceum** Mitt., J. Linn. Soc., Bot. 12: 109. 1896. Lectotype (Florschütz 1955): Peru, San Carlos, Spruce 75 (NY).   – Fig. 61

Lightgreen to whitish plants growing in low tufts; stems short. Leaves erect, rigid, 6-12(-15) mm long and 0.2-0.4 mm wide, moderately fragile, apex obtuse, apiculate or acute; costa in cross-section rounded-triangular nearly to apex, with 2-4 layers of leucocysts on either side of the chlorocysts; in surface view exterior leucocysts irregular-rectangular with large round pores; lamina broad at base, in upper part rounded to the costa, bluntly serrate; central laminal cells rectangular, narrower to the costa, in upper lamina elongate-hexagonal, pores large, about as wide as the cell walls. Perichaetial leaves small, with a short costa. Seta ca. 1 cm long; capsule erect, ovoid (operculum and peristome not seen).

Distribution: Tropical S America, usually at higher altitudes.

Ecology: Epiphytic in low savanna forest and shrubsavanna, also terrestrial on wet humus in lowland rainforest; in the Guianas below 1000 m alt., rare.

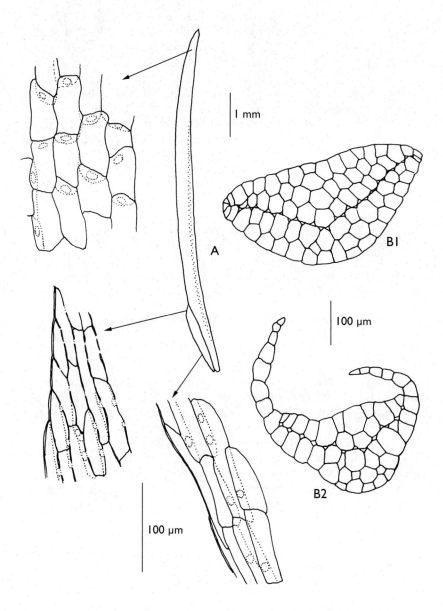

Fig. 61. *Octoblepharum ampullaceum* Mitt.: A, leaf; B, cross-sections: B1, at midleaf; B2, at leaf base (A-B, Maas *et al.* 4222).

Specimens examined: Guyana: Pakaraima Mts., Mt. Latipú, alt. 900 m, Maas *et al.* 4222 (L). Suriname: Emma Range, alt. 1025 m, Jonker & Daniëls 1025B. French Guiana: Station des Nouragues, Arataye, alt. 390 m, Cremers 10965 (CAY, L).

Note: This species is close to *O. erectifolium*; both species have a slender, rounded-triangular costa; the main differences indicated so far are length and fragility of the leaves, but the leaf length is variable in both species and overlap. The shape of the laminal cells as studied in the type collections seems to be variable (Florschütz 1955). A useful character to distinguish *O. ampullaceum* is the presence of large pores in the laminal cells and in the leucocysts of the costa.

3. **Octoblepharum cocuiense** Mitt., J. Linn. Soc., Bot. 12: 109. 1869. Type: Venezuela, Mt. Cocui, Spruce s.n. (NY).

Octoblepharum pellucidum Müll. Hal., Genera Musc. Frond. 89. 1901. Type: Brazil, Mosén 196 (H, PC).

Description see Musci I in Fl. Suriname 6: 108. 1964.

Distribution: C and tropical S America.

Ecology: On tree trunks, logs and moist sand or rocks; common in savanna forest and scrub savanna, also near creeks in lowland rainforest, occasionally at higher altitudes.

Note: *O. pellucidum* is identical with *O. cocuiense* as already suggested in a preliminary study of the types (Florschütz 1955). The shape of the leaf apex is variable and not useful to separate the species. The best character to distinguish *O. cocuiense* from all other species in the Guianas is the dark pseudocostal band created by elongated chlorocysts in the central costa.

4. **Octoblepharum cylindricum** Mont., Ann. Sci. Nat., Bot. sér. 2, 4: 349. 1840. Type: French Guiana, Leprieur 282 (PC).

Description see Musci I in Fl. Suriname 6: 114. 1964.

Distribution: West Indies, C and tropical S America.

E c o l o g y : Terrestrial on sand or debris, occasionally epiphytic; rather common in the vegetation of open rock plateaus and in open areas of savanna forest.

N o t e : The wide-spreading, often recurved leaves make the plants look similar to *O. albidum*; the leaf apices are variable in shape and form no reliable character to separate the species; the main differences are in the sporophyte with longer seta and a cylindric capsule.

5. **Octoblepharum erectifolium** Mitt. ex R.S. Williams, N. Amer. Fl. 15: 162. 1913. Type: Trinidad, Crüger s.n. (NY).

Description see Musci I in Fl. Suriname 6: 107. 1964.

D i s t r i b u t i o n : West Indies, Mexico, C and tropical S America.

E c o l o g y : Epiphytic on tree trunks in lowland rainforest; rare.

N o t e : This species is characterized by the long and narrow leaves, very fragile and seldom intact, in cross-section near midleaf rounded-triangular. The extreme long leaves could not be observed in the collections from French Guiana, the intact leaves were not over 8 mm long. The only collection cited from Suriname (Jonker & Daniëls 1025B) proved to belong to *O. ampullaceum*; for differences with *O. ampullaceum* see under that species.

6. **Octoblepharum pulvinatum** (Dozy & Molk.) Mitt., J. Linn. Soc., Bot. 12: 109. 1869. – *Arthrocormus pulvinatus* Dozy & Molk., Prod. Fl. Bryol. Sur.: 6. 1854. Type: Suriname, Splitgerber 1214 (L, NY).

Description see Musci I in Fl. Suriname 6: 110. 1964.

D i s t r i b u t i o n : West Indies, Mexico, C and tropical S America.

E c o l o g y : Epiphytic on tree trunks and rotten logs, occasionally terrestrial on humus; in the understory of lowland rainforest very common.

N o t e : This species is easily recognized by the very fragile leaves and the short, broad lamina, in upper part abruptly rounded to the costa; the upper laminal cells are short and irregular quadrate or hexagonal, with many small pores.

7. **Octoblepharum stramineum** Mitt., J. Linn. Soc., Bot. 12: 110. 1869. Type: Venezuela, Spruce s.n. (NY).

Description and synonymy see Musci I in Fl. Suriname 6: 113. 1964. Description of the capsule (after Salazar Allen 1991): Seta 6-9 mm long, capsule erect to subinclined, ovoid to cylindric, 1-1.8 mm long, operculum long-rostrate, to 1 mm long; peristome teeth 16, lanceolate, trabeculate, erect when dry, reflexed when wet.

Distribution: C and tropical S America.

Ecology: Epiphytic in the canopy of lowland rainforest and in shrub savannas at higher altitudes (to 900 m).

Note: This species is quickly recognized by the metallic luster of the stramineous leaves, dark red towards base; the laminal cells are rather uniform, elongate-rectangular or hexagonal, slightly shorter in upper lamina, firm-walled, coloured and conspicuously porose with large oval pores.

8. **Octoblepharum tatei** (R.S. Williams) E.B. Bartram, Mem. New York Bot. Gard. 10: 4. 1960. – *Carinafolium tatei* R.S. Williams, Bull. Torrey Bot. Club 58: 502. 1931. Type: Venezuela, Tate 1054 (NY).   – Fig. 62

Pale green plants in dense cushions. Stems to 3 cm high, sparingly branched. Leaves erect-spreading, rather soft and flexuose, linear, 5-8 (-11) mm long and 0.3-0.5 mm wide, keeled at back in the basal part, apex acute; lamina narrow, gradually tapering to the costa in upper part, cells in central lamina rectangular, 80-140 μm long and ca. 40 μm wide, towards margin narrower, conspicuously porose, pores oval, as wide as the cell walls; cross-section in basal leafhalf triangular, keeled at dorsal side, with 2-3 layers of swollen leucocysts on either side of the chlorocysts, cross-section in upper leaf biconvex with 1-2 layers of leucocysts at ventral and 1 layer at dorsal side of the chlorocysts. Gemmae fusiform, produced at leaf apex. Perichaetial leaves with a short costa, at apex acuminate and sharply apiculate. Seta orange-red, to 2 cm long, capsule erect, ovoid, to 2 mm long, operculum long-rostrate, peristome (description after Salazar Allen 1993): 16 hyaline teeth with vertical and semi-circular thickenings on the dorsal plates.

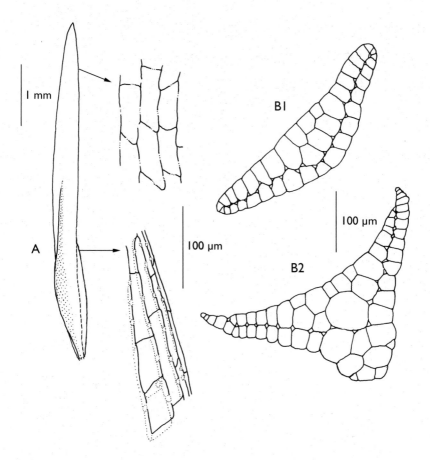

Fig. 62. *Octoblepharum tatei* (R.S. Williams) E.B. Bartram: A, leaf; B, cross-sections: B1, in upper leaf; B2, at leaf base (A-B, Aptroot 17091).

Distribution: C and tropical S America.

Ecology: Epiphytic on branches and tree trunks; in Guyana and Suriname only collected at higher altitudes.

Specimens examined: Guyana: Pakaraima Mts., Mt. Membaru, alt. 600 m, Maas *et al.* 4323 (L); N slope of Mt. Roraima, alt. 700 m, Aptroot 17091 (L). Suriname: Tafelberg, along Geyskes Cr., alt. 565 m, Allen 23259 (L, MO).

7. LEUCOPHANACEAE

Distribution: Pantropical; 1 genus.

Notes: Fleischer (1904) was the first to place *Leucophanes* (together with *Octoblepharum*) in a separate family, the LEUCOPHANACEAE. Later it was placed alternately in the families CALYMPERACEAE and LEUCOBRYACEAE.

In *Leucophanes* the cross-section of the upper leaf is multistratose, composed of a central layer of chlorocysts and 1 or 2 layers of leucocysts on either side. This is comparable with the cross-section in the LEUCOBRYACEAE, where the upper leaf is interpreted as an expanded costa and the hyaline borders at base as a reduced lamina. However, in *Leucophanes* a midstereid band takes the place of the costa and the lamina is unistratose in the basal part of the leaf but becomes gradually multistratose in upper part.

Robinson (1985) compared the hyaline unistratose part of the lamina with the cancellinae of the CALYMPERACEAE. Also the border of incrassate, linear cells and the production of propagules at leaf apex are characters which indicate a close relationship with the CALYMPERACEAE.

Salazar Allen (1986) placed *Leucophanes* in a separate family. She considered the leaf as composed of a costa (the midstereid band), a multistratose lamina and a basal hyaline lamina.

LITERATURE

Fleischer, M. 1904. Leucophanaceae. In: Musc. Buitenzorg 1: 166-193.
Robinson, H. 1985. The structure and significance of the leucobryaceous leaf. Monogr. Syst. Bot. Missouri Bot. Gard. 11: 111-120.
Salazar Allen, N. 1986. A revision of the pantropical moss genus *Leucophanes*. PhD dissertation Univ. of Alberta (314 pp.).
Salazar Allen, N. 1993. Leucophanaceae. Flora Neotrop. Monogr. 59: 1-10.

1. **LEUCOPHANES** Brid., Bryol. Univ. 1: 763. 1827.
   Type: L. octoblepharioides Brid.

Whitish-green to glaucous plants. Stems erect, little branched. Leaves lanceolate, apex acuminate, acute, obtuse or retuse, margins uni- or multistratose, serrate in upper part; lamina in basal part unistratose or irregularly bistratose, towards midleaf in central part gradually becoming multistratose; cross-section in upper part with a median layer of quadrate chlorocysts and one or more layers of leucocysts on either side; costa (midstereid band) narrow, percurrent or short-excurrent, smooth to strongly spinose at back. Dioicous. Perigonial leaves ovate from a broad base, perichaetial leaves not differentiated; seta elongate, smooth, capsule cylindrical, operculum long-rostrate, peristome with a prostome, teeth lanceolate, papillose or papillose-striate on dorsal surface. Calyptra cucullate. Perigonia and sporophytes unknown in the neotropical populations. (Description after Salazar Allen 1993).

D i s t r i b u t i o n : Pantropical and subtropical; in the Guianas 1 species.

1. **Leucophanes molleri** Müll. Hal., Flora 69: 285. 1886. Type: Africa, São Thomé, Moller s.n. (H, PC, S).

   Leucophanes mittenii Cardot, Index Bryol. ed. 2, 3: 192. 1905. Type: Ecuador, Jameson s.n. (NY).

Small to medium-sized plants growing scattered or in loose tufts; stems to 4 cm long. Leaves erect-spreading, linear-lanceolate, carinate, to 5 mm long, apex broad-acute or obtuse, serrate, margins bordered with incrassate, rectangular cells; laminal cells at leaf base hyaline, quadrate-rectangular in a single layer, in central part multistratose with a median layer of chlorocysts and a layer of hyaline cells on both sides; in distal direction this multistratose part is broadening and at midleaf the unistratose part is absent.

D i s t r i b u t i o n : West Indies, C and tropical S America.

E c o l o g y : Epiphytic on bark of trees in lowland rainforest; in the Guianas not common, thus far not collected in Guyana.

Order **Pottiales**

Plants acrocarpous. Stems erect, simple or forked. Leaves with single costa, well-developed, upper laminal cells small, isodiametric, often papillose, basal laminal cells larger and hyaline. Propagules often present. Sporophytes terminal, setae elongate, capsules erect, symmetric, peristome single, teeth 16, often deeply divided.

KEY TO THE FAMILIES

1   Basal laminal cells strongly enlarged in a conspicuous hyaline group, sharply separated from the small, green upper laminal cells. Plants often with propagules at tips of leaves .... *8. CALYMPERACEAE*
    Basal laminal cells often lax and hyaline but not sharply separated in a conspicuous group. Propagules, if present, developed in leaf axils and on rhizoids, seldom on leaves .... *9. POTTIACEAE*

8. CALYMPERACEAE

Description see Musci I in Fl. Suriname 6: 117. 1964.

Distribution: Pantropical; in the Guianas 2 genera.

LITERATURE

Edwards, S.R. 1980. A revision of west tropical African Calymperaceae I. Introduction and Calymperes. J. Bryol. 11: 49-93.

Orbán, S. & W.D. Reese 1990. Syrrhopodon prolifer: A world view. Bryologist 93: 438-444.

Reese, W.D. 1961. The genus Calymperes in the Americas. Bryologist 64: 89-140.

Reese, W.D. 1975. Calymperes rubiginosum (Mitt.) Reese, comb. nov. Bryologist 78: 92-93.

Reese, W.D. 1977. The genus Syrrhopodon in the Americas I. The elimbate species. Bryologist 80: 2-31.

Reese, W.D. 1978. The genus Syrrhopodon in the Americas II. The limbate species. Bryologist 81: 189-225.

Reese, W.D. 1981. Refinements on American Syrrhopodon. Bryologist 84: 244-248.

Reese, W.D. 1983. American Calymperes and Syrrhopodon; identification key and summary of recent nomenclatural changes. Bryologist 86: 23-30.

Reese, W.D. 1984. Calymperes and Syrrhopodon in the Serra do Cachimbo, Brazil and Rhacopilopsis trinitensis (C.Müll.) Britt. & Dix. new to Brazil. Lindbergia 10: 123-126.

Reese, W.D. 1990. Calymperes othmeri reïnstated. Bryologist 93: 37-38.

Reese, W.D. & W.R. Buck. 1991. A sample of Calymperes and Syrrhopodon from French Guiana. Bryologist 94: 298-300.

Reese, W.D. 1993. Calymperaceae. Flora Neotropica Monograph 58: 1-101.

Reese, W.D. 1994. Per Dusén's African species of Syrrhopodon subgenus Pseudo-calymperes. Bryologist 97: 371-376.

Reese, W.D. 1995. Synopsis of Syrrhopodon subgenus Pseudocalymperes. Bryologist 98: 141-145.

Reese, W.D. 1997. Asexual reproduction in Calymperaceae, with special reference to functional morphology. J. Hattori Bot. Lab. 82: 227-244.

Notes: In this family sporophytes are infrequently produced, on the other hand asexual reproduction is rather common. The many different modes of asexual reproduction are examined by Reese (1997). Occasionally dispersal by fragile leaves occurs (*Calymperes venezuelanum*) but most common is the production of propagules. These are sometimes produced along the costa or leaf margins (*Calymperes lonchophyllum, Syrrhopodon helicophyllus, S. lycopodioides*) but more often at the end of the costa. In most species at the ventral side of the leaf tip, sometimes on conspicuous receptacles (*C. palisotii, C. smithii*). In other species the propagules are produced all around the naked tip of the excurrent costa (*C. erosum, C. othmeri*). Also highly modified propaguliferous leaves are often seen in this family: leaves with the upper lamina strongly reduced to a more or less naked costa, bearing propagules at the tip (*S. hornschuchii, S. rigidus*) or leaves with widened lamina, arranged in comae at the end of the stem, bearing propagules at midleaf (*S. parasiticus* and other representatives of subgenus *Pseudocalymperes*). All these forms can be considered as adaptations to successful dispersal. Though not important in taxonomic sense they are very useful as additional characters in identification.

KEY TO THE GENERA

1 Sporophytes present 2
  Sporophytes absent 3

2 Calyptra campanulate, persistent, clasping the seta below the capsule; peristome absent . . . . *1. Calymperes*
  Calyptra cucullate, deciduous; peristome present or lacking . . . .
  . . . . *2. Syrrhopodon*

3 Leaves with a border of elongate, hyaline cells along upper lamina (sometimes incomplete or lacking) . . . . *2. Syrrhopodon*
  Leaves without hyaline border along upper lamina 4

4 Leaves without a band of elongate cells at shoulders 5
  Leaves with a marginal or intramarginal band of elongate cells at shoulders 8

5 Margin of upper lamina thickened 6
   Margin of upper lamina not thickened 7

6 Upper lamina narrowly linear with quadrate, often transversely elongate
   cells. . . . . 1. *Calymperes*
   Upper lamina ovate, oblong or lanceolate, lacking transversely elongate
   cells. . . . . 2. *Syrrhopodon*

7 Leaf margins crenulate. Propaguliferous leaves little differentiated, bearing
   fusiform propagules at ventral side of the broadened apex . . . .
   . . . . 1. *Calymperes (rubiginosum)*
   Leaf margins entire. Propaguliferous leaves enlarged, forming comae at end
   of the stem; propagules filamentous, borne on midleaf along ventral side of
   the costa . . . . .2. *Syrrhopodon (parasiticus)*

8 Band of elongate cells intramarginal . . . . 1. *Calymperes*
   Band of elongate cells marginal, forming a border 9

9 Border at shoulders 2-6 cells wide . . . . 1. *Calymperes*
   Border at shoulders more than 6 cells wide 10

10 Thickened margin of upper lamina smooth or finely serrulate. . . . .
   . . . . 1. *Calymperes*
   Thickened margin of upper lamina double-dentate . . . .
   . . . . 2. *Syrrhopodon (lanceolatus)*

1. **CALYMPERES** Sw. in F. Weber, Tab. Calyptr. Oper. Musc. Frond.
   Gen. 2. 1813.
   Lectotype (Williams 1920): C. lonchophyllum Schwägr.

Description see Musci I in Fl. Suriname 6: 117. 1964.

Distribution: Pantropical; in the Guianas 14 species.

KEY TO THE SPECIES

1 Upper lamina narrowly linear, laminal cells transversely elongate 2
   Upper lamina ovate, lanceolate or broadly linear, laminal cells isodiametric
   to vertically elongate 4

2 Leaves rigid, not contorted when dry; upper lamina in basal part interrupted
   and restricted to the bare costa . . . . 14. *C. venezuelanum*
   Leaves soft, contorted when dry; upper lamina not interrupted at base 3

3　Cells in upper lamina obscure and densely papillose; leaves to 8 mm long
　. . . . . *4. C. levyanum*
　Cells in upper lamina smooth; leaves to 16 mm long . . .*5. C. lonchophyllum*

4　Leaves without elongate cells in shoulder region; costa reddish . . . .
　. . . . . *12. C. rubiginosum*
　Leaves with a band of elongate cells along margin or intramarginal in
　shoulder region (teniolae). 5

5　Teniolae marginal at least at shoulders. 6
　Teniolae intramarginal, sometimes scarcely developed 9

6　Teniolae at shoulders 2-6 cells wide. . . . . *13. C. smithii*
　Teniolae at shoulders 6-16 cells wide. 7

7　Upper laminal cells smooth, thickened margin double-dentate . . . .
　. . . . . .see *11. Syrrhopodon lanceolatus*
　Upper laminal cells papillose, thickened margin smooth or finely serrulate 8

8　Cancellinae extending ca. 3/4 the length of lower lamina, upper laminal cells
　finely papillose dorsally . . . .*6. C. mitrafugax*
　Cancellinae extending ca. 1/2 the length of lower lamina, upper laminal cells
　strongly mammillose-papillose dorsally . . . .*11. C. platyloma*

9　Cancellinae at ventral side strongly mammillose at shoulders. 10
　Cancellinae at ventral side plane 11

10　Teniolae at shoulders 3-8 cells in from margin; propagules borne on all sides
　of the excurrent costa . . . .*2. C. erosum*
　Teniolae at shoulders 1-3 cells in from margin; propagules borne on ventral
　side of the percurrent costa. . . . . *10. C. pallidum*

11　Lower lamina not wider than upper lamina, cancellinae at distal end forming
　very broad angles with the costa; propaguliferous leaves ending in a broad,
　brush-like receptacle at ventral side . . . .*9. C. palisotii*
　Lower lamina wider than upper lamina, cancellinae at distal end acute;
　propagules borne all around costa or at ventral side only, but not in a
　conspicuous receptacle. 12

12　Teniolae at shoulders conspicuous in an uninterrupted band to the upper
　lamina border 13
　Teniolae at shoulders inconspicuous, interrupted or sometimes absent . . . 14

13　Teniolae little incrassate, well-separated from the undifferentiated cells
　towards margins . . . . *1. C. afzelii*
　Teniolae strongly incrassate and porose, not distinctly separated from the
　(also incrassate) cells towards margins . . . .*3. C. guildingii*

14 Leaves narrowly lanceolate, subtubulose, acuminate at apex; teniolae at shoulders indistinctly separated from the cells towards margin; propagules borne on ventral side of the percurrent costa . . . . *7. C. nicaraguense*
Leaves broadly lanceolate, acute at apex; teniolae at shoulders well-separated from the cells towards margin; propagules borne on all sides of the excurrent costa . . . . *8. C. othmerii*

1. **Calymperes afzelii** Sw., Jahrb. Gewächsk. 1: 3. 1818. Type: Africa, Afzelius s.n. (BM).

   Calymperes donnellii Austin, Bot. Gaz. (London) 4: 151. 1879. Type: Florida, Caloosa, Smith & Austin s.n. (NY).

   Description and synonymy see Musci I in Fl. Suriname 6: 138. 1964, as *C. donnellii*.

   D i s t r i b u t i o n : Pantropical.

   E c o l o g y :On rocks in shaded places, on logs, on stems of living trees; common in lowland rainforest, preferably in moist areas.

   N o t e s : *C. donnellii* proved to be identical with the African species *C. afzelii* (Edwards 1980).
   This species is characterized by the clearly delimited teniolae, at shoulders well-separated from the cells along the margins. The plane distal cells of the cancellinae, the thickened margins of the upper lamina and the propagules borne on the ventral side of the costa distinguish it from *C. erosum*.

2. **Calymperes erosum** Müll. Hal., Linnaea 21: 182. 1848. Lectotype (Reese 1961): Suriname, Kegel 539 (GOET, PC).

   Description and synonymy see Musci I in Fl. Suriname 6: 134. 1964.

   D i s t r i b u t i o n : Pantropical.

   E c o l o g y : On decaying wood and epiphytic in all forest types in not too shady places and in the canopy, also on rock and soil; common.

   N o t e : The mammillose distal cells of the cancellinae are not always easy to observe. Good additional characters are the conspicuous teniolae, at shoulders 4-8 cells in from the margins and distinctly separated from the marginal cells; the teniolae extend far up along the upper margins, which for the greater part are not thickened. The propagules are produced on all sides of the excurrent costa, in *C. afzelii* and *C. pallidum* only on ventral side of the costa.

3. **Calymperes guildingii** Hook. & Grev., Edinburgh J. Sci. 3: 223. 1825. Type: Antilles, St. Vincent, Guilding s.n. (BM, NY).

Description see Musci I in Fl. Suriname 6: 144. 1964.

D i s t r i b u t i o n : West Indies, tropical S America.

E c o l o g y : On rock and bark at higher altitudes; in the Guianas rare, only represented by a single collection from each of the countries.

N o t e : The strongly incrassate, porose teniolae are conspicuous and extending to the broadly thickened margins of the upper lamina. The teniolae at shoulders are not sharply separated from the marginal cells, which are also incrassate.

4. **Calymperes levyanum** Besch., Ann. Sci. Nat., Bot. sér. 8, 1: 290. 1896. Type: Nicaragua, Levy s.n. (NY, PC).

Description see Musci I in Fl. Suriname 6: 121. 1964.

D i s t r i b u t i o n : West Indies, Mexico, C America, tropical S America, southern Africa, southern China.

E c o l o g y : On bark of trees; rare, not collected in French Guiana.

5. **Calymperes lonchophyllum** Schwägr., Sp. Musc. Frond. Suppl. 1(2): 333. 1816. Type: French Guiana, Richard s.n. (G, PC).

Description see Musci I in Fl. Suriname 6: 119. 1964.

D i s t r i b u t i o n : Pantropical.

E c o l o g y : Common in lowland rainforest, often in extensive mats on the stems of trees.

6. **Calymperes mitrafugax** Florsch., Fl. Suriname 6: 129. 1964. Type: Suriname, Wane Cr., Lanjouw & Lindeman 521 (LAF, L).

Description see Musci I in Fl. Suriname 6: 129. 1964.

D i s t r i b u t i o n : Venezuela, the Guianas, Brazil, Peru.

E c o l o g y : On bark of living trees; rare, only known from the type locality and a single collection from each of the Guianas.

7. **Calymperes nicaraguense** Renauld & Cardot, Bull. Soc. Roy. Bot. Belgique 33: 117. 1894. Type: Nicaragua, Escondido R., Richmond s.n. (NY, PC).   – Fig. 63 E-D

Small to medium-sized plants growing in tufts or loose mats. Stems usually simple, to 2 cm high. Leaves with broad, appressed lower lamina and involute, lanceolate upper lamina, strongly contorted when dry, spreading and subtubulose when moist, 4-5 mm long; upper lamina with thickened margins nearly to apex, serrulate to distantly double-dentate, apex acuminate, costa percurrent in the very narrow apex, rough-papillose at dorsal side towards apex, in propaguliferous leaves bearing propagules at ventral side; lower lamina much broader than upper lamina, 1/3-1/4 of leaf length, entire to serrulate at shoulders, cancellinae nearly filling the lower lamina, acute or rounded at distal end; upper laminal cells rounded quadrate to oval, more or less collenchymatous, 7-12 μm long, bulging mammillose ventrally, smooth to unipapillose dorsally; teniolae variable, not distinctly separated from the marginal cells which are also elongate. Seta 3-5 mm long, capsule cylindric, ca. 2 mm long, operculum rostrate, 0.8 mm long. Calyptra 3-4 mm long, plicate, rough at apex.

D i s t r i b u t i o n : West Indies, C and tropical S America.

E c o l o g y : Epiphytic in lowland rainforest; rare, represented only by a single collection from each of the Guianas.

S p e c i m e n s e x a m i n e d : Suriname: Sara Cr., beyond Dam, Florschütz 187 (L). French Guiana: Mountain ridge 40 km SW of Cayenne, alt. 100 m, Aptroot 15632 (CAY, L); Arataye, Piste des Cascades, alt. 125 m, Cremers 11056 (CAY, L).

N o t e s : The description of this species in Musci I in Fl. Suriname 6: 141. 1964 is partly based on characters of *C. othmeri*, at that time considered to be identical. Since *C. othmeri* is recognized as a distinct species (Reese 1990), the description of *C. nicaraguense* had to be adjusted.

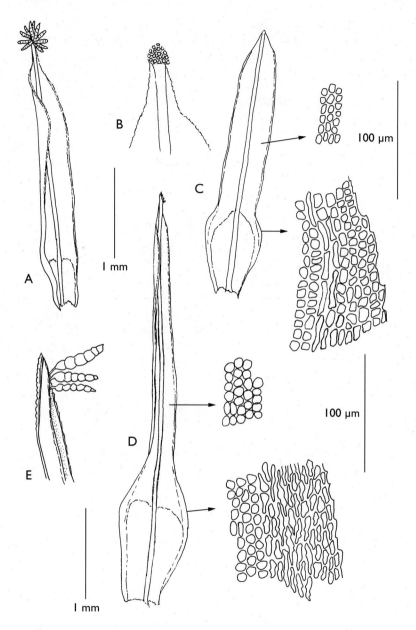

Fig. 63. A-C. *Calymperes othmeri* Herzog: A, leaf with propagula; B, leaf apex with empty propagula-receptacle; C, vegetative leaf (A-C, Florschütz-de Waard & Zielman 5753). D-E. *Calymperes nicaraguense* Renauld & Cardot: D, propaguliferous leaf with empty receptacle at ventral side of the costa; E, leaf apex with propagula (D-E, Aptroot 15632).

The main characters to distinguish this species from *G. othmeri* are the quickly narrowed, subtubulose upper lamina, the acuminate apex and the place of the propagules, borne on ventral side of the costa.

8. **Calymperes othmeri** Herzog, Arch. Bot. São Paulo 1: 60. 1925. Type: Venezuela, Guayana, Caroni, Othmer 443 (JE, M).

– Fig. 63 A-C

Calymperes rupicola P.W. Richards, Bull. Misc. Inf. 8: 323. 1934. Type: Guyana, Essequibo R., Richards 357 (BM, NY).

Medium-sized plants growing scattered or in loose mats. Stems little divided, to 3 cm long. Leaves with whitish, appressed lower lamina and involute, contorted upper lamina when dry, flat and spreading when moist, 3-5 mm long; upper lamina with margin thickened nearly to apex, irregularly serrulate to serrate, apex acute, costa percurrent, towards apex rough-papillose at dorsal side, in propaguliferous leaves excurrent with propagules borne on all sides; lower lamina broader than upper lamina, ca. 1/3 of total leaf length, margins at shoulders irregularly serrulate, cancellinae acute or rounded at distal end, nearly filling lower lamina; upper laminal cells obscure, quadrate, 5-9 µm long, bulging on ventral side, minutely papillose on dorsal side; teniolae indistinct, often interrupted, well-separated from the cells along margin. Sporophyte not seen.

Distribution: Tropical S America.

Ecology: Terrestrial on stones and humus; frequently collected on exposed rocks in lowland rainforest.

Selected specimens: Guyana: Essequibo R. near falls, Smith 2105 (L). Suriname: Bakhuis Mts., 10 km NNW of base camp on W bank of Coppename R., alt 100-600 m, Florschütz & Maas 2838 (L); Kabalebo R. area, rock plateau km 117 of road to Amatopo, Florschütz de Waard & Zielman 5613 (L).

Note: *C. othmeri*, originally placed in synonymy of *C. nicaraguense*, was reïnstated as a separate species by Reese (1990). It is distinguished by a broader leaf apex and in propaguliferous leaves by an excurrent costa with propagules formed on all sides. The teniolae are incomplete or weak, but well separated from the undifferentiated cells towards margin.

9. **Calymperes palisotii** Schwägr., Sp. Musc. Frond. Suppl. 1(2): 334. 1816. Lectotype (Edwards 1980): Africa, Oware, Palisot s.n. (S).

Calymperes richardii Müll. Hal., Syn. Musc. Frond. 1: 524. 1849. – *C. palisotii* Schwägr. subsp. *richardii* (Müll. Hal.) S.R. Edwards, J. Bryol. 11: 81. 1981. Lectotype (Reese 1961): Brazil, Pernambuco, Gardner 47 (BM, PC).

Description and synonymy see Musci I in Fl. Suriname 6: 132. 1964, as *C. richardii*.

D i s t r i b u t i o n : southern U.S.A., West Indies, C and tropical S America, tropical Africa, western Asia.

E c o l o g y : Epiphytic on solitary trees and in savanna forest, occasionally on rocks; very common.

N o t e s : This species first described as *C. richardii* is identical with the African species *C. palisotii*. It was first treated as a subspecies (Edwards 1980), but Reese (1993) concluded that the distinctions between the two subspecies are negligible.
C. palisotii is easily recognized by the broad, blunt cancellinae in a rather narrow lower lamina. Also the conspicuous brush-shaped receptacles at ventral side of the apex form a good decisive character, in which it can only be confused with *C. smithii*.

10. **Calymperes pallidum** Mitt., Philos. Trans. Roy. Soc. London 168: 388. 1879. Lectotype (Reese 1993): Africa, Rodriguez, Balfour s.n. (BM).

Calymperes uleanum Broth., Hedwigia 51: 124. 1912. Type: Brazil, Goiás, Ule 1556 (H, NY).

Description see Musci I in Fl. Suriname 6: 137. 1964, as *C. uleanum*.

D i s t r i b u t i o n : Pantropical.

E c o l o g y : In the Neotropics not common, collected only a few times on tree trunks and logs.

N o t e s : *C. uleanum* was earlier described from Africa as *C. pallidum*, which name has priority (Reese 1993).
A useful character to distinguish this species from *C. erosum* is the distance of the teniolae to the margin at shoulders and the place of the propagules on ventral side of the leaf tip.

11. **Calymperes platyloma** Mitt., J. Linn. Soc. Bot. 12: 128. 1869. Type:
Guyana, Appun 819 (BM, NY).

Description see Musci I in Fl. Suriname 6: 127. 1964.

D i s t r i b u t i o n : Brazil, the Guianas.

E c o l o g y : Epiphytic on tree trunks and in the lower canopy, in savanna
forest and open areas of lowland rainforest, to 800 m high; not common.

N o t e : The cancellinae are short and truncate as in *C. smithii*, but *C.
platyloma* can be distinguished by the sharply serrate margins at shoulders
and the broader teniolae.

12. **Calymperes rubiginosum** (Mitt.) Reese, Bryologist 78: 92. 1975.
– *Syrrhopodon rubiginosus* Mitt., J. Linn. Soc., Bot. 12: 125. 1869.
Lectotype (Reese 1975): Venezuela, Maypures, Spruce s.n. (BM, NY).

Calymperes rufum Herzog, Beih. Bot. Centralbl. Abt. 1, 61: 585. 1942. Type:
Brazil, Amazonia, Franck s.n. (JE).

Description see Musci I in Fl. Suriname 6: 123. 1964, as *C. rufum*.

D i s t r i b u t i o n : C America, tropical S America.

E c o l o g y : Epiphytic on bark of trees in lowland rainforest; rare in
Guyana and Suriname, more often collected in French Guiana.

N o t e s : The older name *C. rubiginosum* replaces *C. rufum* (Reese 1975).
The non-broadened shoulders, the absence of teniolae and the crenulate
margins in upper lamina are characteristic for this species. In habit it
could be confused with *Hyophila tortula*, but the distinct cancellinae and
the rounded, incrassate upper laminal cells are decisive.

13. **Calymperes smithii** E.B. Bartram, Bull. Torrey Bot. Club 66: 223.
1939. Type: Brazil-Guyana boundary, Akarai Mts., Smith 2985 (BM,
FH, L, NY).

Description see Musci I in Fl. Suriname 6: 125. 1964.

D i s t r i b u t i o n : Trinidad, the Guianas.

Ecology: Epiphytic, growing at somewhat higher altitudes (to 700 m); rare.

Note: This species is easily recognized by the conspicuous brush-like receptacles on propaguliferous leaves, only to be confused with *C. palisotii*.

14. **Calymperes venezuelanum** (Mitt.) Broth. ex Pittier, Bol. Soc. Venez. Ci. Nat. 3: 360. 1936. – *Syrrhopodon venezuelanus* Mitt., J. Linn. Soc., Bot. 12: 125. 1869. Type: Venezuela, Schomburgk s.n. (BM, H, NY).

Description and synonymy see Musci I in Fl. Suriname 6: 121. 1964.

Distribution: C America, West Indies, Venezuela, the Guianas.

Ecology: Epiphytic, usually at somewhat higher altitudes; not common.

Note: When dry this species is easily recognized by the stiff, non-contorted leaves. The leaves are very long (to 2.5 mm) and fragile, often broken.

2. **SYRRHOPODON** Schwägr., Sp. Musc. Frond. Suppl. 2(1): 110. 1824. Type: S. gardneri (Hook.) Schwägr. (Calymperes gardneri Hook.)

Description see Musci I in Fl. Suriname 6: 144. 1964.

Distribution: Pantropical; in the Guianas19 species.

Note: In Musci I (p. 152) *Syrrhopodon parasiticus* was treated in a broad species concept with a great variability in cell ornamentation and leaf borders. Meanwhile this species has been divided in a complex of taxa all belonging to subgenus *Pseudocalymperes*. This subgenus is characterized by filiform propagules produced on differentiated propaguliferous leaves, forming comae at the end of the stem. In a worldwide revision of this subgenus Reese (1995) recognized 5 Neotropical species: *S. africanus* subsp. *graminicola*, *S. cymbifolius*, *S. disciformis* Dusén, *S. flexifolius* and *S. parasiticus* s.s. The position of *S. disciformis* is not clear. Reese (1995) commented that this mainly African species is rare in the Neotropics and is not different in any significant ways from *S. parasiticus*.

174

KEY TO THE SPECIES

1 Leaves with a conspicuous border of linear, usually hyaline cells reaching
  nearly to apex 2
  Leaves without such border or border weak, seldom reaching beyond
  midleaf 17

2 Border toothed or ciliate at shoulders 3
  Border smooth or irregularly serrate at shoulders 7

3 Leaves 2-3(-4) mm long, not flexed at shoulders, basal lamina hardly
  wider than upper lamina . . . .*18. S. simmondsii*
  Leaves over 3 mm long, flexed at shoulders, basal lamina distinctly wider
  than upper lamina 4

4 Upper lamina lanceolate, 1-1.5 times the length of lower lamina; cancellinae
  extending to shoulders or not, broad and rounded at distal end 5
  Upper lamina linear, 1.5-3 times the length of lower lamina; cancellinae
  extending beyond shoulders, tapering to the costa 6

5 Upper laminal cells highly bulging pluripapillose at both surfaces . . . .
  . . . . *5. S. elongatus* var. *glaziovii*
  Upper laminal cells with 1-2 conspicuous papillae at both surfaces, spinelike
  towards apex, occasionally smooth or indistinstly pluripapillose and plane
  . . . . *12. S. leprieurii*

6 Leaves remote along stem, on dry plants straight, spreading at an angle of ca.
  45°, often spirally twisted . . . .*16c. S. prolifer* var. *cincinnatus*
  Leaves crowded along stem, variously twisted when dry . . . . .*19. S. tortilis*

7 Upper laminal cells smooth 8
  Upper laminal cells pluripapillose 10

8 Plants large, to 10 cm or more high, leaves 7-14 mm long, border along upper
  lamina coarsely toothed . . . .*14. S. lycopodioides*
  Plants smaller, leaves generally not over 5 mm long, border entire or bluntly
  serrulate 9

9 Leaves oblong, often shorter and broader towards end of stem forming comal
  leaves; upper laminal cells rounded-quadrate, collenchymatous . . . .
  . . . . *6. S. flexifolius*
  Leaves linear-lanceolate, spirally twisted when dry, no comal leaves at end
  of stem; upper laminal cells quadrate-rectangular, incrassate and porose
  . . . . *8. S. helicophyllus*

10 Leaves oblong-lingulate, broader in comal leaves, border weak, irregularly denticulate, costa smooth or slightly rough at back . . . .
. . . . *1. S. africanus* subsp. *graminicola*
Leaves linear or lanceolate, no comal leaves present, border strong, coarsely serrate towards apex, costa serrate or toothed near apex 11

11 Lower lamina obovate, cancellinae filling lower lamina or not reaching shoulders; upper lamina lanceolate 12
Lower lamina oblong, cancellinae extending beyond shoulders; upper lamina linear 13

12 Upper lamina spirally twisted when dry; cancellinae at distal end incrassate and porose, not reaching shoulders . . . . *5. S. elongatus* var. *glaziovii*
Upper lamina crispate and irregularly twisted when dry; cancellinae thinwalled, filling lower lamina . . . . *7. S. gaudichaudii*

13 Leaves remote along stem, on dry plants patent spreading at an angle of ca. 45°, often spirally twisted . . . . *16c. S. prolifer* var. *cincinnatus*
Leaves crowded along stem, variously twisted when dry 14

14 Upper laminal cells highly bulging and pluripapillose 15
Upper laminal cells plane and pluripapillose 16

15 Border slender, smooth and often revolute at shoulders; cancellinae blunt or rounded at distal end . . . . *16d. S. prolifer* var. *scaber*
Border at shoulders broad and strong, with at least at some leaves a few sharp teeth, cancellinae tapering acute to the costa at distal end . . . . *19. S. tortilis*

16 Upper lamina linear, tapering only at apex, slightly flexed at shoulders . . . .
. . . . *16a. S. prolifer* var. *prolifer*
Upper lamina lanceolate, gradually tapering to apex, strongly flexed at shoulders . . . . *16b. S. prolifer* var. *acanthoneuros*

17 Margin of upper lamina not thickened 18
Margin of upper lamina thickened 20

18 Leaf apex round or retuse; cancellinae rounded at distal end; upper laminal cells pluripapillose . . . . *13. S. ligulatus*
Leaf apex acute; cancellinae acute at distal end; upper laminal cells unipapillose or nearly smooth 19

19 Plants to 0.5 mm high, rarely with comal leaves at the end of the stem; border of hyaline cells along the margin extending to half the leaf length . . . .
. . . . *3. S. cymbifolius*
Plants to 2 cm high, often with comae of broad, propaguliferous leaves at the end of the stem; border of hyaline cells along the margin absent . . . .
. . . . *15. S. parasiticus*

20 Upper lamina "winged" with more than 2 rows of irregular teeth. Capsule
immersed . . . . 2. *S. cryptocarpos*
Upper lamina regularly double-serrate. Capsule exserted 21

21 Cancellinae partly coloured, fragile at base; basal lamina hardly broader than
upper lamina. 22
Cancellinae hyaline, persistent; basal lamina in most leaves broader than
upper lamina, forming distinct shoulders. 23

22 Leaves 3-6 mm long; cancellinae restricted to the lower 1/4-1/3 of the lower
lamina, gradually merging into supra-cancellinar cells. . . *9. S. hornschuchii*
Leaves 5-12 mm long; cancellinae extending halfway the middle of the
lower lamina, at distal end coloured, but well-distinguished from the supra-
cancellinar cells . . . . *17. S. rigidus*

23 Lower lamina at shoulders bordered with a broad band of elongate, incrassate
cells. . . . . *11. S. lanceolatus*
Lower lamina not bordered. 24

24 Upper lamina plane when dry, upper laminal cells pellucid, rounded-
quadrate, incrassate . . . . *4. S. elatus*
Upper lamina involute when dry; upper laminal cells obscure, thin walled
. . . . 25

25 Lower lamina with flaring shoulders, quickly narrowed to the subtubulose
upper lamina, apex acuminate; upper laminal cells elongate. . . . .
. . . . *10b. S. incompletus* var. *luridus*
Lower lamina more gradually narrowed to the flat upper lamina, apex broad-
acute; upper laminal cells quadrate . . *10a. S. incompletus* var. *incompletus*

1. **Syrrhopodon africanus** (Mitt.) Paris subsp. **graminicola** (R.S.
Williams) Reese, Bryologist 98: 143. 1995. – *Syrrhopodon
graminicola* R.S. Williams, Bull. Torrey Bot. Club 47: 379. 1920.
Type: Jamaica, Woodstock near Beaufort, Britton 579 (NY).
– Fig. 64 A-D

Calymperes disciforme Müll. Hal., Linnaea 21: 183. 1848., nom. illeg. –
Syrrhopodon parasiticus (Brid.) Besch. var. *disciformis* (Müll. Hal.) Florsch.,
Fl. Suriname 6: 154. 1964. Type: Suriname, Kegel 505 (GOET).

Small plants growing scattered or in small tufts. Stems simple, seldom
divided, to 2 cm high, often densely tomentose at base. Leaves, when
dry, loosely appressed to variously flexuose with undulate margins, when
moist wide-spreading and more or less plane, 2-3 mm long, oblong-
lingulate, bordered nearly to apex with linear, hyaline cells, margin
distantly denticulate throughout, apex acute to slightly acuminate, costa

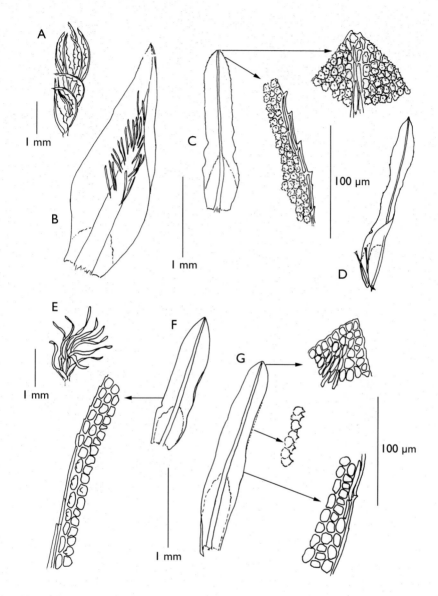

Fig. 64. A-D. *Syrrhopodon africanus* (Mitt.) Paris subsp. *graminicola* (R.S. Williams) Reese: A, habit (dry); B, comal leaf with propagula; C, vegetative leaf; D, perichaetial leaf (A-B, Florschütz 4636A; C-D, Florschütz-de Waard & Zielman 5499). E-G. *Syrrhopodon cymbifolius* Müll. Hal.: E, habit (dry); F-G, leaves (E-F, Florschütz-de Waard 5435; G, Florschütz 1372).

percurrent, firm, near apex rough at dorsal side, propagules septate, borne on ventral side of costa on comal or normal leaves; lower lamina not broader than upper lamina, cancellinae extending 1/4-1/2 the leaf length, at distal end broad-acute; comal leaves broader, oval-ovate, with sometimes indistinct border and cancellinae not clearly delimited; upper laminal cells obscure, pluripapillose, rounded quadrate, 7-12 μm in longest diameter. Sporophyte not seen, description after Reese (1993): seta 2-2.5 mm long, capsule ovoid-cylindric, 1-1.5 mm long, operculum 1 mm long, peristome teeth imperfect, smooth, segmented, bases fused into a pale membrane. Calyptra 2.5 mm long.

Distribution: C America, West Indies, tropical S America.

Ecology: On solitary trees and in the canopy of lowland rainforest; not frequent.

Selected specimens: Guyana: Kamarang, along trail to Waramadan, alt. 500 m, Aptroot 17063 (L). Suriname: area of Kabalebo Dam project, border of granitic rock slab near road to Amatopo, Florschütz-de Waard & Zielman 5621 (L); Brownsberg plateau, alt. 500 m, Florschütz 4636A (L). French Guiana: Saül, Sentier Limonade, alt. 180-210 m, Montfoort & Ek 518 (L).

Notes: In Musci I (p. 154) *Calymperes disciforme* Müll. Hal., the oldest name for this species, was treated as a variety of *Syrrhopodon parasiticus*. Reese (1993) restored it to species level as *S. graminicola*, because the epithet disciforme was preoccupied by the name *S. disciformis* Dusén. Later Reese (1995) changed it in a subspecies of *S. africanus*.
S. africanus var. *graminicola* is one of the 5 Neotropical species belonging to subgenus *Pseudocalymperes*, characterized by comal leaves at the end of the stem (see Note under the genus *Syrrhopodon*). It is distinguished by the undulate leaf margins in dry plants, the weak, denticulate border, ending just below apex and the pluripapillose upper laminal cells.

2. **Syrrhopodon cryptocarpos** Dozy & Molk., Prodr. Fl. Bryol. Surinamensis 14. 1854. Type: Suriname, van Beek Vollenhoven s.n. (L, NY).

Description and synonymy see Musci I in Fl. Suriname 6: 159. 1964.

Distribution: C America, Trinidad, tropical S America, tropical Africa.

Ecology: On tree trunks (often palms) and in the lower canopy of various types of lowland rainforest; rather common.

N o t e : This species is easily recognized by the oblong-ligulate upper lamina usually broader than the basal lamina and winged with at least three rows of irregular serrations nearly to apex. The apex varies from broad-acute to blunt and apiculate.

3. **Syrrhopodon cymbifolius** Müll. Hal., Hedwigia 39: 262. 1900. Lectotype (Reese 1993): Brazil, Goyas, Fl. Curumba, Ule 1571 (NY).
– Fig. 64 E-G

Small plants growing in low mats. Stems short, to 5 mm long, at base often prostrate and densely tomentose. Leaves when dry variously twisted with involute margins, moist wide-spreading, 1.5-2.5(-3.5) mm long, oblong, leaf apex acute to rounded-acute; costa smooth, ending a few cells below apex, propagules filamentous, septate, borne on ventral side of costa, but not on special comal leaves; margin of upper lamina minutely serrulate by projecting cell angles, in basal lamina with a narrow hyaline border, extending 1/4-1/2 (occasionally to 3/4) of leaf length; basal lamina not broader than upper lamina, cancellinae ca. 1/3 of leaf length, at distal end irregular in outline and tapering to costa; upper laminal cells irregularly quadrate, 7-12 µm in diam., unipapillose at dorsal surface, mammillose at ventral side. Monoicous. Perigonia on short branches of the prostrate, basal part of stem, budlike with 3-5 antheridia. Capsule not seen (description after Reese 1981): seta 1.5-2.2 mm long, capsule cylindric, 15-20 mm long, with stomata at base, operculum rostrate, peristome teeth delicate, ca. 85 µm long, spores smooth 16.8-21.6 µm; calyptra cucullate, lobed at base, papillose in upper part.

D i s t r i b u t i o n : Colombia, Brazil, the Guianas.

E c o l o g y : Epiphytic on branchlets and palm leaves in open forest; not common, but lately more frequently collected in the canopy.

S e l e c t e d   s p e c i m e n s : Guyana: Mabura Hill, 180 km SSE of Georgetown, Cornelissen & ter Steege 537 (L). Suriname: area of Kabalebo Dam project, Florschütz-de Waard & Zielman 5435 (L). French Guiana: Saül, "Sentier Limonade", Montfoort & Ek 790 (L).

N o t e : *S. cymbifolius* belongs to subgenus *Pseudocalymperes* (see Note under *Syrrhopodon*) but comal leaves are rarely seen. It was included in the broad species concept of *S. parasiticus* but later it is recognized as a distinct species (Reese 1981). The small size of the plants and the narrow, hyaline border extending halfway the leaf distinguish this species from *S. parasiticus*. Also the narrower leaves with a costa ending a few cells below the non-apiculate apex are distinguishing characters.

4. **Syrrhopodon elatus** Mont., Ann. Sci. Nat., Bot. sér. 2, 3: 198. 1835.
– *Syrrhopodon incompletus* Schwägr. var. *elatus* (Mont.) Florsch., Fl.
Suriname 6: 163. 1964. Type: French Guiana, Leprieur s.n. (PC).
– Fig. 65 A-B

Small to medium sized plants, growing scattered or in small tufts. Stems to 3 cm long, little divided. Leaves when dry erect or secund with plane lamina, 2-5 mm long, lanceolate, margins thickened to just below apex, with coarse double teeth (in young plants indistinct), apex acute or short-acuminate, serrulate; costa percurrent, smooth, in propaguliferous leaves with short propagules on ventral side on apex; basal lamina broad, obovate, cancellinae filling 1/2 to 2/3 of basal lamina, well-delimited and rounded at distal end; upper laminal cells pellucid, incrassate, round or rounded quadrate, 7-15 µm in diam. Sporophytes unknown.

Distribution: Brazil, the Guianas.

Ecology: Epiphytic on tree trunks; only known from a few collections in each of the Guianas.

Specimens examined: Guyana: Mabura Hill, Cornelissen & ter Steege 557 (L). Suriname: Railroad traject km 121, Florschütz 1813 (LAF, L).

Note: In Musci I (p. 163) this species was treated as a variety of *Syrrhopodon incompletus*, but it is restored to specific rank (Reese 1984). Yet the resemblance to *S. incompletus* cannot be ignored, especially in the leaf shape with broad, obovate basal lamina in *S. incompletus* var. *luridus*. The main differences are in the plane, non-involute upper lamina (when dry) and in the thick-walled, rounded-quadrate upper laminal cells. Also the cancellinae are sharply separated from the adjacent cells at distal end.

5. **Syrrhopodon elongatus** Sull. var. **glaziovii** (Hampe) Reese, Fl. Neotrop. Monogr. 58: 36. 1993. – *Syrrhopodon glaziovii* Hampe, Vidensk. Meddel. Naturhist. Foren. Kjøbenhavn ser. 3, 6: 133. 1875. Type: Brazil, Rio de Janeiro, Glaziou 7134 (BM, NY). – Fig. 66

Medium-sized plants growing in dense cushions. Stems little divided, to 4 cm long, green in upper part, brown and densely tomentose below (even with tomentum at leaf apex). Leaves 3-5 mm long, flexed at shoulders, upper lamina lanceolate, spirally twisted when dry, contorted and crispate when moist, strongly bordered, at apex spinose-toothed, at shoulders ciliate to dentate or smooth, apex acute, upper laminal cells round, 10-18 µm in diam., incrassate, bulging and pluripapillose at both surfaces;

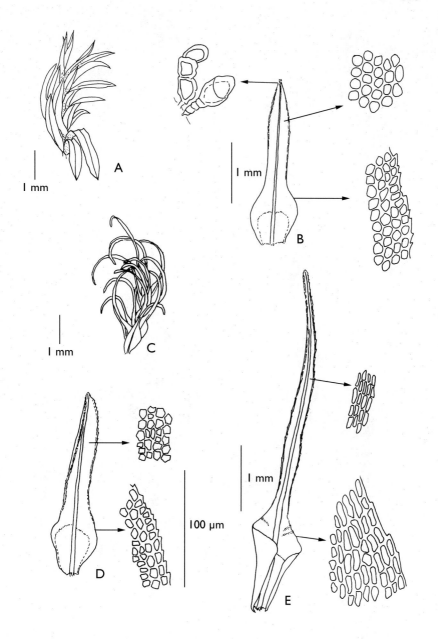

Fig. 65. A-B. *Syrrhopodon elatus* Mont.: A, end of stem (dry); B, leaf with propagula. (A-B, Cornelissen & ter Steege 557). C-D. *Syrrhopodon incompletus* Schwägr. var. *incompletus*: C, end of stem (dry); D, leaf (C-D, Florschütz & Maas 2328). E. *Syrrhopodon incompletus* Schwägr. var. *luridus* (Paris & Broth.) Florsch.: E, leaf (E, Richards 220).

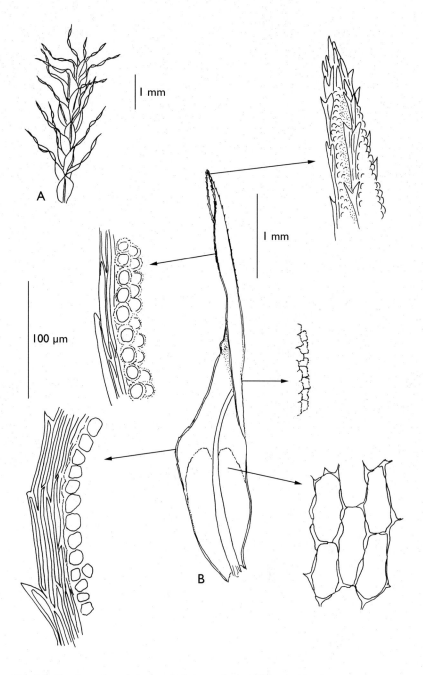

Fig. 66. *Syrrhopodon elongatus* Sull. var. *glaziovii* (Hampe) Reese: A, end of stem (dry); B, leaf (A-B, Gradstein 5632).

costa percurrent, spinose at back towards apex; propagules borne on ventral side of the leaf apex; lower lamina 1/2-1/3 of leaf length, broadest at shoulders, cancellinae not reaching shoulders, broad and rounded at distal end, cells elongate, incrassate and porose. Sporophyte not seen (description from Vital 348, Brazil): seta reddish, 2 mm long; capsule 2-3 mm long, operculum slenderly rostrate, 1.5 mm long, peristome teeth papillose, to 150 μm long. Calyptra 3 mm long.

Distribution: Tropical S America.

Ecology: Terrestrial and on rotten logs in open vegetation; only collected at higher altitudes in Guyana.

Selected specimens: Guyana: Pakaraima Mts., Mt. Membaru, Maas & Westra 4288, alt. 730 m (L); Mt. Latipú, alt. 1000 m, Gradstein 5632 (L).

Note: The collections from Guyana have only faintly dentate or smooth shoulders. These specimens could easily be mistaken for *Syrrhopodon gaudichaudii*, but are different in the round upper laminal cells and in the broad, rounded cancellinae not extending to the shoulders. Also the porose cancellinae cells form a good distinguishing character.

6. **Syrrhopodon flexifolius** Mitt., J. Linn. Soc., Bot. 12: 118. 1869. – *S. parasiticus* (Brid.) Besch. var. *flexifolius* (Mitt.) Reese, Bryologist 84: 245. 1981. Lectotype (Reese 1981): Brazil, São Gabriel, Spruce s.n. (NY). – Fig. 70 D-F

Small plants growing scattered or in loose tufts. Stems simple or divided, to 2 cm high but usually less, sometimes at end of stem comose with gemmiferous leaves. Leaves when dry variously flexuose or crispate, spreading when moist, oblong, 2-4 mm long, bordered with several rows of linear, hyaline cells nearly to apex, margin entire or bluntly serrulate, apex acute, sometimes slightly acuminate, upper laminal cells pellucid, smooth, rounded quadrate or rectangular, to 20 μm long, often collenchymatous; costa smooth, percurrent or ending a few cells below apex; lower lamina not broader than upper lamina, cancellinae extending 1/3-1/2 of leaf length, at distal end acute or rounded. Comal leaves oval to broad-triangular with short upper lamina and reduced border, the inner ones sometimes very small, bearing a cluster of filamentous, septate propagules. Perichaetial leaves little differentiated, more slenderly acute than normal leaves. Seta ca. 3 mm long, reddish-brown; capsule cylindric, 2 mm long and 0.5 mm wide, peristome teeth imperfect, smooth; operculum and calyptra not seen.

Distribution: C America and tropical S America.

Ecology: Epiphytic in savanna forest and in the canopy of moist rainforest; not common.

Selected specimens: Guyana: trail from Kamarang R. to Pwipwi Mts., alt. 800 m, Gradstein 5712 (L); Mabura Hill, 180 km SSE of Georgetown, alt. 0-50 m, Cornelissen & ter Steege 505 (L). Suriname: Nassau Mts., plateau A, alt. 550 m, Lindeman 6954B (L).

Notes: In Musci I (p. 152) *S. flexifolius* was treated as a synonym of *S. parasiticus*. Reese (1981) first recognized it as a variety of *S. parasiticus* but later restored it at species level (Reese 1993).
The flat leaves, not involute when dry, the smooth, collenchymatous upper laminal cells and the complete, (nearly) smooth border of hyaline cells are good characters to distinguish this species from *S. africanus* subsp. *graminicola*.

7. **Syrrhopodon gaudichaudii** Mont., Ann. Sci. Nat., Bot. sér. 2, 2: 376. 1834. Type: Brazil, Gaudichaud s.n. (BM, G, NY).

Description see Musci I in Fl. Suriname 6: 155. 1964. Addition to this description: propagules observed on the ventral side of the costa at leaf apex.

Distribution: Neotropics, tropical Africa.

Ecology: Terrestrial on rock and humus; not common, but in Suriname locally frequent in the rock vegetation around the Voltzberg.

Note: For differences with *Syrrhopodon elongatus* var. *glaziovii* see note under that species.

8. **Syrrhopodon helicophyllus** Mitt., J. Linn. Soc., Bot. 12: 119. 1869. Lectotype (Reese 1993): Brazil, Rio Negro, Spruce 8 (NY).
– Fig. 67

Medium sized plants often growing in dense, bronze-green mats with light-green innovations. Stems to 3 cm high, divided several times, densely tomentose in lower parts. Leaves when dry tightly spirally twisted, when moist flat-spreading, linear, 3-5 mm long, bordered to apex with a conspicuous belt of hyaline, elongated cells, smooth or bluntly serrulate at apex, apex acute; upper laminal cells smooth, strongly incrassate and porose, rectangular or hexagonal, 25-50 μm long and ca. 25 μm wide;

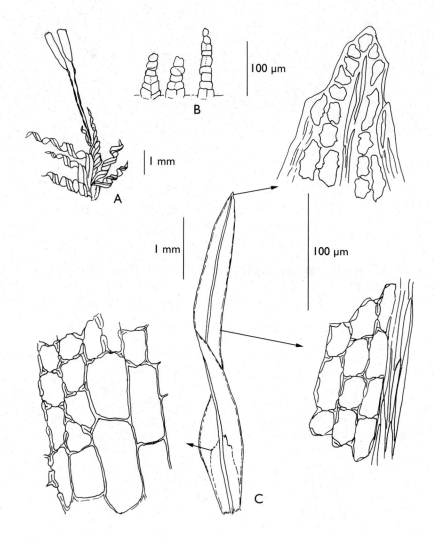

Fig. 67. *Syrrhopodon helicophyllus* Mitt.: A, part of plant with capsules (dry); B, peristome teeth; C, leaf (A-B, Aptroot & Sipman 4802; C, Gradstein 5700).

costa percurrent, propagules filamentous, septate, borne along costa on old, brown, lower leaves (Reese 1993); lower lamina not broader than upper lamina, cancellinae extending 1/5-1/4 of leaf length, at distal end acute or rounded. Perichaetial leaves slender, little differentiated. Seta reddish, ca. 4 mm long, capsule cylindric, 1.5 mm long, operculum ca. 1 mm long, peristome well-developed, trabeculate at outer surface, papillose at inner surface, ca. 100 μm long.

Distribution: Brazil (Amazonas), Guyana.

Ecology: Epiphytic on tree trunks; in Guyana collected only twice, at higher altitudes.

Specimens examined: Guyana: Pakaraima Mts., 2 km N of Kamarang, alt. 500 m, Aptroot & Sipman 4802 (L); Trail from Kamarang to Pwipwi Mt., alt. 800 m, Gradstein 5700 (L).

Notes: This remarkable species is easily recognized by the spirally twisted leaves when dry, the conspicuous border and the large, incrassate cells in upper lamina.
The occurrence of this species in Guyana constitues an expansion of the distribution, which was thus far limited to the Amazon Basin.

9. **Syrrhopodon hornschuchii** Mart., Fl. Bras. 1(2): 6. 1840. – *S. rufus* Hornsch. in Mart., Fl. Bras. 1(2): 6. 1840, nom. nud. in synon. Type: Brazil, Minarum, Amazon R., Martius s.n. (BM, M). – Fig. 68 A-B

Syrrhopodon miquelianus Müll. Hal., Syn. Musc. Frond. 1: 535. 1849. Type: Suriname, hb. Miquel.
Syrrhopodon surinamensis Dozy & Molk., Prod. Fl. Bryol. Surinamensis 12: 1854. Type: Suriname, van Beek Vollenhoven s.n. (L).

Medium sized plants growing in dense mats. Stems to 2 cm high, simple or sparingly branched, usually densely tomentose in basal part. Leaves crispate and strongly involute when dry, plane and spreading when moist, 4-7 mm long, oblong-linear, more or less curved, with thickened margins, bluntly double-serrate nearly to apex, apex broad-acute, sometimes slightly acuminate at the tip; costa percurrent, smooth; upper laminal cells rounded-quadrate to oval, 8-15 µm long, mammillose on ventral surface, smooth or minutely papillose on dorsal surface; lower lamina not broader than upper lamina, margins irregularly serrulate or denticulate, cancellinae short, confined to the basal 1/4-1/3 of the lower lamina, at base hyaline and very fragile, at distal end coloured and incrassate, not clearly separated from the surrounding cells; in propaguliferous leaves upper lamina strongly reduced and costa long-excurrent, densely papillose, with a cluster of propagules on ventral surface. Perigonia small, leaves ca. 1 mm long, ovate, not bordered, serrulate at apex. Perichaetial leaves slightly differentiated, upper lamina shorter and narrower. Seta 1-1.5 cm long, capsule erect, ovoid, ca. 1.5 mm long, operculum slenderly rostrate, 1-1.5 mm long, peristome absent or rudimentary. Calyptra cucullate, split at base.

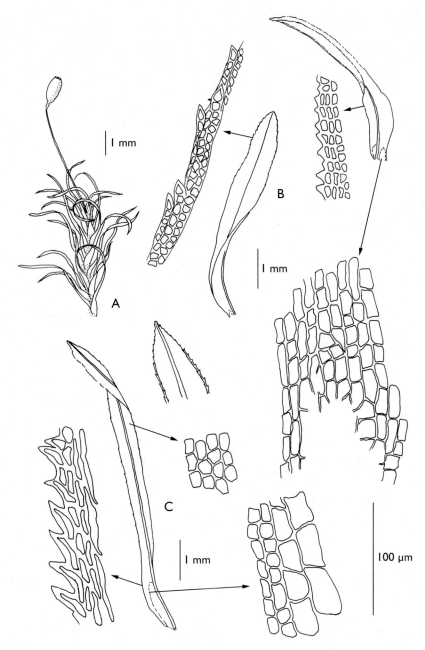

Fig. 68. A-B. *Syrrhopodon hornschuchii* Mart.: A, end of stem with capsule(dry); B, leaves (A-B, Richards 90). C. *Syrrhopodon rigidus* Hook. & Grev.: C, leaf (C, Gradstein 5423).

Distribution: Tropical S America.

Ecology: On roots and logs in lowland rainforest; not common.

Selected specimens: Guyana: Kaieteur Falls, Korume Cr., alt. 150 m, Newton *et al.* 3377 (US, L). Suriname: Zanderij, Cola Cr., Kramer & Hekking 2730 (L); Kabalebo Dam area, marsh forest, Florschütz-de Waard & Zielman 5230 (L). French Guiana: Mt. de la Trinité, alt. 440 m, Cremers 12980 (CAY, L).

Note: The first collections of *S. hornschuchii* from Suriname were described as *S. miquelianus* and *S. surinamensis*. The latter was treated as a synonym of *S. rigidus* in Musci I (p. 164), but Reese (1977) recognized it as a distinct species, identical with *S. hornschuchii*, an earlier name. The main differences with *S. rigidus* are in the cancellinae, which in *S. hornchuchii* are poorly differentiated from the adjacent cells and restricted to the lower half of the basal lamina. Also the size of the leaves, the broader and shorter upper lamina and the broad-acute apex are good distinguishing characters.

10. **Syrrhopodon incompletus** Schwägr., Sp. Musc. Frond. Suppl. 2(1): 119. 1824. Type: Cuba, Poeppig s.n. (BM, PC).

10a. var. **incompletus**   – Fig. 65 C-D

Small to medium-sized plants growing scattered or in tufts. Stems to 3 cm high, but usually much shorter. Leaves variously curved with involute margins when dry, 2-5 mm long, with a broad lower lamina gradually narrowing to a linear-lanceolate upper lamina, margins of upper lamina thickened and double-serrate, apex acute or rounded-acute, costa percurrent, serrate at back in upper half, in propaguliferous leaves bearing a cluster of propagules on ventral side of apex; upper laminal cells small, quadrate-hexagonal, 5-9 μm long, at shoulders not differentiated, bulging at ventral side, mammillose or finely papillose at dorsal side; lower lamina obovate, margin irregularly serrulate at shoulders, cancellinae distinct, filling 1/2 to 3/4 of lower lamina. Perichaetial leaves not differentiated. Seta to 7 mm long, capsule erect, elliptic, operculum long-rostrate, peristome lacking or rudimentary.

Distribution: Tropical and subtropical America, tropical Africa.

Ecology: On tree bases, palm stems and logs, occasionally on rocks in lowland rainforest; common.

N o t e : In the description given in Musci I (p. 161) *Syrrhopodon elatus* was included in this species as a variety. The above given description is an adaptation.

10b. var. **luridus** (Paris & Broth.) Florsch., Fl. Suriname 6: 163. 1964. – *Syrrhopodon luridus* Paris & Broth., Rev. Bryol. 33: 56. 1906. Type: French Guiana, Michel s.n. (L, NY, REN). – Fig. 65 E

Var. *luridus* differs from the typical variety in the following aspects: leaves 4-7 mm long, lower lamina very broad, flaring at shoulders, abruptly contracted to a narrow, subtubulose upper lamina; costa smooth at dorsal side; upper laminal cells elongate, oval-oblong, to 16 µm long, lower laminal cells at shoulders elongate and incrassate, forming teniolae-like features. Cancellinae at distal end not sharply separated from the adjacent cells.

D i s t r i b u t i o n : Tropical S America.

E c o l o g y : On tree trunks in lowland rainforest, also at higher altitudes; rather common.

11. **Syrrhopodon lanceolatus** (Hampe) Reese, Moss Fl. Centr. America 1: 215. 1994. – *Calymperes lanceolatum* Hampe, Vidensk. Meddel. Naturhist. Foren. Kjøbenhavn ser. 4, 1: 78. 1879. – *Syrrhopodon incompletus* Schwägr. var. *lanceolatus* (Hampe) Reese, Bryologist 80: 13. 1977. Type: Brazil, Glaziou 9278 (H, PC).

Description see Musci I in Fl. Suriname 6: 126. 1964, as *Calymperes lanceolatum*.

D i s t r i b u t i o n : Brazil, the Guianas.

E c o l o g y : Epiphytic on branches in lowland rainforest, also at higher altitudes; not common.

N o t e s : Reese (1977) transferred this species to *Syrrhopodon* as a variety of *S. incompletus*, based on the resemblance to *S. incompletus* var. *perangustifolius* Reese. Later a collection with capsules from French Guiana established its place in *Syrrhopodon* and in 1994 Reese recognized it as a distinct species.
Syrrhopodon lanceolatus is distinguished by the broad band of elongate, incrassate cells at shoulders. In this character it resembles *Calymperes platyloma* and *C. mitrafugax*, but it is different in the coarsely double-serrate margin and the smooth cells of the upper lamina.

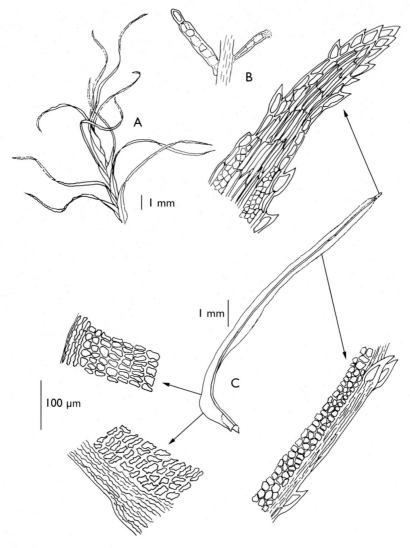

Fig. 69. *Syrrhopodon lycopodioides* (Brid.) Müll. Hal.: A, end of stem (dry); B, propagula along costa; C, leaf (A-C, Gradstein 5218).

12. **Syrrhopodon leprieurii** Mont., Ann. Sci. Nat., Bot. sér. 2, 2: 379. 1834. Type: French Guiana, Leprieur s.n. (BM, L, NY, PC).

Description and synonymy see Musci I in Fl. Suriname 6: 146. 1964.

Distribution: C America, West Indies, tropical S America.

Ecology: Growing in moist areas in lowland rainforest, epiphytic or on logs and rocks; not common.

Note: The most striking character of this species is the apex with spinose-toothed costa and margin and the upper laminal cells with often spinelike papillae, especially towards apex.

13. **Syrrhopodon ligulatus** Mont., Syll. Gen. Sp. Crypt. 47. 1856. Type: French Guiana, Leprieur 1384 (BM, NY, PC).

Description and synonymy see Musci I in Fl. Suriname 6: 149. 1964.

Distribution: Subtropical U.S.A., West Indies, C and tropical S America.

Ecology: Epiphytic on tree trunks and in the canopy of lowland rainforest; not common.

Note: The round leaf apex and the crenulate, not bordered margin in the upper lamina are the typical characters of this small moss. The border of hyaline cells at shoulders can be totally absent but is usually short and occasionally extending to midleaf.

14. **Syrrhopodon lycopodioides** (Brid.) Müll. Hal., Syn. Musc. Frond. 1: 538. 1849. – *Bryum lycopodioides* Brid., Muscol. Recent. 2(3): 54. 1803. – *Dicranum lycopodioides* (Brid.) Sw., Fl. Ind. Occ. 3: 1766. 1806. Type: Jamaica, Swartz s.n. (S).   – Fig. 69

Robust plants growing in loose tufts, blackish in lower parts. Stems to 10 cm or more long, sparingly divided, tomentum scarce. Leaves twisted when dry, wide-spreading when moist, 7-14 mm long, upper lamina linear, bordered with strongly incrassate, elongate cells, often coloured and conspicuously toothed with large, often paired teeth, apex acuminate; costa percurrent or excurrent, propagules filamentous, brown, on both surfaces of the costa; upper laminal cells smooth, incrassate, rounded-quadrate to transversely elongate, 10-16 μm in largest diam., towards shoulders gradually more strongly incrassate and transversely elongate with a broad border of porose, hyaline cells; lower lamina ca. 1/6 of total length, slightly broader than upper lamina, cancellinae extending about 1/2 the length of lower lamina. Perichaetial leaves not differentiated. Seta 8-11 mm long; capsule cylindric, 3-4 mm long, operculum long-rostrate, 1.5-2 mm long, peristome reduced, teeth imperfect, fragile. Calyptra smooth.

Distribution: Mexico, C America, West Indies, tropical S America.

Ecology: Epiphytic on trunks and roots at higher altitudes; in the Guianas only known from Mt. Roraima.

Selected specimen: Guyana: N slope of Mt. Roraima, alt. 1200-1600 m, Gradstein 5218 (L).

Note: This species is readily recognized by the large size, the conspicuous border with prominent teeth and the strongly incrassate, transversely elongate cells at shoulders.

15. **Syrrhopodon parasiticus** (Brid.) Besch., Ann. Sci. Nat., Bot. sér. 8, 1: 298. 1895. – *Bryum parasiticum* Brid., Muscol. Recent. 2(3): 54. 1803. Type: Jamaica, Swartz s.n. (BM, NY).   – Fig. 70 A-C

Syrrhopodon martinicensis Broth. in Urban, Symb. Antill. 3: 422. 1903, nom. illeg.. – *Calymperopsis martinicensis* (Broth.) Broth. in Engler, Nat. Pflanzenfam. ed. 2, 10: 235. 1924. Type: Martinique, Morne Rouge, Duss 372 (H, NY).

Small to medium sized plants growing scattered or in loose tufts. Stems usually simple, to 2 cm high but often much shorter, tomentose in lower part. Leaves loosely appressed and curved with involute margins when dry, flexed and wide-spreading when moist, 2-4 mm long, lanceolate-lingulate, upper leaves often broader, ovate-triangular, forming comae at the end of the stem, leaf apex acute or broad-acute, apiculate, costa smooth, ending in the apiculate apex, propagules filamentous, septate, borne scattered on ventral side of the costa, usually on comal leaves; margins entire, usually bordered with a short row of elongate hyaline cells at leaf base, sometimes indistinct; basal lamina not broader than upper lamina, cancellinae extending ca. 1/3 of leaf length, distinct and well delimited at distal end, merging in acute angles to the costa, in comal leaves broader and less distinct; upper laminal cells rounded quadrate to rectangular, to 10 μm in longest diam., smooth or unipapillose on dorsal surface. Perichaetial leaves little differentiated, perigonial leaves smaller, ovate. Seta smooth, to 4 mm long; capsule ovoid, 1-2 mm long, operculum ca. 1 mm long, peristome teeth irregular, short.

Distribution: Pantropical.

Ecology: Epiphytic in sun exposed habitats; rather common on solitary trees, in the canopy of lowland rainforest, on palm-leaf roofs etc.

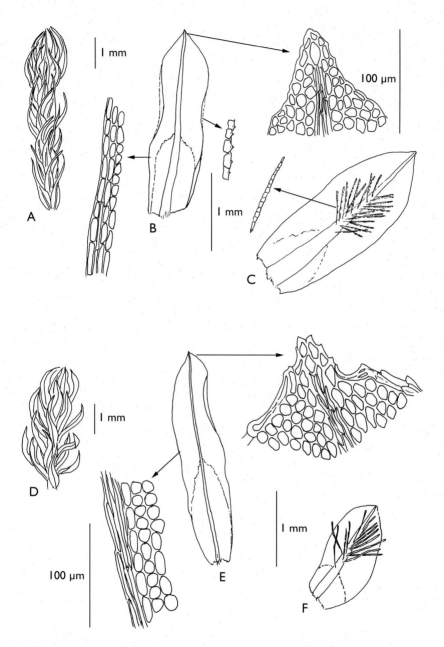

Fig. 70. A-C. *Syrrhopodon parasiticus* (Brid.) Besch.: A, habit (dry); B, leaf; C, comal leaf with propagula (A-C, Florschütz-de Waard 5960). D-F. *Syrrhopodon flexifolius* Mitt.: D, habit (dry); E, leaf; F, comal leaf with propagula (D-F, Cornelissen & ter Steege 505).

194

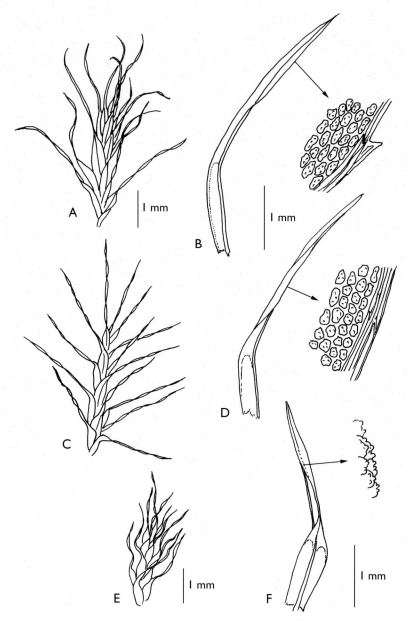

Fig. 71. A-B. *Syrrhopodon prolifer* Schwägr. var. *acanthoneuros* (Müll. Hal.) Müll. Hal.: A, end of stem (dry); B, leaf (A-B, Florschütz & Maas 3036). C-D. *Syrrhopodon prolifer* Schwägr. var. *cincinnatus* (Hampe) Reese: C, end of stem (dry); D, leaf (C-D, Gradstein 5447). E-F. *Syrrhopodon prolifer* Schwägr. var. *scaber* (Mitt.) Reese: E, end of stem (dry); F, leaf (E-F, Cremers *et al.* 5925).

Selected specimens: Guyana: Mabura Hill, 180 km SSE of Georgetown, Cornelissen & ter Steege 532 (L). Suriname: Brownsberg plateau, alt. 500 m, Florschütz 4796 (L). French Guiana: Tour de l'Isle, 20 km S of Cayenne, Florschütz-de Waard 5850 (L); Saül, "Sentier Limonade", alt. 180-210 m, Montfoort & Ek 793 (L).

Note: The description given in Musci I (p. 152) of *S. parasiticus* was based on a broad species concept with a great variation in leaf border and cell ornamentation. A new species description is given here (see also note under the genus *Syrrhopodon*).

16. **Syrrhopodon prolifer** Schwägr., Sp. Musc. Frond. Suppl. 2(2): 99. 1827. Type: Brazil, Serra dos Orgãos, Beyrich s.n. (NY).

16a. var. **prolifer**

Description see Musci I in Fl. Suriname 6: 156. 1964.

Distribution: Pantropical, but most abundant in the Neotropics.

Ecology: Predominantly epiphytic, but sometimes on rock or soil, usually at higher altitudes.

Selected specimens: Guyana: Trail from Kamarang R. to Pwipwi Mt., alt. 650 m, Gradstein 5687. Suriname: Brownsberg plateau, alt. ca. 500 m, Florschütz 4684 (L). French Guiana: Mt. de l'Inini, alt. 670-800 m, Cremers 9076 (CAY, L).

Notes: Worldwide 9 varieties are recognized in *S. prolifer* (Orbán & Reese 1990) of which 5 have been recognized for the Neotropics (Reese 1978). In the collections of the Guianas 4 varieties can be dstinguished, but it requires some experience to see the differences. As some varieties seem to intergrade, it is not always possible to determine the variety with certainty (see key).
S. prolifer var. *prolifer* is determined by the leaves, little or not flexed at shoulders and with linear upper lamina and flat, densely papillose cells. For differences with *S. prolifer* var. *acanthoneuros* see under that variety.

16b. var. **acanthoneuros** (Müll. Hal.) Müll. Hal., Syn. Musc. Frond. 1: 542. 1849. – *Syrrhopodon acanthoneuros* Müll. Hal., Bot. Zeit. 2: 727. 1844. Type: Brazil, Serra de Navidade, Gardner 50 (BM).

– Fig. 71 A-B

S. prolifer var. *acanthoneuros* is very similar to var. *prolifer* in the flat, densely papillose cells; it is only different in the more gradually tapering leaves, flexed at shoulders. These 2 characters are not always linked, which makes the determination often insecure. The darker color and the preference for growing on rocks are 2 additional characters to recognize this variety.

D i s t r i b u t i o n : C America, West Indies, tropical S America, tropical E Africa.

E c o l o g y : Preferably on rocks, but sometimes on tree trunks, log or soil; not common.

S e l e c t e d s p e c i m e n s : Suriname: Bakhuis Mts., alt. 700 m, Florschütz & Maas 3036 (L). French Guiana: Sud du Massif des Emerillons, alt. 200-250 m, Cremers 6858 (CAY, L).

16c. var. **cincinnatus** (Hampe) Reese, Bryologist 81: 200. 1978. – *Syrrhopodon cincinnatus* Hampe, Vidensk. Meddel. Naturhist. Foren. Kjøbenhavn ser. 3, 6: 131. 1875. Type: Brazil, Rio de Janeiro, Glaziou 7136 (BM). – Fig. 71 C-D

S. prolifer var. *cincinnatus* is usually easily recognized by the elongated, unbranched stems (to 5 cm long) with widely spaced, spreading leaves, often spirally twisted when dry. The border of elongate cells is rather heavy and sometimes toothed or denticulate at shoulders.

D i s t r i b u t i o n : C America, West Indies, tropical S America.

E c o l o g y : Epiphytic or on rock; not common, usually at higher altitudes.

S e l e c t e d s p e c i m e n s : Guyana: N slope of Mt. Roraima, alt. 1200-1600 m, Gradstein 5447 (L). Suriname: Emmaketen, summit of Grote Hendrik Mt., alt. 1025 m, Jonker & Daniëls 1026 (L). French Guiana: R. Nana, Saut Fracas, Cremers 7638 (CAY, L).

16d. var. **scaber** (Mitt.) Reese, Fl. Neotrop. Monogr. 58: 13. 1993. – *Syrrhopodon scaber* Mitt., J. Linn. Soc., Bot. 12: 119. 1869. Type: Cuba, Wright 46 (NY). – Fig. 71 E-F

Syrrhopodon papillosus Müll. Hal., Bot. Zeitung (Berlin) 6: 766. 1848. – *S. prolifer* Schwägr. var. *papillosus* (Müll. Hal.) Reese, Bryologist 81: 196. 1978, nom. illeg. Lectotype (Reese 1978): Venezuela, Prov. Caracas, Galipán, Funck & Schlim 358 (NY).

S. prolifer var. *scaber* is mainly distinguished by the thick-walled, highly bulging upper laminal cells. The upper lamina is linear and slightly contorted when dry. For differences with *S. tortilis* see under that species.

Distribution: Widespread in tropical and subtropical America.

Ecology: Terrestrial and occasionally epiphytic on rock savannas; not common, not seen from Guyana, in Suriname and French Guiana only at higher altitudes.

Selected specimens: Suriname: Tafelberg, Maguire 24598 (L). French Guiana: Saül, Piste de Carbet Maïs, alt. 500 m, Cremers 5931 (CAY, L).

17. **Syrrhopodon rigidus** Hook. & Grev., Edinburgh J. Sci. 3: 226. 1825. Type: St. Vincent, Mt. St. Andrew, Guilding s.n. (BM). – Fig. 68 C

Syrrhopodon androgynus (Mont.) Besch., Ann. Sci. Nat., Bot. sér 6, 3: 197. 1876. – *Calymperes androgynum* Mont., Ann. Sci. Nat., Bot. sér. 2, 3: 195. 1835. Type: French Guiana, Leprieur s.n. (PC).
Syrrhopodon longisetaceus Müll. Hal., Syn. Musc. Frond. 1: 535. 1849, nom. illeg.

Medium sized plants, growing in loose tufts. Stems to 2 cm high, simple or sparingly branched, tomentose in basal part. Leaves when dry with strongly incurved margins and twisted, moist erect-spreading, linear to ligulate, 6-11 mm long, margins in upper lamina thickened, with 2 rows of coarse teeth in upper half, apex acute or acuminate; costa percurrent or excurrent, smooth at back or nearly so, in propaguliferous leaves with reduced upper lamina costa sharp-papillose; upper laminal cells quadrate to oval, 8-15 μm long, mammillose at ventral surface, along shoulders more elongate and incrassate, sometimes forming an inconspicuous border; lower lamina hardly broader than upper lamina, margins at shoulders strongly serrate-dentate, cancellinae filling 1/2 to 3/4 of lower lamina, coloured, at apical end well-separated from the surrounding cells. Perigonia inconspicuous, budlike. Perichaetial leaves little differentiated, basal lamina longer and cancellinae less distinctly delimited as in normal leaves. Seta 1.5-2.5 cm long, capsule erect, ovoid, 1.5-2 mm long (including the long-rostrate operculum ca. 4 mm long), peristome absent. Calyptra smooth, slender, to 4 mm long.

Distribution: Mexico, C America, West Indies, tropical S America.

Ecology: On tree trunks and rocks; rather common in lowland rainforest, also at higher altitudes.

Selected specimens: Guyana: Mabura Hill, alt. 100-200 m, Maas *et al*. 5907 (L); North slope of Mt. Roraima, alt. 1200-1600 m, Gradstein 5297 (L). Suriname: Emmaketen, alt. 800 m, Jonker & Daniëls 1230 (L). French Guiana: Mt. de l'Inini, alt. 700 m, Cremers 8977 (CAY, L).

Note: In the description given in Musci I (p.164) *S. rigidus* was treated in a broad species concept, including *S. surinamensis* Dozy & Molk. as a synonym. Reese (1977) considered the latter as a different species synonymous with *S. hornschuchii*. For the differences see under that species.

18. **Syrrhopodon simmondsii** Steere, Bryologist 49: 8. 1946. Type: Trinidad, St. George, Simmonds 69 (MICH).

> *Syrrhopodon brevisetus* Florsch., Fl. Suriname 6: 148. 1964. Type: Suriname, Tibiti savanna, Lanjouw & Lindeman 1780 (LAF, L).

Description see Musci I in Fl. Suriname 6: 148. 1964, as *S. brevisetus*.

Distribution: Tropical S America.

Ecology: Epiphytic and on logs; regularly collected in the canopy of lowland rainforest or in open savanna forest.

Note: The small size and the narrow basal lamina, with coarse cilia in the shoulder region are characteristic for this species. The cell ornamentation on the dorsal surface is variable and can be with single papillae, with multifid papillae or bulging and pluripapillose. The hyaline margin is serrate at apex, but not spinose-toothed as in *S. leprieurii*.

19. **Syrrhopodon tortilis** Hampe, Vidensk. Meddel. Naturhist. Foren. Kjøbenhavn ser. 3, 4: 38. 1872. Type: Brazil, Rio de Janeiro, Glaziou 5188 (BM). – Fig. 72

Medium sized plants growing in loose mats. Stems to 5 cm long, little divided, densely tomentose in basal part. Leaves contorted when dry, erect spreading when moist, flexed at shoulders, 4-5 mm long, bordered from base to apex with a broad band of elongate, hyaline cells; upper lamina linear-lanceolate, apex acute, margin dentate to spinose-toothed towards apex and at shoulders, sometimes dentate throughout; costa percurrent or excurrent, often ending in a sharp, unicellular tooth, serrate at back in upper half, spinose at apex; upper laminal cells rounded-quadrate or

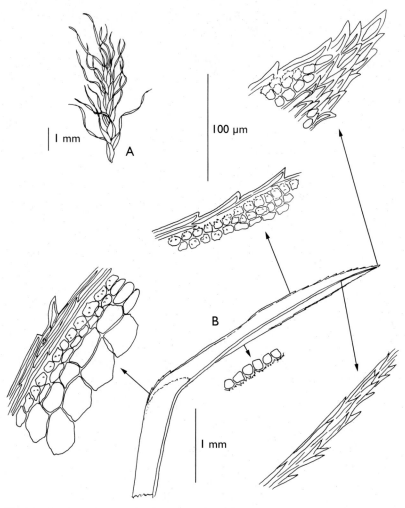

Fig. 72. *Syrrhopodon tortilis* Hampe: A, end of stem (dry); B, leaf (A-B, Cremers *et al.* 8362).

rectangular, bulging and pluripapillose at both surfaces, 10-18 µm long and ca. 10 µm wide; lower lamina 1/3 to 1/4 of leaf length, cancellinae extending beyond shoulders, at distal end tapering with sharp angles to the costa. Sporophyte not seen.

Distribution: C America, West Indies, tropical S America.

Ecology: On rock and soil in creeks and open rock vegetations; not common, not known from Suriname, collected only once in Guyana, in French Guiana more frequent.

Selected specimens: Guyana: Kamoa R., Toucan Mt., alt. 260-360 m, Jansen-Jacobs *et al.* 1635 (L). French Guiana: Carbet Mitan, 12 km E of Saül, Cremers 6214 (L).

Note: This species is distinguished by the strong border, at shoulders usually dentate or spinose-toothed and by the bulging pluri-papillose upper laminal cells. In specimens with inconspicuously dentate shoulders it may be mistaken for *S. prolifer* var. *scaber*, also with bulging, pluripapillose upper laminal cells, but in that species the cancellinae are rounded at the distal end and the border is weaker and often recurved at the shoulders.

9. POTTIACEAE

Description see Musci I in Fl. Suriname 6: 168. 1964.

Distribution: Worldwide; in the Guianas 4 genera.

Note: The genus *Leptodontium* cited in Musci I (p. 168) does not occur in the Guianas, the mentioned collection McConnell & Quelch 511 originates from the Venezuelan side of Mt. Roraima.

LITERATURE

Sollman, Ph. 1984. Notes on Pottiaceous mosses 2. Lindbergia 10: 53-56.
Zander, R.H. 1979. Notes on Barbula and Pseudocrossidium in North America and an annotated key to the taxa. Phytologia 44: 177-214.
Zander, R.H. 1995. Phylogenetic relationships of Hyophiladelphus gen. nov. and a perspective on the cladistic method. Bryologist 98: 363-374.

KEY TO THE GENERA

1 Upper laminal cells densely papillose, obscure 2
  Upper laminal cells bulging mammillose, pellucid 3

2 Leaves not over 1.5 mm long, apex obtuse or rounded acute, mucronate; costa percurrent, ending in the mucro . . . . *1. Barbula*
  Leaves 1.5-3 mm long, apex acute, costa excurrent as a sharp, hyaline tooth
  . . . . *4. Trichostomum*

3   Leaf apex obtuse or rounded-acute, upper cells rounded-quadrate, usually
     not over 10 µm in largest diam. . . . . 2. *Hyophila*
     Leaf apex acute, upper cells quadrate-rectangular, usually over 10 µm in
     largest diam. . . . . *3. Hyophiladelphus*

1. **BARBULA** Hedw., Sp. Musc. Frond. 115. 1801, nom. cons.
    Lectotype (Steere in Grout, Moss Fl. N. Amer. 1: 149. 1938): B.
    unguiculata Hedw.

Description see Musci I in Fl. Suriname 6: 169. 1964.

D i s t r i b u t i o n : Worldwide; in the Guianas 1 species.

1. **Barbula indica** (Hook.) Spreng. ex Steud., Nomencl. Bot. 2: 72.
    1824. – *Tortula indica* Hook., Musci Exot. 2: 135. 1819. Type: India,
    Madras, Röttler s.n. (NY).

Barbula cruegeri Sond. ex Müll. Hal., Syn. Musc. Frond. 1: 618. 1849. Type:
Trinidad, Crüger s.n. (BM).

Description see Musci I in Fl. Suriname 6: 169. 1964, as *B. cruegeri*.

D i s t r i b u t i o n : Pantropical and subtropical.

E c o l o g y : Terrestrial on calcareous rock and soil; in the Guianas
not common.

N o t e : The extensive synonymy of this species was studied by
Zander (1979).

2. **HYOPHILA** Brid., Bryol. Univ. 1: 760. 1826.
    Type: H. javanica (Nees & Blume) Brid. (Gymnostomum javanicum
    Nees & Blume)

Description see Musci I in Fl. Suriname 6: 171. 1964.

D i s t r i b u t i o n : Worldwide; in the Guianas 1 species.

1. **Hyophila involuta** (Hook.) A. Jaeger, Ber. St. Gallischen Naturwiss.
    Ges. 1871-72: 354. 1873. – *Gymnostomum involutum* Hook., Musci
    Exot. 2: tab. 154. 1819. Type: Nepal, Gardner s.n.

Hyophila tortula (Schwägr.) Hampe, Bot. Zeitung (Berlin) 4: 267. 1846. –
Gymnostomum tortula Schwägr., Sp. Musc. Frond. Suppl. 2(2): 78. 1826.
Type: Cuba, Pöppig s.n. (G).

Description see Musci I in Fl. Suriname 6: 173. 1964, as *H. tortula*.

Distribution: N America, Mexico, C America, West Indies, tropical S
America, C Europe, tropical Africa, SE Asia.

Ecology: Epilithic on wet rocks, concrete and brick walls; in the
Guianas rather common on regularly inundated stones along riversides
and in rapids.

Note: The small, round upper laminal cells are bulging- mammillose on
both surfaces but occasionally also minutely papillose on dorsal surface.

3. **HYOPHILADELPHUS** (Müll. Hal.) Zander, Bryologist 98: 363. 1995.
Lectotype (Zander 1995): H. agrarius (Hedw.) Zander (Barbula
agraria Hedw.)

Barbula sect. *Hyophiladelphus* Müll. Hal., Syn. Musc. Frond. 1: 604. 1849.
Barbula subgenus *Hyophiladelphus* (Müll. Hal.) Zander, Phytologia 44:
201. 1979.

Distribution: Mexico, C America, West Indies, tropical S America;
1 species.

Note: *Barbula agraria* was placed in a separate section *Hyophiladelphus*
by Müller (1849), was later raised to subgenus level by Zander (1979),
but was always seen as an incongruous element in the genus. A cladistic
analysis resulted in the recognition of *Hyophiladelphus* as a distinct genus
(Zander 1995).

1. **Hyophiladelphus agrarius** (Hedw.) Zander, Bryologist 98: 372. 1995.
– *Barbula agraria* Hedw., Sp. Musc. Frond. 116. 1801. Lectotype
(Zander 1995): "Jamaica et Domingo", Swartz s.n. (BM, G).

Description see Musci I in Fl. Suriname 6: 170. 1964, as *Barbula agraria*.

Ecology: Terrestrial on calcareous rock and soil; in the Guianas
not common.

4. **TRICHOSTOMUM** Bruch, Flora 12: 295. 1829, nom. cons.
Type: T. brachydontium Bruch

Small to medium-sized plants. Stems erect, little branched. Leaves strongly contorted when dry, erect-spreading when moist, lanceolate to linear-lanceolate, acute, apiculate; costa short-excurrent, smooth to densely papillose at back, in cross-section with 2 stereid bands; upper laminal cells rounded-quadrate to hexagonal, pluripapillose on both surfaces, obscure, basal cells enlarged, rectangular to oblong-hexagonal, smooth. Perichaetial leaves little differentiated. Seta elongate, straight, capsule erect, cylindric, operculum obliquely long-rostrate; peristome teeth inserted below the mouth, split nearly to the base into 2 filiform divisions. Calyptra cucullate.

Distribution: Worldwide; in the Guianas 1 species.

1. **Trichostomum duidense** E.B. Bartram, Mem. New York Bot. Gard. 9: 237. 1957. Type: Venezuela, Cerro Duida, Maguire, Cowan & Wurdack 22652 (FH). – Fig. 73

Dull green plants growing in loose tufts. Stems to 2.5 cm long, densely tomentose in lower part. Leaves tubulose, contorted and incurved when dry, wide-spreading and concave in upper part when moist, oblong-lanceolate, 1.5-3 mm long, apex acute, margins plane to broadly inflexed; costa stout, excurrent as a stout, hyaline mucro, often reflexed; upper laminal cells rounded-quadrate, bulging and pluripapillose on both surfaces, 5-7 μm in diam., in the basal part of the leaf more elongate, smooth, often incrassate, central cells at leaf base wider, thin-walled and more or less inflated. Dioicous. Seta smooth, reddish, to 3 cm long, capsule cylindric, to 3 mm long, operculum ca. 1 mm long, conic; peristome teeth often rudimentary, 30-300 μm, entire or 2-3 forked. Calyptra to 3 mm long, twisted.

Distribution: Worldwide.

Ecology: Epiphytic on tree trunks and logs, also on moss-covered rocks; in the Guianas in open forest, also at higher altitudes; not common.

Selected specimens: Guyana: Pakaraima Mts., Mt. Latipu, alt. 900 m, Maas et al. 4223 (L). Suriname: Tafelberg, Lisa Cr., Grace Falls trail, alt. 690 m, Allen 20698 (MO). French Guiana: Route de l'Est, après l'Inini, Cremers 4323 (CAY, L).

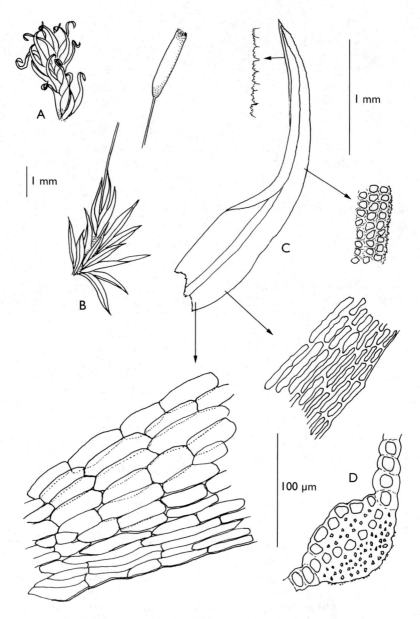

Fig. 73. *Trichostomum duidense* E.B. Bartram: A, upper part of plant (dry); B, upper part of plant with capsule; C, leaf; D, cross- section of costa (A-C, Aptroot 17071; D, Gradstein 5240).

N o t e s : In the Guianas this species may be confused with *Barbula indica*, another POTTIACEAE with densely papillose upper laminal cells. It can be distinguished by the acute, leaf apex with excurrent costa ending in a sharp, hyaline tooth; the basal laminal cells are thin-walled and more or less inflated (in *Barbula indica* firm-walled and quadrate-rectangular). According to Sollman (pers. comm.) *Trichostomum duidense* is closely related to *T. brachydontium* Bruch and only different in the papillae at the back of the costa; previously he considered both species synonymous (Sollman 1984).

Order **Funariales**

Plants acrocarpous. Stems erect, leaves mostly broad and obovate, larger and crowded at the stem tip, costa single (occasionally absent), laminal cells large, lax, thin-walled. Sporophytes terminal. Setae usually elongate, capsules erect and symmetric or inclined and asymmetric, peristome single or double, teeth 16, endostome segments opposite the exostome teeth.

KEY TO THE FAMILIES

1   Plants minute, growing from a persistent protonema . . .*10. EPHEMERACEA*
       Plants small to medium-sized, protonema not persistent 2

2   Leaves acute, bordered with elongate cells or unbordered. Archegonia clustered in perichaetia. . . . . *11. FUNARIACEAE*
       Leaves rounded-obtuse, apex bordered by a single row of short cells. Archegonia solitary at leaf axils, scattered along the stem . . . .
       . . . . *12. SPLACHNOBRYACEAE*

10. **EPHEMERACEAE**

Description see Musci I in Fl. Suriname 6: 174. 1964.

D i s t r i b u t i o n : Worldwide; in the Guianas 1 genus.

LITERATURE

Crosby, M.R. 1968. Micromitrium Aust., an earlier name for Nanomitrium Lindb., Bryologist 71: 114-117.

1. **MICROMITRIUM** Austin, Musci Appalach. 10. 1870.
Type: M. austinii Sull. ex Austin

Nanomitrium Lindb., Not. Sällsk. Fauna Fl. Fenn. Förh. 13: 408. 1874, nom. nud.

Description see Musci I in Fl. Suriname 6: 174. 1964, as *Nanomitrium*.

Distribution: Neotropical, also in temperate regions; in the Guianas 1 species.

Note: The generic name *Nanomitrium* as used in Musci I is replaced by *Micromitrium* (Crosby 1968).

1. **Micromitrium thelephorothecum** (Florsch.) Crosby, Bryologist 71: 116. 1986. – *Nanomitrium thelephorothecum* Florsch., Fl. Suriname 6: 175. 1964. Type: Suriname, Florschütz 2253 (L).

Description see Musci I in Fl. Suriname 6: 175. 1964, as *Nanomitrium thelephorothecum*.

Distribution: Amazon Basin, Guyana, Suriname.

Ecology: Terrestrial on wet sand and clay of river banks; seldom collected, probably overlooked.

11. FUNARIACEAE

Description see Musci I in Fl. Suriname 6: 177. 1964.

Distribution: Worldwide; in the Guianas 2 genera.

LITERATURE

Fife, J.A. 1987. Taxonomic and nomenclatural observations on the Funariaceae 5. A revision of the Andean species of Entosthodon. Mem. New York Bot. Gard. 45: 301-325.

KEY TO THE GENERA

1   Leaves bordered with 1-2 rows of elongate cells. Capsule symmetric, annulus absent, peristome rudimentary or absent . . . . *1. Entosthodon*
Leaves unbordered. Capsule curved-asymmetric with oblique mouth, annulus present, peristome teeth lanceolate, trabeculate at back, often united at the tip . . . . *2. Funaria*

1. **ENTOSTHODON** Schwägr., Sp. Musc. Frond. Suppl. 2(1): 44. 1823. – *Funaria* subgenus *Entosthodon* (Schwägr.) Lindb., Musci Scand. 18. 1879.
Type: E. templetonii (Sm.) Schwägr. (Funaria templetonii Sm.)

Stems short, usually simple; upper leaves crowded, oblong-ovate or obovate, acute or acuminate, entire or serrulate above, costa ending in apex or below. Upper laminal cells lax, hexagonal to oblong-hexagonal, sometimes linear along the margin forming a border. Autoicous, sometimes polygamous. Seta elongate, capsule erect, symmetric, pyriform with a distinct neck, shrivelled when dry, operculum flat to convex-conic, exothecial cells incrassate, annulus absent, peristome single or (rarely) double, well-developed or rudimentary, inserted well below the mouth, exostome teeth papillose-striate, little or not trabeculate, endostome at most half the length of the teeth, often rudimentary. Calyptra cucullate, long-rostrate, inflated at base.

Distribution: Worldwide; in the Guianas 1 species.

1.   **Entosthodon bonplandii** (Hook.) Mitt., J. Linn. Soc., Bot. 12: 245. 1869. – *Gymnostomum bonplandii* Hook., Pl. Crypt. 1b. 1816.
Type: Colombia, Cundinamarca, Humboldt & Bonpland s.n. (NY).
– Fig. 74

Stems ca. 5 mm high, sparingly branched. Leaves contorted when dry, spreading when moist, small and distant below, larger and crowded towards end of stem, to 3 mm long and 1.5 mm wide, apex acute or short-acuminate, margin bluntly serrate in upper part, costa firm, extending 3/4 of leaf length to a few cells below apex, lamina bordered with 1-2 rows of linear cells; upper laminal cells hexagonal to oblong-hexagonal, 50-90 µm long and ca. 30 µm wide, more elongate towards

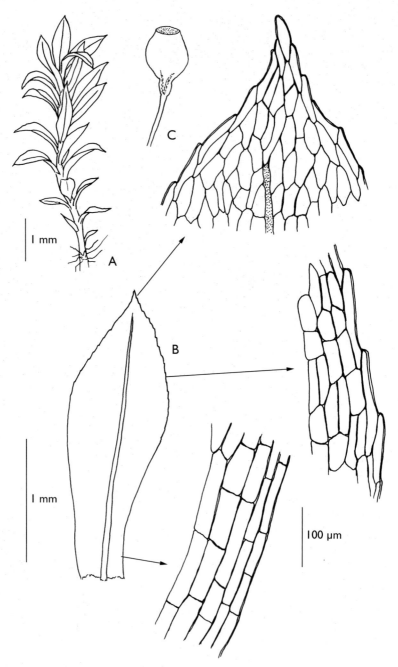

Fig. 74. *Entosthodon bonplandii* (Hook.) Mitt.: A, plant (moist); B, leaf; C, capsule (A-B, Gradstein 5073; C, Buck 3522, Puerto Rico).

midleaf, basal cells rectangular, 100-180 μm long. Autoicous. Seta smooth, 5-15 mm long, capsule erect, symmetric, globose-pyriform, with age becoming urceolate with wide mouth and short neck, operculum flat; peristome (after Fife 1987) usually absent, occasionally weakly developed, single. Calyptra cucullate-rostrate.

Distribution: Mexico, C America, West Indies, S America (Andes, Brazil, Paraguay, Guyana).

Ecology: Terrestrial on wet sand and in rock crevices at higher altitudes; rare.

Specimen examined: Guyana: Upper Mazaruni Distr., Waruma R., alt. 550 m, Gradstein 5073 (L).

2. **FUNARIA** Hedw., Sp. Musc. Frond. 172. 1801.
Type: F. hygrometrica Hedw.

Description see Musci I in Fl. Suriname 6: 177. 1964.

Distribution: Worldwide; in the Guianas 1 species.

1. **Funaria calvescens** Schwägr., Sp. Musc. Frond. Suppl. 1(2): 77. 1816. – *F. hygrometrica* Hedw. var. *calvescens* (Schwägr.) Mont., Ann. Sci. Nat., Bot. sér. 2, 12: 54. 1839. Type: Dominican Republic, Thouin s.n.; syntype?: Nova Belgia, La Billardière s.n. (G).

Description see Musci I in Fl. Suriname 6: 178. 1964, as *F. hygrometrica* var. *calvescens*.

Distribution: Pantropical.

Ecology: Terrestrial on disturbed (often burnt) soil; rare, not collected in French Guiana.

Note: The variety *calvescens* differs from the typical *Funaria hygrometrica* by the erect seta and only slightly inclined capsule. Allen (2002) concluded that these differences are sufficient to recognize *F. calvescens* as a distinct species.

210

12. SPLACHNOBRYACEAE

Small plants growing scattered or in thin mats. Stems erect, simple or sparingly divided. Leaves elliptic, ovate, spathulate or linear, apex obtuse, seldom acute, margin entire below, crenulate at apex, costa ending just below apex, rarely percurrent; upper laminal cells hexagonal, rhomboidal or rectangular, lax, gradually smaller towards margins. Dioicous. Archegonia solitary, scattered along the stem, seta elongate, thin, capsule ovoid to cylindric, usually short-necked, annulus persistent, operculum small, conic, umbonate to rostrate; peristome double, exostome reduced, hyaline, endostome papillose, consisting of 16 often rudimentary segments. Calyptra cucullate, long and narrow.

Distribution: Worldwide; in the Guianas 1 genus.

Notes: *Splachnobryum* was previously placed in the families SPLACHNACEAE or POTTIACEAE. Koponen (1981) concluded that in sporophytic and in gametophytic aspects it is remarkably different from other genera in the SPLACHNACEAE. It did not fit in the POTTIACEAE either, so she created the monogeneric family SPLACHNOBRYACEAE for this genus.
In a revision of this family Arts (2001) erected a second genus.

LITERATURE

Arts, T. 2001. A revision of the Splachnobryaceae. Lindbergia 26: 77-96.
Koponen, A. 1981. Splachnobryaceae, a new moss family. Ann. Bot. Fenn. 18: 123-132.

1. **SPLACHNOBRYUM** Müll. Hal., Verh. Zool.-Bot. Ges. Wien 19: 503. 1869.
   Type: S. obtusum (Brid.) Müll. Hal. (Weisia obtusa Brid.)

Description see Musci I in Fl. Suriname 6: 179. 1964.

Distribution: Worldwide; in the Guianas 1 species.

1. **Splachnobryum obtusum** (Brid.) Müll. Hal., Verh. Zool.-Bot. Ges. Wien 19: 504. 1869. – *Weisia obtusa* Brid., Muscol. Recent. Suppl. 1: 118. 1806. Type: S. Domingo, Poiteau s.n. (BM).

Description and synonymy see Musci I in Fl. Suriname 6: 180. 1964.

Distribution: Tropical and subtropical regions of all continents, also introduced in glasshouses in temperate regions.

Ecology: Terrestrial on loamy sand or clay, occasionally on bricks; in Suriname not common, not collected in Guyana or French Guiana.

Order **Bryales**

Plants acrocarpous. Stems generally erect. Leaves variable in shape, costa single. Sporophytes terminal, sometimes basal, setae elongate, capsules often inclined to horizontal, peristome double, exostome teeth 16, endostome with high basal membrane and perforate segments, alternating with exostome teeth.

KEY TO THE FAMILIES

1 Stems prostrate. Leaves complanate, strongly asymmetric, lamina divided in 2 unequal parts by the costa .... *15. PHYLLODREPANIACEAE*
   Stems erect. Leaves spreading, subsymmetric 2

2 Laminal cells often prorulose by projecting cell ends. Capsules subglobose or ovoid............................... *13. BARTRAMIACEAE*
   Laminal cells smooth or with central papillae. Capsules elongate........3

3 Margins, if serrate, not bistratose. Perichaetia at end of stem or branches...
   .... *14. BRYACEAE*
   Leaves double-serrate, with bistratose margins. Perichaetia at base of stem
   .... *16. RHIZOGONIACEAE*

13. **BARTRAMIACEAE**

Description see Musci I in Fl. Suriname 6: 201. 1964.

Distribution: Worldwide; in the Guianas 3 genera.

Note: The family BARTRAMIACEAE is characterized by globose and often furrowed capsules, leaves with a strong costa and cells often prorulose by projecting cell ends. Griffin & Buck (1989) emphasized the importance of the axillary hairs to divide the family in 3 subfamilies: BARTRAMIOIDEAE including *Leiomela*, BREUTELIOIDEAE including *Breutelia* and *Philonotis*, and CONOSTOMOIDEAE without representatives in the Guianas.

LITERATURE

Allen, B. 1999. The genus Philonotis in Central America. Häussknechtia 9: 19-36.

Griffin, D. & W.R. Buck 1989. Taxonomy and phylogenetic studies on the Bartramiaceae. Bryologist 92: 368-380.

KEY TO THE GENERA

1   Plants small; leaves not over 2 mm long, acute . . . . *3. Philonotis*
    Plants more robust; leaves over 2 mm long, slenderly acuminate 2

2   Leaves plicate at base, laminal cells uniform, alar cells differentiated . . . . .
    . . . . *1. Breutelia*
    Leaves not plicate, basal cells linear and nearly smooth, upper laminal cells rectangular with double papillae, alar cells not differentiated. . . . . .
    . . . . *2. Leiomela*

1. **BREUTELIA** (Bruch & Schimp.) Schimp., Coroll. Bryol. Eur. 85. 1856. – *Bartramia* sect. *Breutelia* Bruch & Schimp., Bryol. Europ. 4: 1. 1851.
    Type: B. arcuata (Sw.) Schimp. (Mnium arcuatum Sw.)

Description see Musci I in Fl. Suriname 6: 206. 1964.

Distribution: Worldwide; in the Guianas 1 species.

1. **Breutelia tomentosa** (Brid.) A. Jaeger, Ber. St. Gallischen Naturwiss. Ges. 1873-74: 93. 1875. – *Mnium tomentosum* (Brid.) Sw. ex Brid., Muscol. Recent. 2(3): 78. 1803. Type: Jamaica, Swartz s.n.   – Fig. 75

Plants in loose tufts, 2-15 cm high. Stems prostrate, densely tomentose in basal part, simple or irregularly branched, branches ascending. Leaves erect to wide-spreading, 2-4 mm long, lanceolate, plicate at base, apex slenderly acute, margin serrulate to sharply serrate in upper part, costa excurrent, serrate; laminal cells linear, 50-70 μm long and ca. 5 μm wide, thick-walled, prorulose at upper or lower end or both, alar cells slightly differentiated, shorter and wider, usually 1-2 inflated. Dioicous. Sporophytes not seen from the Guianas, description from Cleef 10227 (Colombia). Perichaetial leaves not differentiated. Seta 10-20 mm long, smooth, capsule inclined, ovoid-globose, furrowed, to 4 mm long,

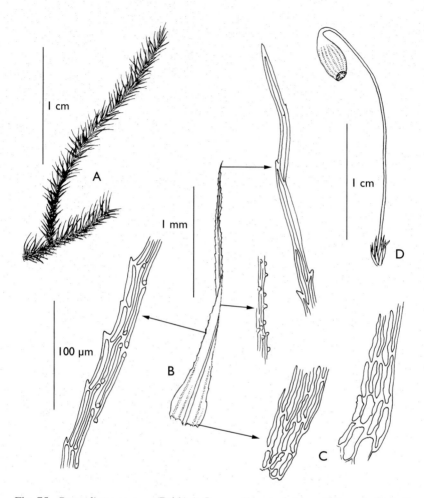

Fig. 75. *Breutelia tomentosa* (Brid.) A. Jaeger: A, part of stem with branch; B, leaf; C, inflated alar cells; D, perichaetium with sporophyte (A-C, Gradstein 5385; D, Cleef 10227, Colombia).

operculum plano-convex, exostome teeth to 500 μm long, papillose at upper end, endostome segments of same length, keeled, bifid, densely papillose at upper end.

D i s t r i b u t i o n : Mexico, West Indies, C and S America.

E c o l o g y : On wet rock or soil at higher altitudes; in the Guianas only known from Mt. Roraima.

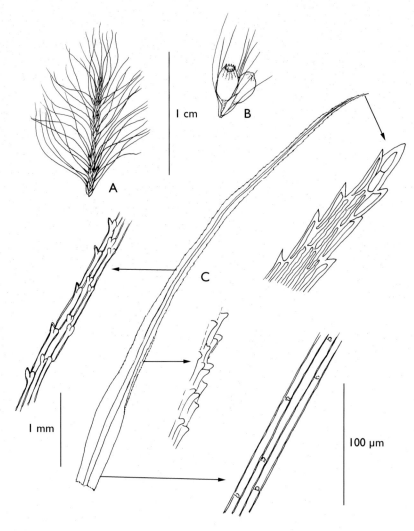

Fig. 76. *Leiomela bartramioides* (Hook.) Paris: A, end of stem; B, perichaetia with capsules; C, leaf (A, C, Gradstein 5313; B, Cleef 4835, Colombia).

Specimen examined: Guyana: N slope Mt. Roraima, alt. 1200-1600 m, Gradstein 5385 (L).

Note: In Musci I (p. 206) *Breutelia scoparia* (Schwägr.) A. Jaeger is recorded for Guyana, based on the collection McConnell & Quelch 514 (not seen); the latter collection came from the Venezuelan side of Mt. Roraima and may belong to the same species.

215

2. **LEIOMELA** (Mitt.) Broth. in Engler & Prantl, Nat. Pflanzenfam.
1(3): 634. 1904. – *Bartramia* subsect. *Leiomela* Mitt., J. Linn. Soc.,
Bot. 12: 253. 1869.
Type: not designated.

Robust plants in loose tufts. Stems simple or sparingly branched,
tomentose below. Leaves linear-lanceolate, apex subulate-acuminate,
serrate, costa percurrent to long-excurrent; upper laminal cells quadrate
to rectangular, prorulose at upper ends, basal cells longer and laxer, alar
cells little differentiated. Synoicous or dioicous. Capsules immersed,
erect or nodding, ovoid to subglobose, operculum plano-convex or plane;
peristome double (or lacking), exostome teeth 16, inserted below the
mouth, lanceolate, endostome usually rudimentary.

Distribution: Tropical and subtropical, mainly in America and Africa;
in the Guianas 1 species.

1. **Leiomela bartramioides** (Hook.) Paris, Index Bryol. 2, 3: 132.
1905. – *Leucodon bartramioides* Hook., Icon. Pl. 1: 71. 1836. Type:
Ecuador, Surucucho, Jameson s.n.   – Fig. 76

Plants in large tufts to 5(-7) cm high. Stems simple or with innovations
below the sporophytes, densely tomentose in lower part. Leaves wide-
spreading from an erect basal part, slenderly lanceolate, 5-10 mm
long and 0.3-0.5 mm wide at base, gradually narrowed to a long, often
flexuose, subulate apex, margin double-serrate by projecting cell ends,
serrate-dentate in upper part; costa excurrent, papillose at both sides,
sharply serrate in apex; basal laminal cells pellucid, linear, 90-180 μm
long and 8-10 μm wide with small prorulae at cell ends, upper laminal
cells rectangular, 15-35 μm long and 5-8 μm wide, with broad, 1-3
lobed prorulae protruding at upper cell ends. Synoicous or dioicous.
Perichaetial leaves erect, to 10 mm long, serrate, setaceous from a short,
oblong base. Seta 0.6-1 mm long, capsules immersed, ovoid-globose,
2-4 mm long and 1-2 mm wide, indistinctly furrowed below the mouth,
operculum plano-convex, exostome ca. 200 μm long, smooth, perforate,
endostome rudimentary.

Distribution: Pantropical.

Ecology: Epiphytic or terrestric on humus in montane forest; in the
Guianas only collected on Mt. Roraima.

Specimen examined: Guyana: N slope of Mt. Roraima, alt. 1200-
1600 m, Gradstein 5313 (L).

216

3. **PHILONOTIS** Brid., Bryol. Univ. 2:15. 1827.
Type: P. fontana (Hedw.) Brid. (Mnium fontanum Hedw.)

Description see Musci I in Fl. Suriname 6: 202. 1964.

D i s t r i b u t i o n : Worldwide; in the Guianas 3 species.

KEY TO THE SPECIES

1 Leaves ovate-oblong, apex rounded acute, costa ending well below
   apex; upper laminal cells rounded hexagonal, prorulae indistinct.....
   .... *1. P. hastata*
   Leaves lanceolate, apex sharp-acute, costa percurrent or excurrent; upper
   laminal cells rectangular with bulging prorulae at upper cell ends 2

2 Leaves ovate-lanceolate, costa excurrent, toothed in excurrent part ....
   ....*2. P. sphaerocarpa*
   Leaves triangular-lanceolate, costa percurrent to short-excurrent, but not
   toothed in excurrent part............................*3. P. uncinata*

1. **Philonotis hastata** (Duby) Wijk & Margad., Taxon 8: 74. 1959. –
   Hypnum hastatum Duby in Moritzi, Syst. Verz. 132. 1846. Type:
   Java, Fl. Tjapus, Zollinger s.n.   – Fig. 77

   Philonotis gracillima Ångstr., Öfvers. Kongl. Vetensk.-Akad. Förh. 33: 17.
   1876. – *P. uncinata* (Schwägr.) Brid. var. *gracillima* (Ångstr.) Florsch., Fl.
   Suriname 6: 205. 1964. Type: Brazil, Regnell 38 (BM, S).

Slender plants growing in low tufts. Stems ca. 1 cm high, erect or
prostrate, sparingly branched, tomentose at base. Leaves erect to wide-
spreading, distant to imbricate, 0.4-0.8 mm long, oblong-lanceolate, apex
rounded, broad-acute, margin bluntly serrate with a single or double row
of protruding marginal cells; costa ending well below apex (occasionally
percurrent), distantly serrate at back; cells in upper lamina rounded
hexagonal or oblong-rectangular, 15-25 µm long and 8-14 µm wide,
smooth or with one prorula at distal end, marginal cells oval, bulging
mammillose at distal end, inner basal cells larger, smooth. Sporophyte
not seen, description after Allen 1999: Dioicous. Seta 15-25 mm long,
flexuose. Capsule inclined, ovoid to subglobose, 1-2 mm long, operculum
conic-mammillate; peristome double, exostome teeth triangular,
trabeculate, outer surface finely papillose, coarsely papillose at the tip,
inner surface smooth, endostome not seen.

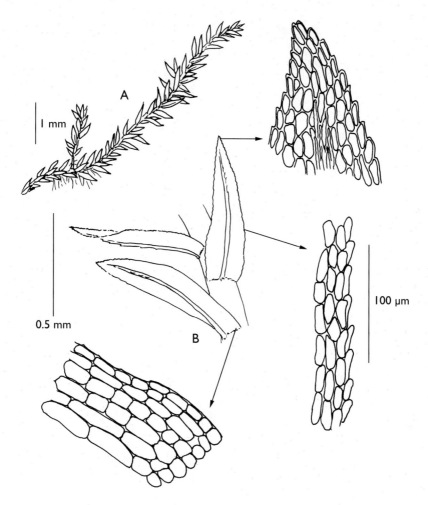

Fig. 77. *Philonotis hastata* (Duby) Wijk & Margad.: A, part of stem with young branch; B, leaves (A-B, Allen 19314A).

Distribution: Pantropical.

Ecology: Terrestric on clay or sandy soil; not common, only known from Suriname.

Specimens examined: Suriname: Paramaribo, Bakboord lake property, Allen 19314A (MO); Paramaribo, Cultuurtuin, Florschütz 561 (L).

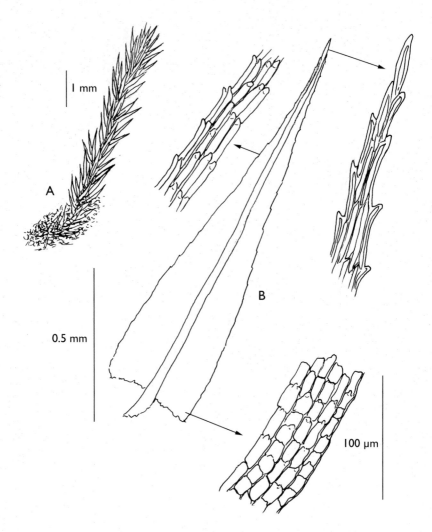

Fig. 78. *Philonotis sphaerocarpa* (Hedw.) Brid.: A, part of stem; B, leaf (A-B, Allen 23648).

2. **Philonotis sphaerocarpa** (Hedw.) Brid., Bryol. Univ. 2: 25. 1827. – *Mnium sphaerocarpon* Hedw., Sp. Musc. Frond. 197. 1801. – *Bartramia sphaerocarpa* (Hedw.) P. Beauv., Prodr. Aethéogam. 44. 1805. Type: Jamaica, sin coll. – Fig. 78

Small, yellowish green plants growing in loose tufts. Stems erect to prostrate, to 3 cm long, densely tomentose at base, sparingly branched; leaves erect-spreading, straight or slightly falcate, 1.2-1.5(-2) mm long, apex acuminate, margin double-serrate, often revolute, costa excurrent, toothed at the tip; laminal cells strongly papillose at apical ends, rectangular, 30-45 µm long and ca. 9 µm wide, in basal lamina gradually shorter and wider, alar cells quadrate. Dioicous. Seta 15-25 mm long, capsule 1.5-2 mm long, subglobose, inclined to pendent, operculum 0.5 mm long, conic-mammillate. Peristome double, exostome teeth densely papillose above, finely papillose below, endostome segments about as long as exostome, finely papillose, cilia 2-3. (Sporophyte description after Allen 1999).

Distribution: SE U.S.A., Mexico, West Indies, C America, northern and western S America.

Ecology: Terrestric, on soil over boulder; in the Guianas not common, only collected in Suriname.

Specimen examined: Suriname, Tafelberg, trail along south rim through burned savanna, Allen 23648 (MO).

3. **Philonotis uncinata** (Schwägr.) Brid., Bryol. Univ. 2: 22. 1827. – *Bartramia uncinata* Schwägr., Sp. Musc. Frond. Suppl. 1(2): 60. 1816. Type: Guadeloupe, Richard s.n. (PC).

Philonotis uncinata (Schwägr.) Brid. var. *glaucescens* (Hornsch.) Florsch., Fl. Suriname 6: 205. 1964. – *Bartramia glaucescens* Hornsch. in Mart., Fl. Bras. 1(2): 40. 1840. Type: Brazil, Beyrich s.n. (BM).

Description see Musci I in Fl. Suriname 6: 202. 1964.

Distribution: Neotropics and subtropics.

Ecology: Terrestric in moist areas, also on wet rocks and bricks, occasionally on decaying wood.

Note: The varieties of *P. uncinata* described in Musci I in Fl. Suriname 6: 205. 1964 are not maintained: var. *glaucescens* is difficult to separate from the species by many intergradations. Var. *gracillima* is identical with *P. hastata*, which is recognized as a distinct species (Allen 1999).

14. BRYACEAE

Description see Musci I in Fl. Suriname 6: 182. 1964.

Distribution: Worldwide; in the Guianas 2 genera.

LITERATURE

Frahm, J-P. 2002. The taxonomic status of Bryum arachnoideum C. Müll. and B. lanatum (P. Beauv.) Brid.. Trop. Bryol. 21: 53-56.
Ochi, H. 1971. What is true Bryum truncorum? Bryologist 74: 503-506.
Ochi, H. 1980. A revision of the Neotropical Bryoideae 1. J. Fac. Educ. Tottori Univ., Nat. Sci. 29: 49-154.
Ochi, H. 1981. A revision of the Neotropical Bryoideae 2. Ibid. 30: 21-55.
Steere, W.C. 1948. Contribution to the bryogeography of Ecuador. 1. A review of the species of Musci previously reported. Bryologist 51: 65-167.
Syed, H. 1973. A taxonomic study of Bryum capillare Hedw. and related species. J. Bryol. 7: 265-326.

KEY TO THE GENERA
(for specimens without capsules see key under *Bryum*)

1 Capsule erect, endostome incomplete, without segments or cilia . . . .
. . . . *1. Brachymenium*
   Capsule inclined or pendulous, endostome well-developed. . . . . . *2. Bryum*

1. BRACHYMENIUM Hook. ex Schwägr., Sp. Musc. Frond. Suppl. 2(1): 131. 1824.
Lectotype (A.L. Andrews in Grout, Moss Fl. N. Amer. 2: 208. 1935): B. nepalense Hook. ex Schwägr.

Description see Musci I in Fl. Suriname 6: 192. 1964.

Distribution: Pantropical and subtropical; in the Guianas 3 species.

KEY TO THE SPECIES

1 Leaves lanceolate, apex acute-acuminate, margin double-serrate in upper part . . . . *2. B. speciosum*
   Leaves ovate or obovate-orbicular, apex cuspidate, margin serrulate in upper part 2

2   Border indistinct, consisting of 1(-2) rows of thin-walled, elongate cells in
upper leaf half. . . . . *1. B. klotzschii*
Border distinct, consisting of 3-4 rows of incrassate, elongate cells . . . .
. . . . *3. B. wrightii*

1. **Brachymenium klotzschii** (Schwägr.) Paris, Index Bryol. 123. 1894.
– *Didymodon klotzschii* Schwägr., Sp. Musc. Frond. Suppl. 4: 310a.
1842. Type: not designated; locality Brazil.   – Fig. 79 A

Small plants in dense to loose mats. Stems erect, to 1 cm long, branched
by subfloral innovations, rhizoids reddish. Leaves twisted when dry, wide-
spreading when moist, ovate or ovate-oblong, to 1 mm long and 0.5 mm
wide, apex acute or rounded-acute, cuspidate, margins smooth or serrulate at
apex, recurved at base; costa strong, reddish, protruding at back, excurrent,
hyaline at the tip; upper laminal cells hexagonal or elongate-rhomboidal,
along the margin 1 row of elongate cells, forming an indistinct border
extending halfway down the leaf, basal laminal cells quadrate-rectangular,
along the basal margin smaller, quadrate. Dioicous. Seta 1-2.3 mm long;
capsule erect, oblong or oblong-clavate, operculum conic; peristome teeth
fragile, sublinear, hyaline, separate at base, sometimes bifid at the tip,
endostome hyaline, reduced to a basal membrane. (Sporophyte description
after Ochi in Sharp, Crum & Eckel, 1994).

Distribution: SE U.S.A., Mexico, West Indies, C and tropical S
America.

Ecology: On rocks and terrestric in open savanna areas.

Specimen examined: Guyana: S Rupununi Savanna, granitic
outcrop 5 km N of Dadanawa Ranch, Florschütz-de Waard 6026 (L).

2. **Brachymenium speciosum** (Hook. & Wilson) Steere, Bryologist 51:
98. 1948. – *Leptotheca speciosa* Hook. & Wilson in Hook., Icon. Pl.
8: tab. 748. 1845. Type: Venezuela, Maracaibo, Purdie, sept. 1844.

Brachymenium sipapoense E.B. Bartram, Mem. New York Bot. Gard. 10(2):
7. 1960. Type: Venezuela, Maguire & Politi 27811 (NY).

Description see Musci I in Fl. Suriname 6: 192. 1964, as *B. sipapoense*.

Distribution: Jamaica, Venezuela, Andes, the Guianas.

Ecology: Epiphytic; only at higher altitudes.

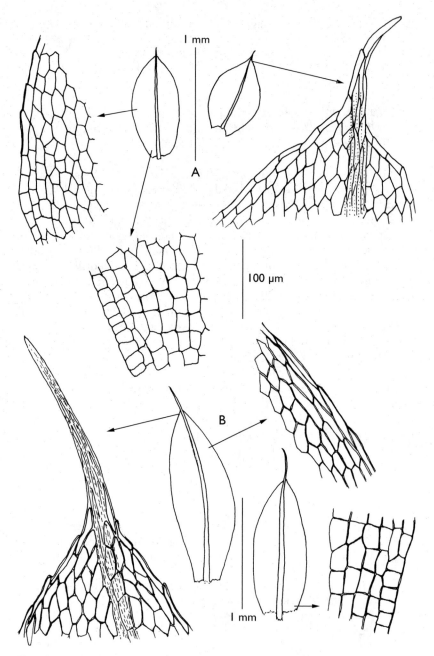

Fig. 79. A. *Brachymenium klotzschii* (Schwägr.) Paris: A, leaves. (A, Florschütz-de Waard 6026). B. *Bryum capillare* Hedw.: B, leaves (B, Florschütz-de Waard & Zielman 5515).

3. **Brachymenium wrightii** (Sull.) Broth. in Engler & Prantl, Nat.
Pflanzenfam. 1(3): 559. 1903. – *Leptotheca wrightii* Sull., Proc.
Amer. Acad. Arts 5: 281. 1861. Type: Cuba, Wright 53 (NY).

Brachymenium wrightii (Sull.) Broth. var. *mnioides* (Besch.) Florsch., Fl.
Suriname 6: 195. 1964. – *Brachymenium mnioides* Besch., J. Bot. 16: 7.
1902. Type: Guadeloupe, Duss 1018 (H).

Description see Musci I in Fl. Suriname 6: 193. 1964, as *B. wrightii*
var. *mnioides*.

D i s t r i b u t i o n : Mexico, West Indies, C and tropical S America.

E c o l o g y : Epiphytic on logs and living trees; rare, collected only once
in Suriname and French Guiana.

2. **BRYUM** Hedw., Sp. Musc. Frond. 178. 1801.
Type: B. argenteum Hedw.

Description see Musci I in Fl. Suriname 6: 183. 1964.

D i s t r i b u t i o n : Worldwide; in the Guianas 6 species.

KEY TO THE SPECIES
(including sterile *Brachymenium* species)

1 Plants whitish when dry; upper part of leaf hyaline 2
  Plants not whitish; upper part of leaf green 3

2 Leaf apex obtuse or rounded-acute, costa not reaching apex . . . .
  . . . .*2a. B. argenteum* var. *argenteum*
  Leaf apex acuminate to filiform, costa excurrent or percurrent . . . .
  . . . .*2b. B. argenteum* var. *lanatum*

3 Leaves not or indistinctly bordered 4
  Leaves distinctly bordered with incrassate, elongate cells 6

4 Leaves ovate-oblong, apex rounded, cuspidate; small border of 1 row of
  elongate cells present along upper part of leaf . . . .*Brachymenium klotzschii*
  Leaves lanceolate, apex acute, no distinct border. 5

5   Costa percurrent (occasionally short-excurrent); upper laminal cells elongate-
    rhomboidal, often flexuose, 5-10 times as long as wide. Capsule pyriform
    with long-tapering neck ....*1. B. apiculatum*
    Costa long-excurrent; upper laminal cells hexagonal, 2-4 times as long as
    wide. Capsule cylindric with truncate neck .... *6. B. coronatum*

6   Robust plants, stems to 10 cm long; leaves 6-12 mm long ....
    ....*3. B. beyrichianum*
    Plants much smaller; leaves not over 6 mm long 7

7   Leaves obovate to orbicular with cuspidate apex ... *Brachymenium wrightii*
    Leaves obovate-oblong, spathulate or lanceolate with acute-acuminate apex
   8

8   Border narrow, 1-2 cells broad at midleaf, margin crenulate or bluntly serrate
    in upper half .... *5. B. capillare*
    Border 2-4 cells broad at midleaf, margin coarsely dentate in upper half. ... 9

9   Leaves lanceolate, border double-dentate .... *Brachymenium speciosum*
    Leaves spathulate or obovate-oblong, border serrate-dentate ....
    .... *4. B. billardieri*

1. **Bryum apiculatum** Schwägr., Sp. Musc. Frond. Suppl. 1(2): 102.
    1816. – *Pohlia apiculata* (Schwägr.) H.A. Crum & L.E. Anderson,
    Mosses E. N. Amer. 1: 534. 1981. Type: America meridional, Richard
    s.n. (G).

    Bryum cruegeri Hampe in Müll. Hal., Syn. Musc. Frond. 1: 300. 1848.
    Syntypes: Trinidad, Crüger s.n. (BM); Suriname, Kegel 503 (GOET, PC).

    Description see Musci I in Fl. Suriname 6: 188. 1964, as *B. cruegeri*.

    D i s t r i b u t i o n : Pantropical, extending to temperate areas.

    E c o l o g y : Terrestrial, occasionally on rocks and tree trunks.

2. **Bryum argenteum** Hedw., Sp. Musc. Frond. 181. 1801. Type: not
    designated; locality: Europe.

2a. var. **argenteum**   – Fig. 80 C

    Bryum candicans Taylor in Mitt., J. Linn. Soc., Bot. 12: 303. 1869. Type:
    Ecuador, Andes Quitenses, Jameson 108 (NY).

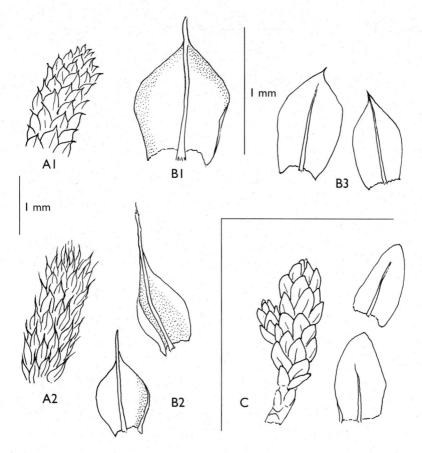

Fig. 80. A-B. *Bryum argenteum* Hedw. var. *lanatum* (P. Beauv.) Hampe: A, end of stems; B, leaves (A1, B1, Jansen-Jacobs 4829; A2, B2, Aptroot 17054; B3, Jansen-Jacobs 5605). C. *Bryum argenteum* Hedw. var. *argenteum*: C, end of stem (C, Florschütz 4568).

Description see Musci I in Fl. Suriname 6: 183. 1964, as *B. candicans* (reduced to synonymy by Ochi 1980).
The typical variety has an obtuse or rounded-acute apex and a costa extending 2/3 of leaf length.

D i s t r i b u t i o n : Worldwide.

E c o l o g y : Terrestrial on bare rocks and rock crevices; in the Guianas only collected on the Voltzberg in Suriname.

2b. var. **lanatum** (P. Beauv.) Hampe, Linnaea 13: 44. 1839. – *Mnium lanatum* P. Beauv., Prodr. Aethéogam. 75. 1805. Type: not designated; locality North America.   – Fig. 80 A-B

Var. *lanatum* is different from the typical variety in the acuminate, sometimes long-filiform apex and excurrent costa.

D i s t r i b u t i o n : Worldwide; in the Guianas only collected in Guyana.

S p e c i m e n s   e x a m i n e d : Guyana: Upper Mazaruni Distr., Kamarang, Aptroot 17054 (L); Rupununi savanna, Florschütz-de Waard 6027, Jansen-Jacobs *et al.* 4829, 5605 (L).

N o t e s : Ochi (1980) synonymized var. *lanatum* with the species because many intermediate forms occur. Frahm (2002) studied African collections of *B. argenteum* var. *lanatum* and concluded that the differences with the species are important. He proposed to treat it on species level as *B. lanatum*.
In the collections from Guyana the acuminate apex is a constant character but the length of the costa is variable. The different aspect of the plants with excurrent costae, however, makes it useful to maintain at least the variety.

3. **Bryum beyrichianum** (Hornsch.) Müll. Hal., Syn. Musc. Frond. 1: 249. 1848. – *Mnium beyrichianum* Hornsch. in Mart., Fl. Bras. 1(2): 45. 1840. – *Rhodobryum beyrichianum* (Hornsch.) Müll. Hal. in Hampe, Vidensk. Meddel. Naturhist. Foren. Kjøbenhavn ser. 3, 6: 146. 1875. Type: Brazil, Beyrich s.n. (BM).

Description see Musci I in Fl. Suriname 6: 191. 1964, as *Rhodobryum beyrichianum*.

D i s t r i b u t i o n : Neotropical.

E c o l o g y : Terrestric or epiphytic in the undergrowth of moist rainforest at higher altitudes; not common, not seen from Guyana and French Guiana.

N o t e : Ochi (1980, 1981) treated the genus *Rhodobryum* as a subgenus of *Bryum*, distinguished only by the size, the absence of stereïd tissue in the costa and the presence (occasionally) of subterranean stolons.

4. **Bryum billardierei** Schwägr., Sp. Musc. Frond. Suppl. 1(2): 115. 1816. Type: Tasmania, Billardière s.n. (not located).

Description see Musci I in Fl. Suriname 6: 184. 1964, as *Bryum truncorum* (Brid.) Brid.

Distribution: Pantropical.

Ecology: On wet stones and tree trunks in lowland rainforest; in the Guianas not common.

Note: This species, hitherto erroneously named *B. truncorum*, proves to belong to *B. billardierei* (Ochi 1971). In Buck (2003) this species is reported as *Rosulabryum billardierei* (Schwägr.) Spence.

5. **Bryum capillare** Hedw., Sp. Musc. Frond. 182. 1801. Type: "Ad vias cavas, muros, latera fossarum, ligna putrida Europae", sin. col. (G).
   – Fig. 79 B

Small, green plants in loose or dense tufts. Stems erect, 0.5-2(-5) cm long, repeatedly branched, rhizoids at base orange-brown, finely papillose, tubers present. Leaves more or less crowded above, spirally twisted when dry, erect-spreading when moist, 1-3 mm long and ca. 0.5 mm wide, ovate-oblong, sometimes spathulate, acute or rounded-acute, abruptly mucronate or cuspidate at apex, margin bluntly serrulate or crenulate in upper part; costa strong, brown, ending in the cuspidate apex to long-excurrent as a smooth, rigid point; upper laminal cells thinwalled, hexagonal, 30-60 µm long and 15-20 µm wide, along the margins 1-2 rows of linear cells forming a narrow, but distinct border; basal laminal cells rectangular or quadrate, 20-50 µm long, marginal cells quadrate. Dioicous. Perichaetial leaves narrow-lanceolate with long-piliferous apex, costa excurrent, perigonial leaves broad-ovate, costa percurrent. Seta 10-35 mm long, capsule inclined or subpendulous, symmetrical, pyriform to subcylindric with a distinct neck, 2.5-5 mm long, operculum convex-conic, apiculate, peristome teeth brown, endostome segments fenestrate, cilia 1-4, appendiculate. (Sporophyte description after Syed 1973).

Distribution: Worldwide.

Ecology: On rock, soil and walls, also epiphytic on tree trunks and branches; in the Guianas rare, only one collection.

Specimen examined: Suriname: Kabalebo-Dam project, granite plateau km 114, Florschütz-de Waard & Zielman 5515 (L).

6. **Bryum coronatum** Schwägr., Sp. Musc. Frond. Suppl. 1(2): 103. 1816. Lectotype (Ochi 1980): French Guiana, Richard s.n. (PC).

Description see Musci I in Fl. Suriname 6: 187. 1964.

Distribution: Pantropical, extending to temperate areas.

Ecology: Terrestrial, often on burnt substrates, also collected in the lower canopy of lowland rainforest.

15. PHYLLODREPANIACEAE

Drepanophyllaceae (see Note under *Phyllodrepanium*).

Small to medium-sized plants in flat cushions. Stems prostrate, sparingly branched, often with bundles of propagules at end of stem; leaves in 4 rows, complanate-spreading, ovate-oblong to oblong-lanceolate, curved-asymmetric to falcate, costa strong, ending in the apiculate apex, dividing the lamina in 2 unequal parts; laminal cells incrassate, smooth or unipapillose. Perichaetial leaves elongate, linear-ligulate; sporophyte (only known in *Phyllodrepanium*) terminal, seta elongate, capsule erect, symmetric, peristome single with 16 short teeth.

Distribution: West Indies, C and tropical S America; in the Guianas 2 genera.

LITERATURE

Crosby, M.R. 1970. Some remarks on the genus Drepanophyllum Schwägr. Rev. Bryol. Lichénol. 37: 345-353.

KEY TO THE GENERA

1 Leaves curved, asymmetric but not falcate; laminal cells rounded-quadrate, unipapillose .... *1. Mniomalia*
   Leaves falcate; laminal cells rhomboid-elliptic, smooth ....
   .... *2. Phyllodrepanium*

229

1. **MNIOMALIA** Müll. Hal., J. Mus. Godeffroy 3(6): 60. 1874. Type: M. semilimbata (Mitt.) Müll. Hal. (Drepanophyllum semilimbatum Mitt.)

Small plants, stems little divided, prostrate. Leaves in four rows, spreading laterally, obliquely ovate, unequally divided by the costa; laminal cells rounded, smooth or unipapillose.

Distribution: Pantropical; in the Guianas 1 species.

1. **Mniomalia viridis** (Mitt.) Müll. Hal., J. Mus. Godeffroy 3(6): 61. 1874. – *Drepanophyllum viride* Mitt., J. Linn. Soc. Bot. 12: 318. 1869. Type: Brazil, Pará, Spruce 553 (BM). – Fig. 81

Mniomalia bernouillii Müll. Hal., Bull. Herb. Boissier 5: 176. 1897. Type: Guatemala, Boissier 91.

Slender, dull green plants growing in thin mats. Stems densely tomentose at base, to 2 cm long but usually shorter, prostrate, often bearing propagules at end of stems; leaves in four rows, complanate-spreading or curved to the substrate, asymmetric, ovate-oblong, to 1.7 mm long, smaller at the end of propagulum-bearing stems, apex rounded-acute and apiculate or short-acuminate, margin entire or minutely serrulate, often narrowly revolute; costa at base nearest to one side of the oblique insertion, above midleaf central, at apex ending in the short acumen; laminal cells incrassate, rounded-quadrate, 8-15 µm in diameter, with one coarse papilla excentric over each lumen, basal cells larger, oblong, along the margin one row of small quadrate cells. Propagules reddish brown, septate, in dense clusters at the end of stems or branches, sometimes also in the axils of the small upper leaves. Perichaetia conspicuous, at the end of short branches, inner perichaetial leaves to 3 mm long and 0.3 mm wide, ligulate with acuminate apex, cells smooth, elongate at midleaf, subquadrate at apex. Seta to 1.2 cm long, capsule to 1.4 mm long, erect, radially symmetric or more or less bilaterally symmetric. (Sporophyte description after Pursell & Allen 2002).

Distribution: C and tropical S America.

Ecology: Epiphytic in the canopy or in open areas of lowland rainforest; not common.

Selected specimens: Suriname: Kabalebo Dam project, road km 212, Florschütz & Zielman 5762 (L). French Guiana: Concession CIPRIO, S of Mt. des Chevaux, Cremers 5324 (CAY, L).

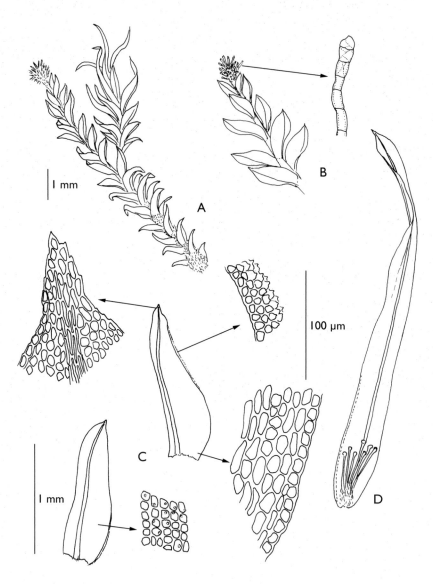

Fig. 81. *Mniomalia viridis* (Mitt.) Müll. Hal.: A, stem with propagula and branch with perichaetium; B, end of stem with propagula; C, leaves; D, perichaetial leaf (A-D, Cremers 5324).

2. **PHYLLODREPANIUM** Crosby, Rev. Bryol. Lichénol. 37: 345. 1970. – *Drepanophyllum* Schwägr., Sp. Musc. Frond. Suppl. 2(1): 84. 1823, nom. illeg.
Type: P. falcifolium (Schwägr.) Crosby (Fissidens falcifolius Schwägr.)

Description see Musci I in Fl. Suriname 6: 197. 1964, as *Drepanophyllum*.

Distribution: West Indies, C and northern S America; 1 species.

Note: The name *Drepanophyllum* Hook. was published without valid description. Schwägrichen's description in 1823 was valid but meanwhile the name was used for a genus of phanerogams. Crosby (1970) proposed the name *Phyllodrepanium* and also described the variation in leaf length, cell shape and differentiation of marginal cells, concluding that the genus is represented by only 1 variable species.

1. **Phyllodrepanium falcifolium** (Schwägr.) Crosby, Rev. Bryol. Lichénol. 37: 346. 1970. – *Fissidens falcifolius* Schwägr., Sp. Musc. Frond. Suppl. 1(2): 9. 1816. – *Drepanophyllum falcifolium* (Schwägr.) Wijk & Margad., Taxon 8: 106. 1959. Lectotype (Crosby 1970): French Guiana, Richard s.n. (G).

Drepanophyllum duidense R.S. Williams, Bull. Torrey Bot. Club 58: 501. 1931. Type: Venezuela, Mt. Duida, Tate 222 (NY).

Description see Musci I in Fl. Suriname 6: 197. 1964, as *Drepanophyllum falcifolium*.

Distribution: West Indies, C and tropical S America.

Ecology: Epiphytic on tree trunks and rotting logs, occasionally on wet rocks; rather common in tropical lowland rainforest.

16. RHIZOGONIACEAE

Description see Musci I in Fl. Suriname 6: 198. 1964.

Distribution: Pantropical; in the Guianas 1 genus.

LITERATURE

Manuel, G. 1980. Classification of Rhizogonium Brid., Miscellanea bryologica 2. Cryptog. Bryol. Lichénol. 1: 67-70.

Notes: In a reclassification of the genus *Rhizogonium* (Manuel, 1980) the section *Pyrrhobryum* was recognized as a separate genus. *Rhizogonium spiniforme*, belonging to this section, was transferred to *Pyrrhobryum*. It is different from *Rhizogonium* by the leaves with multistratose, double-toothed borders, spreading in all directions.
Rhizogonium lindigii (Hampe) Mitt. was recorded for Guyana (Musci I in Fl. Suriname 6: 201) but it was collected on the Venezuelan side of Mt. Roraima. No collections from the Guianas are known.

1. **PYRRHOBRYUM** Mitt., J. Linn. Soc., Bot. 10: 174. 1868.
   Lectotype (Manuel 1980): P. spiniforme (Hedw.) Mitt. (Hypnum spiniforme Hedw.)

   Rhizogonium sect. *Pyrrhobryum* (Mitt.) Mitt., J. Linn. Soc., Bot. 12: 326. 1869.

Description see Musci I in Fl. Suriname 6: 200. 1964, as *Rhizogonium*.

Distribution: Pantropical (also subtropical); in the Guianas 1 species.

1. **Pyrrhobryum spiniforme** (Hedw.) Mitt., J. Linn. Soc., Bot. 10: 174. 1868. – *Hypnum spiniforme* Hedw., Sp. Musc. Frond. 236. 1801. – *Rhizogonium spiniforme* (Hedw.) Bruch in Krauss, Flora 29: 134. 1846. Type: Jamaica, Sloane s.n. (G).

Description see Musci I in Fl. Suriname 6: 200. 1964, as *Rhizogonium spiniforme*.

Distribution: Pantropical, extending into subtropical areas.

Ecology: Usually on wet rock cliffs, also terrestrial or epiphytic on bark of trees; at higher altitudes, not common.

Order **Orthotrichales**

Plants acrocarpous or cladocarpous. Stems erect or creeping with many erect branches. Leaves erect-spreading or contorted when dry, costa single, strong. Sporophytes terminal on stem and branches; capsules erect and symmetric, peristome double, often variously reduced and then single or lacking.

17. MACROMITRIACEAE

Medium-sized plants, cladocarpous. Stems creeping with ascending branches. Leaves often spirally twisted or crispate when dry, oblong-lingulate to linear-lanceolate, apex acute-acuminate, seldom obtuse, costa percurrent to long-excurrent, laminal cells thick-walled, small and rounded in upper leaf, in basal part usually elongate, sometimes tuberculate. Capsule exserted, mostly erect, smooth or furrowed, neck often distinct; peristome double, single or reduced, exostome teeth 16, often papillose, endostome often reduced. Calyptra mitrate or campanulate, seldom cucullate.

N o t e : Traditionally the MACROMITRIACEAE were treated as a subfamily of the ORTHOTRICHACEAE, but phylogenetic research has lead to the separation of the families (De Luna 1995). MACROMITRIACEAE are recognized as a distinct family by Churchill & Linares (1995). The genera *Groutiella*, *Macromitrium* and *Schlotheimia*, treated under ORTHOTRICHACEAE (Musci I in Fl. Suriname 6: 206. 1964) belong to this family. The genus *Zygodon* Hook. & Taylor remained in ORTHOTRICHACEAE, but no collections from the Guianas are known (the collection McConnell & Quelch 540, mentioned in Musci I in Fl. Suriname 6: 207. 1964, is from the Venezuelan side of Mt. Roraima).

D i s t r i b u t i o n : Pantropical and -subtropical; in the Guianas 3 genera.

LITERATURE

Churchill, S.P. & C.E. Linares. 1995. Prodromus Bryologiae Nova Granatensis. Bibl. José Jéronimo Triana 12: 1-924.

Crosby, M.R. 1970. A study of Groutiella apiculata and G. mucronifolia. Bryologist 73: 607-611.

De Luna, E. 1995. The circumscription and phylogenetic relationships of the Hedwigiaceae. Syst. Bot. 20: 347-373.

Grout, A.J. 1944. Preliminary synopsis of the North American Macromitriae. Bryologist 47: 1-22.

Grout, A.J. 1946. Orthotrichaceae. North American Flora 15A: 1-62.

Vitt, D.H. & H. Crum. 1970. Groutiella tomentosa, new to the United States. Bryologist 73: 145-149.

Vitt, D.H. 1979. New taxa and new combinations in the Orthotrichaceae of Mexico. Bryologist 82: 1-19.

Vitt, D.H. 1994. Orthotrichaceae. In A.J. Sharp, H. Crum & P.M. Eckel. Moss Flora of Mexico, 2: Mem. New York Bot. Gard. 69: 590-656.

Vitt, D.H., T. Koponen & D.H. Norris. 1995. Bryophyte flora of the Huon Peninsula, Papua New Guinea. LV. Desmotheca, Groutiella, Macrocoma and Macromitrium (Orthotrichaceae, Musci). Acta Bot. Fenn. 154: 1-94.

KEY TO THE GENERA

1 Basal laminal cells quadrate or transversely elongate (oblate), with a border of elongate cells . . . . *1. Groutiella*
  Basal laminal cells all elongate (or all isodiametric), no border differentiated. 2

2 Leaves crispate or spirally twisted when dry. Calyptra campanulate, often plicate, laciniate at base . . . . *2. Macromitrium*
  Leaves erect, appressed when dry. Calyptra cucullate or mitriform. 3

3 Midlaminal cells elongate. Calyptra cucullate . . . .
  . . . . *2. Macromitrium (pellucidum)*
  Midlaminal cells rounded or transversely elongate. Calyptra campanulate, lobed at base . . . . *3. Schlotheimia*

1. **GROUTIELLA** Steere, Bryologist 53: 145. 1950.
Type: G. tomentosa (Hornsch.) Wijk & Margad. (Macromitrium tomentosum Hornsch.)

Description and synonymy see Musci I in Fl. Suriname 6: 207. 1964.

D i s t r i b u t i o n : Pantropical; in the Guianas 4 species.

N o t e : The species *Groutiella wagneriana* (Müll. Hal.) H.A. Crum & Steere, mentioned in Musci I in Fl. Suriname 6: 215. 1964, was not collected in the Guianas.

KEY TO THE SPECIES

1 Leaf apex acute or rounded acute 2
  Leaf apex obtuse, mucronate 3

2 Leaves often abruptly contracted above midleaf to a long, linear apex, very fragile and usually broken . . . . *4. G. tomentosa*
  Leaves without fragile upper part . . . . *1. G. chimborazensis*

3 Leaves very rugose in upper part; costa ending in a short, blunt mucro . . . .
  . . . . *3. G. obtusa*
  Leaves not or inconspicuously rugose; costa excurrent in a sharp apiculus
  . . . . *2. G. mucronifolia*

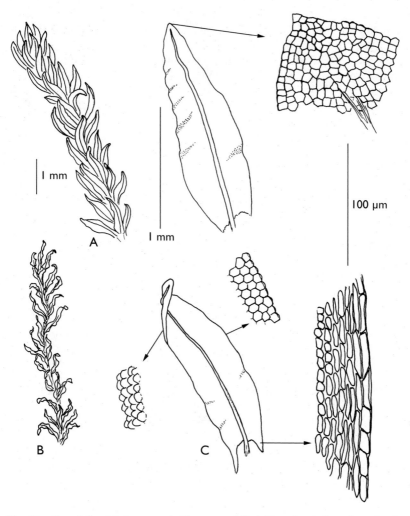

Fig. 82. *Groutiella chimborazensis* (Spruce ex Mitt.) Florsch.: A, branch moist; B, branch dry; C, leaves (A-C, Florschütz 4797).

1. **Groutiella chimborazensis** (Spruce ex Mitt.) Florsch., Fl. Suriname 6: 215. 1964. – *Macromitrium chimborazense* Spruce ex Mitt., J. Linn. Soc., Bot. 12: 218. 1869. Type: Ecuador, Spruce 110 (NY).
– Fig. 82

Groutiella undosa (Cardot) H.A. Crum & Steere, Bryologist 53: 147. 1950. – *Macromitrium undosum* Cardot, Rev. Bryol. 36: 108. 1909. Type: Mexico, Jalisco, Pringle 10560 (NY, PC).

Olive-green plants growing in thin mats. Stems elongate, creeping, radiculose; branches distant, erect, to 1 cm high. Stem and branch leaves not different, contorted-crispate when dry, erect-spreading when moist, keeled at base, oblong-lanceolate with acute or rounded-acute apex, 1.5-2(-2.5) mm long, margins smooth or crenulate, often undulate, costa firm, prominent at back, ending in or below apex; laminal cells rounded quadrate or rounded hexagonal, 5-8 μm in diam., thin-walled to slightly incrassate, flat or bulging, uniform throughout the leaf, inner basal cells enlarged; border extending 1/3-1/2 of leaf length, 2-5 cells wide at base. Dioicous. Perigonia terminal, perichaetial leaves not differentiated. Seta 4-13 mm long, smooth, capsule ovoid-cylindric, 1.8-3.4 mm long, operculum slenderly rostrate, peristome membranous, rudimentary. Calyptra mitriform, plicate, naked, split at base. (Sporophyte description after Vitt 1994).

Distribution: Mexico, West Indies, C and tropical S America.

Ecology: Epiphytic on tree trunks and branches; in the Guianas not common, only known from a few collections in lowland rainforest.

Selected specimens: Guyana: Rupununi Distr., Cold Creek Falls, alt. 100 m, Maas *et al.* 3880 (L). Suriname: Brownsberg, alt. 450 m, Florschütz 4797 (L). French Guiana: Saül, Sentier Limonade, alt. ca. 200 m, Montfoort & Ek 673 (L).

Notes: This species is characterized by the uniform rounded-quadrate or hexagonal laminal cells, incrassate towards leaf base.
The type of *G. chimborazensis* (NY) has a more acuminate apex and less bulging cells than in the Guianan collections. The type of *G. undosa* (NY) shows more resemblance: more undulate margins, more bulging upper laminal cells and a costa ending below apex. According to Vitt (1979) there are so many intermediate specimens that *G. undosa* has to be considered synonymous.

2. **Groutiella mucronifolia** (Hook. & Grev.) H.A. Crum & Steere, Bryologist 53: 146. 1950. – *Orthotrichum mucronifolium* Hook. & Grev., Edinburgh J. Sci. 1: 116. 1824. Type: St. Vincent, Guilding s.n. (BM, NY).

Description and synonymy see Musci I in Fl. Suriname 6: 213. 1964.

Distribution: Southern U.S.A., Mexico, West Indies, C and northern S America.

Ecology: Epiphytic on tree trunks and branches; rather common in coastal region and lowland rainforest.

Note: In gametophytic aspects *G. mucronifolia* is difficult to separate from *G. apiculata* (Hook.) H.A. Crum & Steere (not yet reported for the Guianas). Crosby (1970) stated that only the sporophytes are different and Vitt (1979) concluded that both species are synonymous. However, in the Moss Flora of Central America (Allen 2002) *G. mucronifolia* is distuinguished from *G. apiculata* by differences in the sporophytes: shorter setae (3-8 mm long), shorter, obovoid capsules (1-2 mm long) and rudimentary peristomes. Accept for a single longer seta, these characters were all in accordance with the Guianan collections of *G. mucronifolia*. Allen (2002) reported for C America the occurrence of *G. apiculata* above 700 m and of *G. mucronifolia* below 500 m.

3. **Groutiella obtusa** (Mitt.) Florsch., Fl. Suriname 6: 210. 1964. – *Macromitrium obtusum* Mitt., J. Linn. Soc., Bot. 12: 201. 1869. Lectotype (Florschütz 1964): Brazil, Spruce 111 (NY, PC).

Description and synonymy see Musci I in Fl. Suriname 6: 210.1964.

Distribution: West Indies, the Guianas.

Ecology: Epiphytic on branches of solitary trees and in the canopy of lowland rainforest; common.

4. **Groutiella tomentosa** (Hornsch.) Wijk & Margad., Taxon 9: 51. 1960. – *Macromitrium tomentosum* Hornsch. in Mart., Fl. Bras. 1(2): 21. 1840. Type: Uruguay, Sellow s.n. (NY).

Groutiella fragilis (Mitt.) H.A. Crum & Steere, Bryologist 53: 146. 1950. – *Macromitrium fragile* Mitt., J. Linn. Soc., Bot. 12: 218. 1869. Type: Brazil, Spruce 110 C (NY).

Description and synonymy see Musci I in Fl. Suriname 6: 208. 1964, as *G. fragilis*.

Distribution: Pantropical, also extending to subtropical areas.

Ecology: Epiphytic on stem and branches of solitary trees and in the canopy of lowland rainforest, occasionally at higher altitudes; common.

Notes: This species is easily recognized by the long and slender leaves which are usually best visible at the end of the branches; the elongate, linear apex is fragile and quickly broken. Lower branch leaves with acute apices may resemble the leaves of *G. chimborazensis*, but the laminal cells are not uniform throughout the leaf but become transversely elongate and strongly incrassate in the basal part.

The extensive synonymy of this species was studied by Vitt & Crum (1970), who showed that the species is pantropical in distribution. They also described the variation of several characters such as the length of the basal border and the papillosity of the basal cells, ranging from smooth to conical-papillose.

2. **MACROMITRIUM** Brid., Muscol. Recent. Suppl. 4: 132. 1819.
Type: *M. aciculare* Brid., nom. illeg. [= *M. pallidum* (P. Beauv.) Wijk & Margad. (Orthotrichum pallidum P. Beauv.)]

Description see Musci I in Fl. Suriname 6: 221. 1964.
Addition to the description: Stem leaves sometimes different from branch leaves, often smaller; costa percurrent to long-excurrent. Occasionally laminal cells not elongate towards base, but uniform throughout the leaf (*M. orthostichum*).

Distribution: Pantropical; in the Guianas 13 species.

Note: The expression 'secondary stem' is not used in the species descriptions since the difference with branches is not clearly defined.

KEY TO THE SPECIES

1 Laminal cells uniform, isodiametric throughout the leaf; upper laminal cells pluripapillose . . . .*4. M. orthostichum*
  Basal laminal cells elongate and narrow, upper laminal cells smooth, bulging or unipapillose 2

2 Basal cells of leaf tuberculate-papillose 3
  Basal cells of leaf smooth 9

3 Leaf apex obtuse, mucronate . . . . *10. M. stellulatum*
  Leaf apex acute or acuminate 4

4 Costa percurrent, ending in or just below apex (occasionally short-excurrent in *M. cirrosum*) 5
  Costa distinctly excurrent 8

5   Upper laminal cells elongate, upper margins serrate 6
    Upper laminal cells isodiametric, upper margins crenulate or finely
    serrulate 7

6   Leaves less than 5 mm long; upper laminal cells smooth or mammillose . . .
    . . . . *1. M. cirrosum*
    Leaves 5-7 mm long; upper laminal cells unipapillose . . . . *11. M. trinitense*

7   Leaves 1-2 mm long, apex regularly incurved when dry; upper laminal cells
    round, bulging . . . . *7. M. podocarpi*
    Leaves 2-4 mm long, apex keeled and contorted when dry; upper laminal
    cells rounded-quadrate or rounded-hexagonal, flat or mammillose . . . .
    . . . . *13. M. xenizon*

8   Leaves less than 4 mm long, costa excurrent as a firm arista, margins entire
    . . . . *2. M. harrisii*
    Leaves 4-10 mm long, costa long-excurrent as a denticulate hairpoint,
    margins dentate in upper part . . . . *12. M. ulophyllum*

9   Leaves oblong-lingulate, rugose in upper part . . . . *5. M. pellucidum*
    Leaves lanceolate, not rugose 10

10 Upper laminal cells elongate. 11
    Upper laminal cells rounded, bulging-mammillose 13

11 Leaf apex truncate with excurrent costa; perichaetial leaves longer than
    normal leaves, with long, flexuose hairpoint . . . . *6. M. perichaetiale*
    Leaf apex acute or acuminate; perichaetial leaves not longer than normal
    leaves 12

12 Leaves shorter than 5 mm, acute or short-acuminate, costa percurrent . . . . .
    . . . . *3. M. leprieurii*
    Leaves over 5 mm long, gradually long-acuminate, costa excurrent . . . .
    . . . . *9. M. scoparium*

13 Leaves arranged in distinct spiral rows, conspicuous when moist; leaf margin
    denticulate in upper part . . . . *8. M. punctatum*
    Leaves not ranked; margin entire or finely serrulate at apex . . *7. M. podocarpi*

1. **Macromitrium cirrosum** (Hedw.) Brid., Bryol. Univ. 1: 316. 1826.
    – *Anictangium cirrosum* Hedw., Sp. Musc. Frond. 42. 1801. Type:
    Jamaica, Swartz s.n. (NY).

    Macromitrium schwaneckeanum Hampe, Linnaea 25: 360. 1853. Type:
    Puerto Rico, Schwanecke s.n. (NY).

Description and synonymy see Musci I in Fl. Suriname 6: 226. 1964 (and 224, as *M. schwaneckeanum*).

Distribution: Mexico, West Indies, C and tropical S America.

Ecology: Epiphytic on branches, tree trunks and decaying wood, occasionally terrestrial; in the Guianas rather common at higher altitudes, also collected in the canopy of lowland rainforest.

Note: *M. schwaneckeanum* is difficult to separate from *M. cirrosum* as already suggested in Musci I. Great variation can be observed in leaf apex (acute to gradually acuminate), in costa (percurrent to short-excurrent) and in upper laminal cells (rounded-quadrate to elongate). In the type of *M. schwaneckeanum* the upper laminal cells are slightly more elongate and incrassate, but this is variable even in the type collection (NY) and it fits largely within the variation seen in the collections from the Guianas.

2. **Macromitrium harrisii** Paris, Index Bryol. Suppl. 238. 1900. Type: Jamaica, Harris 10033 (not seen).   – Fig. 83

Brownish green plants, growing in dense mats. Stems long, creeping, branches erect, to 2 cm long. Stem leaves falcate-secund, 1-1.5 mm long, smaller than branch leaves but otherwise not different; branch leaves ca. 3 mm long, flexuose when dry, falcate when moist, lanceolate, deeply keeled; costa excurrent as a firm arista, margins entire except for a few blunt teeth at the apex; upper laminal cells arranged in longitudinal rows, incrassate, hexagonal with oval lumen, 10-20 µm long and ca. 12 µm wide, in apex more elongate, basal cells rectangular, strongly incrassate with narrow lumina, tuberculate in central part of basal lamina. Perichaetial leaves abruptly short-acuminate, strongly serrate. Seta smooth, ca. 1 cm long, capsule slightly ribbed, endostome membranous. (Sporophyte description after Grout 1946).

Distribution: West Indies, Guyana.

Ecology: Epilithic in mesophytic rainforest at higher altitude (Grout 1946: mainly epiphytic); apparently rare in the Guianas.

Specimen examined: Guyana: Upper Potaro R., Mt. Wokomung, alt. 1530 m, Boom & Samuels 9219 (NY).

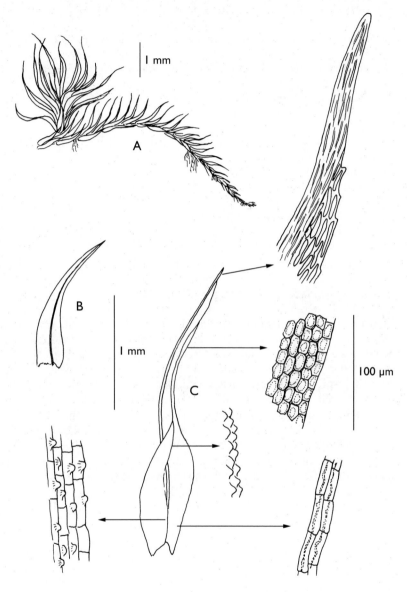

Fig. 83. *Macromitrium harrisii* Paris: A, end of stem with young branch; B, stem leaf; C, branch leaf (A-C, Boom & Samuels 9219).

3. **Macromitrium leprieurii** Mont., Ann. Sci. Nat., Bot. sér. 2, 14: 347. 1848. Type: French Guiana, Leprieur 334 (L, NY).

Description and synonymy see Musci I in Fl. Suriname 6: 233. 1964.

D i s t r i b u t i o n : West Indies, C America, Venezuela, the Guianas.

E c o l o g y : Epiphytic on tree trunks and branches in lowland rainforest, also at higher altitudes; not common.

N o t e s : In collections from higher altitudes the plants are more robust: branches to 4 cm long and leaves to 5 mm long occur in collections from the North side of Mt. Roraima.
The collections Boom & Samuels 9086 and 9100 from Guyana, cited as *M. dubium* Schimp. ex Müll. Hal. (Buck 1990), belong here. The rather abruptly short-acuminate, entire apex is typical for *M. leprieurii*.

4. **Macromitrium orthostichum** Nees ex Schwägr., Sp. Musc. Frond. Suppl. 4: 316. 1842. Lectotype (Vitt, Koponen & Norris 1995): Indonesia, Java, ex. hb. Dozy & Molk. (NY).   – Fig. 84

Stem to 10 cm long, creeping, densely covered with brown tomentum; stem leaves squarrose-recurved, triangular-lanceolate, to 0.8 mm long, apex acute, costa percurrent. Branches erect, ca. 1 cm long, sparingly divided, at base covered with dense tomentum; branch leaves dullgreen, contorted when dry, spreading when moist, oblong-lanceolate, to 1 mm long, keeled at base, apex acute or slightly rounded and mucronate, costa strong, smooth or distantly papillose at back, ending in apex or short-excurrent; laminal cells uniform isodiametric, firm-walled, rounded-quadrate, 8-12 µm in diam., bulging with 2-5 small papillae, occasionally forked; basal cells elongate in a few rows at insertion, bulging, mammillose. Dioicous. Perichaetial leaves little differentiated. Seta 3-4 mm long, strongly papillose; capsule ovoid with a short neck, ca. 1 mm long, operculum conic-rostrate, peristome single, membranous. Calyptra covering the capsule, mitrate, plicate, covered with long, hyaline, dentate hairs.

D i s t r i b u t i o n : SE Asia, Guyana; this is the first record for the Neotropics (see Vitt *et al.* 1995).

E c o l o g y : Epiphytic in mesophytic savanna forest.

S p e c i m e n   e x a m i n e d : Guyana: Upper Mazaruni Distr., trail from Kamarang R. to Pipwi Mt., alt. 800 m, Gradstein 5703 (L).

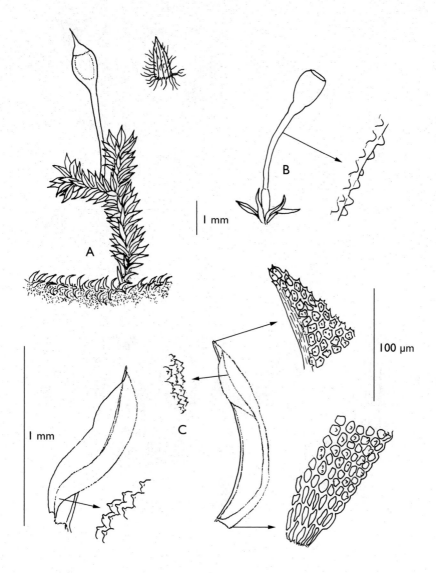

Fig. 84. *Macromitrium orthostichum* Nees ex Schwägr.: A, stem with branch and capsule; B, perichaetium with capsule; C, branch leaves (A-C, Gradstein 5703).

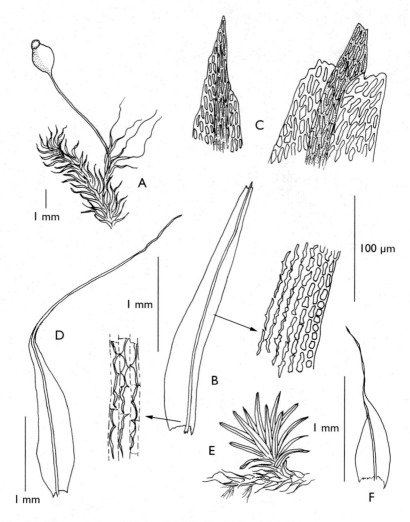

Fig. 85. *Macromitrium perichaetiale* (Hook. & Grev.) Müll. Hal.: A, branch (dry) with perichaetium and capsule; B, branch leaf; C, leaf apices; D, perichaetial leaf; E, part of stem with young branch; F, stem leaf (A, D, Gradstein 5641; B-C, Florschütz & Maas 2968; E-F, Maas & Westra 4195).

5. **Macromitrium pellucidum** Mitt., J. Linn. Soc., Bot. 12: 203. 1869. Type: Brazil, Spruce 80 (NY).

Description and synonymy see Musci I in Fl. Suriname 6: 232. 1964.

Distribution: West Indies, tropical S America.

Ecology: Epiphytic in the canopy of lowland rainforest, on solitary trees and in savanna scrub vegetation; rather common in the Guianas.

Note: *M. pellucidum* with blunt, rugose leaves, erect-appressed when dry, can easily be confused with rugose *Schlotheimia* species. It is different in the cucullate calyptra and the more elongate laminal cells: at midleaf oblong, strongly incrassate and porose, at base linear (not sigmoid as in *Schlotheimia*).

6. **Macromitrium perichaetiale** (Hook. & Grev.) Müll. Hal., Bot. Zeitung (Berlin) 3: 544. 1845. – *Orthotrichum perichaetiale* Hook. & Grev., Edinburgh J. Sci. 1: 127. 1824. Type: St. Vincent, Guilding s.n. (BM, NY). – Fig. 85

Rather robust, rusty brown plants growing in dense mats. Stems creeping, elongate; stem leaves falcate-secund when moist, ca. 1 mm long, ovate-lanceolate, acuminate with long-excurrent costa. Branches numerous, erect, later procumbent and divided; branch leaves contorted or spirally twisted when dry, erect-spreading when moist, 2.5-4 mm long, lanceolate, deeply keeled in basal part, apex truncate or emarginate, entire or sometimes bluntly serrulate, costa strong and short-excurrent as a mucro; laminal cells smooth, elongate and strongly incrassate throughout the leaf, at midleaf rectangular 25-40 μm long and ca. 10 μm wide, porose, towards margins and apex less porose and shorter with irregular-round or oblong lumina, basal cells linear, strongly porose, not tuberculate. Dioicous. Perigonia not seen. Perichaetial leaves conspicuous, 5-8 mm long, acuminate with a long-excurrent costa forming a flexuose hairpoint of the same length as the lamina. Seta smooth, 10-20 mm long, capsule erect, ovoid, smooth, contracted below the mouth; peristome double, exostome teeth truncate, papillose, endostome with low basal membrane and linear segments, longer than the exostome teeth. Calyptra mitrate, hairy.

Distribution: West Indies, Venezuela, Guyana, Suriname.

Ecology: Epiphytic at higher altitudes; not known from French Guiana.

Selected specimens: Guyana: Mt. Latipú, 8 km N of Kamarang, alt. 1000 m, Gradstein 5641 (L). Suriname: Bakhuis Mts., alt. 700 m, Florschütz & Maas 3027 (L).

Note: The truncate leaf apex with excurrent costa makes this species easy to recognize. The strong, excurrent costa can give the apex a tridentate appearance, but occasionally the margins are tapering gradually to the costa which makes the apex more or less acuminate. The perichaetial leaves are very long and conspicuous.

7. **Macromitrium podocarpi** Müll. Hal., Bull. Herb. Boissier 6: 96. 1898. Type: Brazil, Ule s.n. (FH, NY).

Macromitrium portoricense R.S. Williams, Bryologist 32: 69. 1929. Type: Puerto Rico, Britton 2638 (NY).

Description see Musci I in Fl. Suriname 6: 227. 1964.

D i s t r i b u t i o n : C and tropical S America.

E c o l o g y : Epiphytic on branches, tree trunks and decaying wood; in the Guianas rather common in the canopy of lowland rainforest, also at higher altitudes.

N o t e s : The differences between *M. podocarpi* and *M. portoricense* are very small and not clearly defined. The type collections (NY) show no significant differences in leaf length, leaf shape and cell size. In the Guianan collections the variation in leaves run from oblong-lanceolate with blunt, apiculate apex to linear-lanceolate with abruptly acuminate apex.
The collections cited as *M. portoricense* in Musci I, p. 229, belong to *M. xenizon*.

8. **Macromitrium punctatum** (Hook. & Grev.) Brid., Bryol. Univ. 1: 739. 1826. – *Orthotrichum punctatum* Hook & Grev., Edinburgh J. Sci. 1: 119. 1824. Lectotype (Vitt 1979): Brazil, Raddi s.n. (hb. Hooker stems A and B) (BM).

Macromitrium pentastichum Müll. Hal., Linnaea 21: 186. 1848. Type: Suriname, Kegel 1405 (GOET, NY).

Description see Musci I in Fl. Suriname 6: 230. 1964, as *M. pentastichum*. Adjustment of the description: upper laminal cells round to rounded-hexagonal, 10-17 μm in longest diam., bulging-mammillose, in extreme apex more elongate, elliptic, smooth.

D i s t r i b u t i o n : Mexico, West Indies, C and tropical S America.

E c o l o g y : Epiphytic; in the Guianas common in the canopy of lowland rainforest, in low savanna forest and in xeromorphic scrub at higher altitudes.

N o t e : The earlier described *M. punctatum* proved to be identical with *M. pentastichum* (Vitt 1979).

9. **Macromitrium scoparium** Mitt., J. Linn. Soc., Bot. 12: 206. 1869.
Syntypes: Jamaica, Wilson s.n.; Trinidad, Crüger s.n. (NY).
– Fig. 86

Rather robust, golden-brown plants growing in dense mats. Stems creeping, radiculose, stem leaves slender, 1.5-2 mm long, smaller than branch leaves, otherwise not different; branches erect or prostrate and ascending, 2-3 cm long, sparingly divided. Branch leaves crowded, spirally twisted when dry, erect-spreading when moist, 5-7 mm long, linear-lanceolate, keeled at base, gradually narrowed to a long and slender apex, costa excurrent in a dentate arista, margins serrate in upper half; laminal cells incrassate, in upper lamina oval-oblong, 20-40 µm long and ca. 10 µm wide, towards margins narrower and more elongate, forming an indistinct border, basal cells rectangular to linear, strongly incrassate and porose with narrow lumina, smooth. Perichaetial leaves lanceolate, 3-4 mm long, keeled, short-acuminate at apex. Seta ca. 8 mm long, rough in upper part; capsule ovoid, ca. 2 mm long, slightly plicate, not ribbed, operculum long-rostrate, peristome double, exostome teeth truncate, papillose, endostome of the same length, membranous, papillose. Calyptra wide-mitriform, deeply laciniate at base.

Distribution: West Indies, C America, Brazil, Guyana, French Guiana.

Ecology: Epiphytic in savanna forest at higher altitudes.

Specimen examined: Guyana, summit of Mt. Wokomung, alt. 1650 m, Boom & Samuels 9067 (NY).

Note: Buck (1990) cited the collection Boom & Samuels 9067 as *M. trachypodium* Mitt..

10. **Macromitrium stellulatum** (Hornsch.) Brid., Bryol. Univ. 1: 314. 1826. – *Schlotheimia stellulata* Hornsch., Horae Phys. Berol. 61. 1820. Lectotype (Florschütz 1964): Venezuela, coll. unknown, hb. de Thümen (NY).

Description see Musci I in Fl. Suriname 6: 223. 1964.

Distribution: Brazil, Bolivia, Venezuela, Guyana, Suriname.

Ecology: Epiphytic on branches and tree trunks; in the Guianas not common, not known from French Guiana.

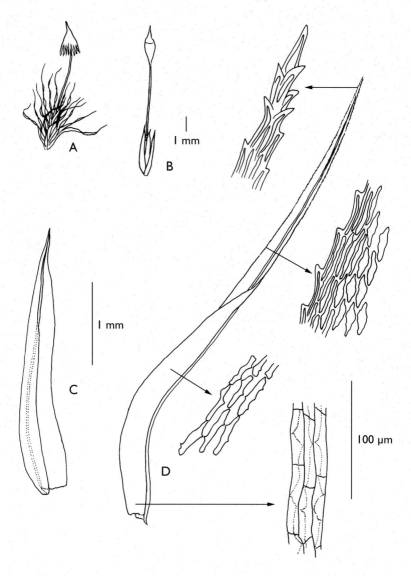

Fig. 86. *Macromitrium scoparium* Mitt.: A, end of branch with capsule; B, perichaetium with capsule; C, perichaetial leaf; D, branch leaf (A-D, Boom & Samuels 9067).

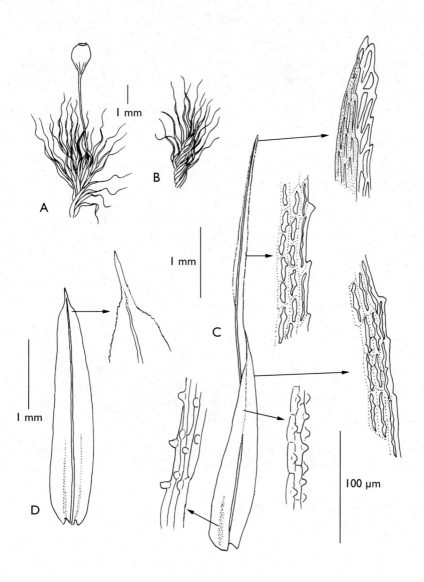

Fig. 87. *Macromitrium trinitense* R.S. Williams: A, end of branch with capsule
(moist); B, end of branch (dry); C, branch leaf; D, perichaetial leaf (A-D, Aptroot
& Sipman 4801).

250

11. **Macromitrium trinitense** R.S. Williams, Bryologist 24: 65. 1922.
Type: Trinidad, Johnston 163 (NY).   – Fig. 87

Medium-sized plants growing in loose mats. Stems elongate, creeping, densely tomentose; stem leaves lanceolate, 2-3 mm long, shorter than branch leaves but otherwise not different. Branches erect, to 2 cm high; branch leaves contorted and spirally twisted when dry, linear-lanceolate, 5-7 mm long, apex slenderly acuminate, margin serrate in upper half, sometimes nearly to base, costa percurrent; laminal cells strongly incrassate, unipapillose, rectangular, 18-30 μm long and ca. 10 μm wide, arranged in longitudinal rows, along the margins more elongate and smooth; basal cells linear, extremely incrassate, tuberculate in central part. Dioicous. Inner perichaetial leaves smooth, ca. 3 mm long, oblong, abruptly short acuminate. Seta 6-10 mm long, smooth, contorted when dry; capsule globose-pyriform, smooth or ribbed at neck when dry, operculum conic, slenderly rostrate; peristome double, exostome teeth truncate, finely striate at outer surface, endostome membranous. Calyptra mitrate, scabrous.

Distribution: Trinidad, Guyana, Suriname.

Ecology: Epiphytic on treelets in open forest at higher altitudes; not common, not collected in French Guiana.

Selected specimens: Guyana: Kaieteur Falls, Noel 89 (NY, FH); Pakaraima Mts., 2 km NW of Kamarang, alt. 500 m, Aptroot & Sipman 4801 (L); Kaieteur Falls, alt. 450 m, Newton *et al*. 3460 (L, US). Suriname: Lely Mts., alt. 650 m, Florschütz 4851 (L).

Note: This species was already reported for Guyana in Musci I in Fl. Suriname 6: 224. 1964 with a short description.

12. **Macromitrium ulophyllum** Mitt., J. Linn. Soc., Bot. 12: 206. 1869.
Type: Ecuador, Abitagua, Spruce 94 (NY).   – Fig. 88

Bronze-green plants, growing in dense mats. Stems creeping, 10 cm or more long, with long rhizoids at leaf bases; stem leaves ca. 2 mm long, appressed at base, recurved in upper part, ovate-lanceolate with long filiform acumen, costa long-excurrent, dentate. Branches short, erect, densely foliate; branch leaves twisted and crispate when dry, erect-spreading when moist, undulate in upper half, 4-10 mm long, linear-lanceolate, gradually acuminate, margins dentate in upper part, denticulate towards base, costa red, excurrent as a long, denticulate hairpoint; laminal

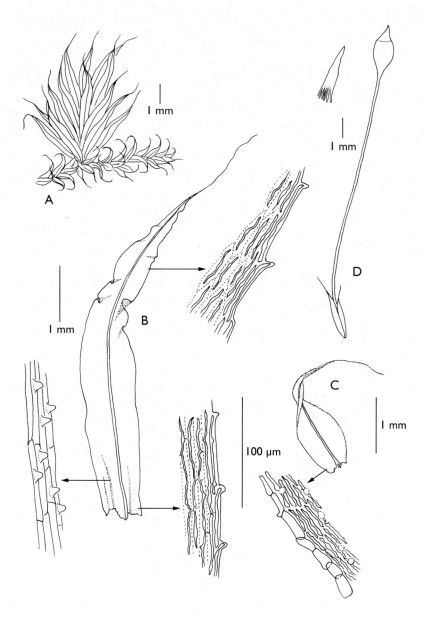

Fig. 88. *Macromitrium ulophyllum* Mitt.: A, end of stem with branch (moist); B, branch leaf; C, stem leaf; D, perichaetium with capsule and calyptra (A-D, Gradstein 5406).

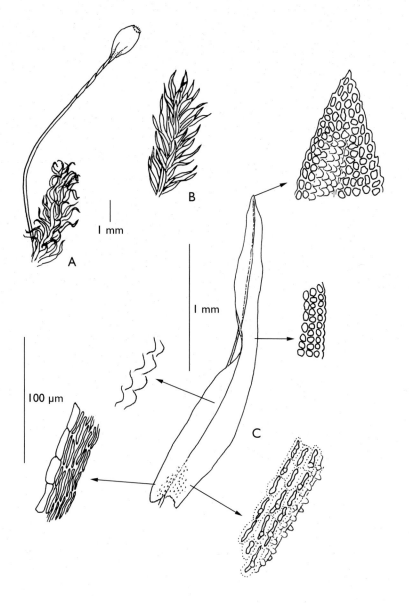

Fig. 89. *Macromitrium xenizon* B.H. Allen & W.R. Buck: A, end of branch with capsule (dry); B, end of branch (moist); C, branch leaf (A, Jansen-Jacobs 2455; B-C, Allen 19265A).

cells arranged in diverging rows, strongly incrassate and porose, elongate-oblong, 40-70 µm long in upper lamina, more elongate towards margins and towards base, basal cells yellowish, strongly tuberculate. Autoicous. Perichaetial leaves little differentiated, shorter than branch leaves. Seta smooth, 15-20 mm long; capsule globose-pyriform with a short neck, smooth, operculun conic, short-rostrate; peristome double, exostome teeth ca. 200 µm high, blunt, finely striate, endostome membranous, densely papillose, as long as the exostome. Calyptra mitrate, smooth, laciniate at base.

Distribution: Costa Rica, Panama, Guyana, Ecuador.

Ecology: Epiphytic on exposed branches, only at higher altitudes; rare.

Specimens examined: Guyana: N slope of Mt. Roraima, alt. 2000-2300 m, Gradstein 5406; ibid., alt. 1200-1600 m, Gradstein 5392 (L); Mt. Roraima, alt. 2800 m, Liesner 23299 A (NY).

13. **Macromitrium xenizon** B.H. Allen & W.R. Buck, Mem. New York Bot. Gard. 76: 82. 2003. Type: French Guiana, Commune de Saül, Buck 38090 (NY). – Fig. 89

Medium sized, dark green plants growing in dense mats. Stems creeping, densely radiculose at lower side, stem leaves not differentiated. Branches erect, 2-15 mm long. Leaves fragile, irregularly twisted when dry, wide-spreading and slightly contorted at apex when moist, 2-4 mm long, ovate-lanceolate or lanceolate, deeply keeled, apex acute or rounded acute with short-acuminate tip, margin crenulate or finely serrulate, costa percurrent; upper laminal cells rounded quadrate or rounded hexagonal, 6-10 µm wide, mammillose, inner basal cells rectangular or hexagonal, strongly incrassate and tuberculate, towards margins linear with narrow lumen, bordered by a short row of hyaline, thinwalled cells (often inconspicous). Autoicous. Perichaetial leaves erect-appressed, 1-2 mm long, ovate-oblong, apex short-acuminate, costa faint. Seta 5-10 mm long, smooth; capsule ovoid, 1.5-2 mm long, ribbed at base, operculum conic-rostrate, peristome double, often incomplete.

Distribution: The Guianas; thus far not known from other countries but probably more wide-spread.

Ecology: Epiphytic on upper tree trunks and branches in canopy of lowland rainforest; not common.

Selected specimens: Guyana: Rupununi Distr., Kuyuwini Landing, alt. 200 m, Jansen-Jacobs *et al.* 2455 (L). Suriname: Sipaliwini, along road NW of Blanche Marie Guest House, Allen 19265A (MO). French Guiana: Saül, Sentier Limonade, alt. 180-210 m, Montfoort & Ek 675 (L).

Notes: The specimens mentioned in Musci I in Fl. Suriname 6: 229. 1964 under *M. portoricense* belong to this species. After seeing more collections it became clear that it was a new species, lately described as *M. xenizon* by Buck (2003).

The longer, often broken leaves with distinct tubercula at base distinguish this species from *M. portoricense*, now synonymized with *M. podocarpi*. The swollen marginal cells at leaf base are characteristic but not always obvious.

3. **SCHLOTHEIMIA** Brid., Muscol. Recent. Suppl. 2: 16. 1812.
Lectotype (Grout 1946): S. torquata (Hedw.) Brid. (Hypnum torquatum Hedw.)

Description see Musci I in Fl. Suriname 6: 215. 1964.

Distribution: Pantropical, also extending to subtropical areas; in the Guianas 4 species.

KEY TO THE SPECIES

1 Leaves distinctly rugose in upper part 2
  Leaves smooth or faintly rugose. 3

2 All cells in upper part of leaf roundish (at most one single row along the costa more elongate); perichaetial leaves hardly differentiated . . . .
  . . . .*3. S. rugifolia*
  Cells in upper part of leaf along the costa elongate in several rows, often diverging from the costa; perichaetial leaves twice as long as vegetative leaves . . . . *4. S. torquata*

3 Leaves in upper half abruptly narrowed to a linear or subulate part about 1/3 of leaf length, in older leaves mostly broken . . . . *1. S. angustata*
  Leaves gradually narrowed to a rounded acute apex . . . . *2. S. jamesonii*

1. **Schlotheimia angustata** Mitt., J. Linn. Soc., Bot. 12: 223. 1869.
Syntypes: Peru, in Monte Campana, Spruce 113; Ecuador, Lligua near Baños, Spruce 113 b (NY).   – Fig. 90

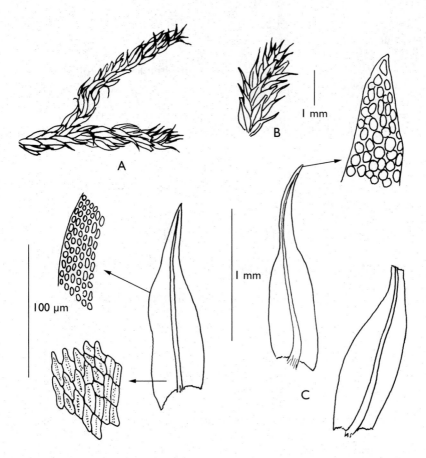

Fig. 90. *Schlotheimia angustata* Mitt.: A, stem with branch (dry); B, end of branch (moist); C, leaves (A-C, Newton *et al*. 3601).

Brownish plants with darkgreen tips, growing in dense mats. Stems elongate, prostrate; branches erect, slender, ca. 1 cm long. Leaves appressed when dry, erect-spreading when moist, 1.5-2 mm long, oblong-lanceolate, abruptly narrowed to a firm, subulate point, 1/3 of leaf length, usually broken, margin entire, costa ending in the thickened apex; laminal cells incrassate, rounded-quadrate to elliptic, 6-9 μm long, smaller towards margins, apical cells in more than one layer; basal cells more elongate, rectangular or sigmoid-rhomboidal with very narrow lumen. Seta 3.5-4 mm long, smooth, capsule erect, cylindric, urn to 2 mm long, slightly furrowed, neck short, operculum rostrate; peristome double, exostome teeth linear, thick, reflexed when dry, endostome segments of equal length, linear. Calyptra smooth, rather deeply lobed. (Sporophytes description after Grout 1946).

Distribution: C America, Andes (Peru, Ecuador, Bolivia, Colombia), Guyana.

Ecology: Epiphytic in canopy at higher altitude; rare, collected only once in Guyana.

Specimen examined: Guyana: Potaro-Siparuni region, Paramakatoi, alt. 720 m, Newton *et al.* 3601 (L, US).

2. **Schlotheimia jamesonii** (Arn.) Brid., Bryol. Univ. 1: 742. 1826. – *Orthotrichum jamesonii* Arn., Mém. Soc. Linn. Paris 1: 349. 1823. Type: Brazil, Jameson s.n.

Description see Musci I in Fl. Suriname 6: 216. 1964.

Distribution: Mexico, West Indies, C and tropical S America.

Ecology: Epiphytic in forest at higher altitudes, occasionally on wet rock; not common, not seen from French Guiana.

3. **Schlotheimia rugifolia** (Hook.) Schwägr., Sp. Musc. Frond. Suppl. 2(1): 150. 1824. – *Orthotrichum rugifolium* Hook., Musci Exot. 128. 1820. Type: Brazil, Swainson s.n. (NY).

Description see Musci I in Fl. Suriname 6: 219. 1964.

Distribution: SE U.S.A., Mexico, West Indies, C and tropical S America.

Ecology: Epiphytic; common in the canopy of lowland rainforest.

4. **Schlotheimia torquata** (Hedw.) Brid., Muscol. Recent. Suppl. 2: 16. 1812. – *Hypnum torquatum* Hedw., Sp. Musc. Frond. 246. 1801. Type: Jamaica, Swartz s.n. (G, NY).

Description and synonymy see Musci I in Fl. Suriname 6: 218. 1964.

Distribution: Mexico, West Indies, C and tropical S America.

Ecology: Epiphytic on tree trunks and branches at higher altitudes, occasionally in the canopy of lowland rainforest; not seen from French Guiana.

N o t e : If no perichaetial leaves are present the best character to distinguish this species from *S. rugifolia* is the presence of several rows of elongated cells along the costa in the upper leaf half.

Order Leucodontales

Plants pleurocarpous. Primary stems usually creeping, stolonlike with often reduced leaves; secondary stems erect, ascending or pendent, simple or branched. Sporophytes lateral, setae often short, capsules erect and symmetric, peristome double, endostome often reduced.

N o t e s : An extensive discussion about the delimitation of the families in this order is on-going. The THAMNOBRYACEAE, originally included in the *Leucodontales* are now placed in the *Hypnales* (Churchill in Gradstein *et al*. 2001).
In recent studies the families of the *Leucodontales* are considered as reduction series and are intercalated among those of the *Hypnales*.

KEY TO THE FAMILIES

1   Laminal cells obscure, covered with a network of fine papillae, marginal cells smooth forming a pellucid border. Capsule without peristome . . . . *24. RHACOCARPACEAE*
    Laminal cells smooth or papillose, marginal cells not differentiated. Capsule with peristome 2

2   Leaves dimorphous, arranged in four rows, lateral leaves wide-spreading, dorsal leaves erect, much smaller. . . . . *23. RACOPILACEAE*
    Leaves different 3

3   Leaves in two opposite rows, strongly asymmetric 4
    Leaves subsymmetric, evenly arranged around the stem 5

4   Leaves plane, asymmetric to cultriform . . . . *20. NECKERACEAE*
    Leaves strongly concave to conduplicate . . . . . . *21. PHYLLOGONIACEAE*

5   Secondary stems elongate, often pendent; leaves with rather uniform aereolation, alar cells little differentiated. Peristome with well-developed endostome. . . . . *19. METEORIACEAE*
    Secondary stems short, dendroid or simple, erect or spreading; leaves with differentiated alar cells. Endostome often rudimentary or absent. 6

6. Laminal cells fusiform to elongate-rhomboidal, basal cells subquadrate in many oblique rows . . . . *18. LEPTODONTACEAE*
    Laminal cells rhomboidal to linear, thick-walled and porose, basal cells coloured, variously differentiated. . . . . *22. PTEROBRYACEAE*

18. LEPTODONTACEAE

Medium-sized plants with creeping primary stem and ascending secondary stems, irregularly pinnately branched. Microphyllous branchlets often present. Leaves ovate to short-lanceolate, apiculate to acuminate, costa single, ending in upper leaf half; laminal cells fusiform or elliptic, smooth or prorulose by projecting cell ends. Dioicous. Capsules immersed or exserted, peristome single or double, exostome teeth 16, endostome reduced.

D i s t r i b u t i o n : Worldwide; in the Guianas 1 genus.

N o t e s : The genera *Leucodontopsis* and *Pseudocryphaea* were previously placed in the LEUCODONTACEAE (Musci I in Fl. Suriname 6: 241). Manuel (1974) placed both genera in a new subfamily FORSSTROEMIOIDEAE. Buck (1980) transferred this subfamily to the LEPTODONTACEAE. In 1989 he transferred *Leucodontopsis* (renamed as *Henicodium*) to the PTEROBRYACEAE.
Akiyama (1994) presented a critical view on the attempts to find a natural classification for the LEUCODONTACEAE: all genera are epiphytic and the morphological features used to define relationships may be influenced by extreme habitat conditions.

LITERATURE

Akiyama, H. 1994. Suggestions for the delimitation of the Leucodontaceae and the infrageneric classification of the genus Leucodon. J. Hattori Bot. Lab. 76: 1-12.

Buck, W.R. 1980. Animadversions on Pterigynandrum with special commentary on Forsstroemia and Leptopterigynandrum. Bryologist 83: 451-465.

Buck, W.R. 1989. Henicodium replaces Leucodontopsis. Bryologist 92: 534.

Churchill, S.P. & N. Salazar-Allen. 2001. Leptodontaceae. In S.R. Gradstein *et al.*, Guide to the Bryophytes of Tropical America. Mem. New York Bot. Gard. 86: 392-393.

Churchill, S.P. & N. Salazar-Allen. 2001. Leucodontaceae. In S.R. Gradstein *et al.*, Guide to the Bryophytes of Tropical America. Mem. New York Bot. Gard. 86: 404-405.

Manuel, M.G. 1974. A revised classification of the Leucodontaceae and a revision of the subfamily Alsioideae. Bryologist 77: 531-550.

1. **PSEUDOCRYPHAEA** E. Britton, Bull. Torrey Bot. Club 32: 261. 1905.
Type: P. flagellifera (Brid.) E. Britton, nom. illeg. [= P. domingensis (Spreng.) W.R. Buck]

Distribution: Neotropics; 1 species.

1. **Pseudocryphaea domingensis** (Spreng.) W.R. Buck, Bryologist 83: 455. 1980. – *Neckera domingensis* Spreng., Syst. Veg. 4: 185. 1827. Type: Cuba, Bertero s.n. (NY).

Pseudocryphaea flagellifera (Brid.) E. Britton, Bull. Torrey Bot. Club 32: 261. 1905. – *Pilotrichum flagelliferum* Brid., Bryol. Univ. 2: 259. 1827, nom. illeg.

Description see Musci I in Fl. Suriname 6: 244. 1964, as *P. flagellifera.*

Distribution: West Indies, southern U.S.A., Mexico, C and tropical S America.

Ecology: Epiphytic on tree trunks and branches in lower canopy; not frequent, not known from French Guiana.

19. METEORIACEAE

Description see Musci I in Fl. Suriname 6: 252. 1964.

Distribution: Pantropical; in the Guianas 7 genera.

Note: METEORIACEAE resemble PTEROBRYACEAE in many aspects, but can often be distinguished by their habit: hanging in loosely tangled mats (METEORIACEAE) or growing upright with stipitate-frondose stems (PTEROBRYACEAE). This does not hold for all genera; for more differences see note under PTEROBRYACEAE.

LITERATURE

Allen, B.H. & M.R. Crosby. 1986. Revision of the genus Squamidium. J. Hattori Bot. Lab. 61: 423-476.
Allen, B.H. 1987. On distinguishing Pterobryaceae and Meteoriaceae by means of pseudoparaphyllia. Bryol. Times 42: 1-3.

260

Allen, B.H. & R.E. Magill. 2007. A revision of Orthostichella. Bryologist 110: 1-45.

Buck, W.R. 1981. The taxonomy of Eriodon and notes on other South American genera of Brachytheciaceae with erect capsules. Brittonia 33: 556-563.

Buck, W.R. 1994a. A new attempt at understanding the Meteoriaceae. J. Hattori Bot. Lab. 75: 51-72.

Buck, W.R. 1994b. The resurrection of Orthostichella. Bryologist 97: 434-435.

Buck, W.R. 1998. Meteoriaceae. In Pleurocarpous mosses of the West Indies. Mem. New York Bot. Gard. 82: 253-273.

Manuel, M.G. 1977a. The genus Meteoridium (Müll. Hal.) Manuel, stat. nov., Lindbergia 4: 45-55.

Manuel, M.G. 1977b. A monograph of the genus Zelometeorium Manuel, gen. nov., J. Hattori Bot. Lab. 43: 107-126.

Robinson, H. 1967. Preliminary studies on the bryophytes of Colombia. Bryologist 70: 1-61.

KEY TO THE GENERA

1   Leaves without costa. 2
     Leaves with single costa 3

2   Leaves lanceolate, acuminate, not spirally arranged . . . . *2. Lepyrodontopsis*
     Leaves ovate, abruptly apiculate, arranged in 5 spiral rows . . . .
     . . . . *4. Orthostichella*

3   Laminal cells papillose 4
     Laminal cells smooth 5

4   Leaves wide-spreading or complanate-spreading; laminal cells linear, not in diverging rows in basal part . . . .*1. Floribundaria*
     Leaves appressed to erect, usually plicate; laminal cells rhomboidal to linear, in basal part in diverging rows to the margins . . . . *5. Papillaria*

5   Alar cells differentiated in a conspicuous group . . . . *6. Squamidium*
     Alar cells little differentiated, inconspicuous 6

6   Stem and branch leaves little differentiated, spreading from the insertion, margin sharply serrate . . . . *3. Meteoridium*
     Stem leaves long-acuminate to piliform, with a clasping base, branch leaves acuminate, squarrose recurved, margin denticulate . . . . . *7. Zelometeorium*

1. **FLORIBUNDARIA** M. Fleisch., Hedwigia 44: 301. 1905.
Type: F. floribunda (Dozy & Molk.) M. Fleisch. (Leskea floribunda Dozy & Molk.)

Papillaria sect. *Floribundaria* Müll. Hal., Linnaea 35: 267. 1867, nom. nud.

Stems elongate, creeping or pendent, irregularly to pinnately branched. Branch leaves spreading, often complanate, lanceolate, apex acute or acuminate, costa narrow, ending below apex; laminal cells linear to elongate rhomboidal, usually seriate-papillose, basal cells wider, smooth, alar cells not clearly differentiated. Dioicous or synoicous. Seta short, smooth; capsule ovoid-cylindric, erect or inclined, operculum conic-rostrate; peristome double, exostome teeth lanceolate, transversely striolate at base, papillose above, endostome segments keeled, slender, from a rather high membrane, cilia lacking or rudimentary. Calyptra cucullate, slightly hairy.

Distribution: Pantropical; in the Guianas 2 species.

KEY TO THE SPECIES

1 Branch leaves to 2.5 mm long; laminal cells with 3-5 papillae on each surface .... *1. F. cardotii*
   Branch leaves to 1.2 mm long; laminal cells with 4-10 papillae in a distinct row on each surface .... *2. F. floribunda*

1. **Floribundaria cardotii** (Broth.) Broth. in Engler & Prantl, Nat. Pflanzenfam. 1(3): 822. 1906. – *Papillaria cardotii* Broth., Bih. Kongl. Svenska. Vetensk.-Akad. Handl. 21 Afd. 3(3): 48. 1895. Type: Brazil, Binot s.n. (hb. Card. 22 (H); hb. Card. 33 (PC)).   – Fig. 91

Slender, lightgreen plants with long, creeping or pendent stems, irregularly branched; branches variable in size: short and erect or elongate and pendent. Leaves on erect branches wide-spreading to squarrose, on elongate branches more or less complanate-spreading; stem leaves lanceolate with clasping base, gradually tapering to a long acumen, often hyaline and piliform at the tip, margins distantly serrulate or crenulate throughout, costa extending ca. half the leaf length; branch leaves ovate-lanceolate, cordate at base, to 2.5 mm long and 0.7 mm wide, apex acute, often twisted, margins serrate at apex, serrulate or crenulate below, costa extending 1/2-3/4 of leaf length; laminal cells linear, flexuose, 35-60 µm long and 3-4 µm wide, with 3-5 small papillae over the lumen on each surface, basal cells shorter and wider, incrassate, smooth. Sporophyte not seen.

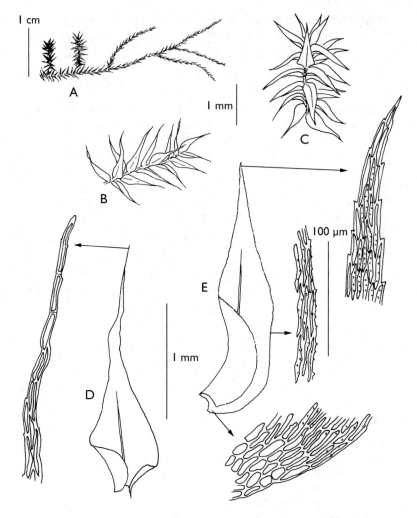

Fig. 91. *Floribundaria cardotii* (Broth.) Broth.: A, end of stem with branches; B, part of stem; C, end of branch; D, stem leaf; E, branch leaf (A-E, Florschütz-de Waard & Zielman 5034).

Distribution: Brazil, Suriname.

Ecology: Epiphytic on branches and lianas in the understory of lowland rainforest; locally abundant, only known from Brownsberg in Suriname.

Selected specimens: Suriname: Brownsberg, alt. 500 m, Florschütz 4632 (L); ibid. Florschütz-de Waard & Zielman 5034 (L).

2. **Floribundaria floribunda** (Dozy & Molk.) M. Fleisch., Hedwigia
44: 302. 1905. – *Leskea floribunda* Dozy & Molk., Ann. Sci. Nat.,
Bot. sér. 3, 2: 310. 1844. Type: not designated. – Fig. 92

Slender lightgreen plants growing in dense, tangled mats. Stems
elongate, creeping or pendent, irregularly to pinnately branched; branches
to 2 cm long, attenuate at ends. Leaves to 1.2 mm long and 0.3 mm
wide, diminishing in size in distal direction; stem leaves distant, ovate-
lanceolate, cordate at base, long-acuminate, costa extending to midleaf
or shorter; branch leaves complanate-spreading, lanceolate, rounded at

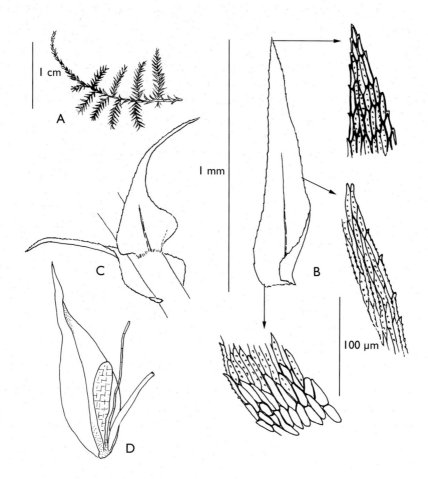

Fig. 92. *Floribundaria floribunda* (Dozy & Molk.) M. Fleisch.: A, end of stem; B,
branch leaf; C, stem leaves; D, perichaetial leaf (A-D, Florschütz 4751).

base, costa extending 1/2-3/4 of leaf length, margins serrulate throughout; laminal cells linear, 35-60 μm long and 3-6 μm wide, with 4-10 papillae in a row on each surface, basal cells shorter and wider, smooth, slightly differentiated in basal angles. Synoicous. Perichaetial leaves ovate, to 1 mm long, acuminate, ecostate, cells elongate rhomboidal, smooth except in apex. No sporophytes seen.

Distribution: Pantropical.

Ecology: Epiphytic on tree trunks and branches, also epiphyllous; not common, not collected in Guyana.

Selected specimens: Suriname: Brownsberg, alt. 450 m, Florschütz 4751 (L); Lely Mts., alt. 550-700 m, Lindeman *et al.* 462 (L). French Guiana: Massif des Emerillons, alt. 500 m, Cremers 6874 (CAY, L).

Notes: Buck (2003) reports *F. flaccida* (Mitt.) Broth. from French Guiana and claims this to be the oldest name for the species. However, *F. floribunda* was already described in 1844 as *Leskea floribunda* by Dozy & Molkenboer. It is not clear which of the two species Buck has seen for he does not mention the papillae. Unfortunately I did not see the collections.
The number and arrangement of the papillae is one of the best characters to recognize the species: in *F. floribunda* 4-10 papillae, distinctly arranged in rows, in *F. cardotii* 3-5 papillae scattered over the lumen.

2. **LEPYRODONTOPSIS** Broth. in Engler, Nat. Pflanzenfam. ed. 2, 11: 358. 1925.
Type: L. trichophylla (Sw. ex Hedw.) Broth. (Hypnum trichophyllum Sw. ex Hedw.)

Distribution: Neotropical; 1 species.

Note: *Lepyrodontopsis*, originally included in the BRACHYTHECIACEAE (Brotherus 1925), was placed in a separate family the LEPYRODONTOPSIDACEAE by Buck (1981). Plicate leaves and flagellate branches, considered as distinctive characters for the family, could not be observed in Guianan collections. Recently the genus was placed in the METEORIACEAE based on peristome characters (Buck 1998).

1. **Lepyrodontopsis trichophylla** (Sw. ex Hedw.) Broth. in Engler, Nat. Pflanzenfam. ed. 2, 11: 358. 1925. – *Hypnum trichophyllum* Sw. ex Hedw., Sp. Musc. Frond. 274. 1801. Type: Jamaica, Swartz s.n.

– Fig. 93

Slender plants growing in glossy, lightgreen mats. Stems elongate, prostrate, freely branched, branches ascending. Stem and branch leaves not different, erect-spreading, slenderly lanceolate, sometimes plicate, variable in size, 1-4 mm long, ecostate, apex slenderly acuminate, margin serrate throughout, in upper part coarsely serrate-dentate; laminal cells

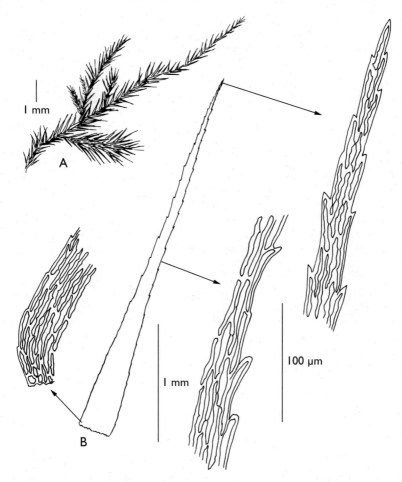

Fig. 93. *Lepyrodontopsis trichophylla* (Sw. ex Hedw.) Broth.: A, end of stem with branches; B, leaf (A-B, Gradstein 5592).

uniform, incrassate and porose, linear, at midleaf 65-85 μm long and 3.5-5 mm wide, in apex shorter and more strongly porose, basal cells shorter at insertion, alar cells not differentiated. Dioicous. Perichaetial leaves erect, costate, apex slenderly acuminate, sharply serrate. Seta to 2 cm long, red, capsule erect, narrowly cylindric, operculum conic-rostrate, peristome double, exostome teeth lanceolate, papillose, endostome segments on a basal membrane, hyaline, papillose. Calyptra cucullate (Buck 1981). (Sporophyte description after Brotherus 1925).

Distribution: West Indies, Venezuela, Colombia, Guyana, French Guiana.

Ecology: On trunks and roots, occasionally on rocks at higher altitudes; in the Guianas rare.

Specimens examined: Guyana: Mt. Latipu, alt. 800 m, Gradstein 5592 (L). French Guiana: Mt. de l'Inini, alt. 700 m, Cremers 9165 (CAY, L).

3. **METEORIDIUM** (Müll. Hal.) Manuel, Lindbergia 4: 45. 1977. Type: M. remotifolium (Müll. Hal.) Manuel (Leskea remotifolia Müll. Hal.)

Neckera subsect. *Meteoridium* Müll. Hal., Syn. Musc. Frond. 2: 276. 1851. *Meteoriopsis* sect. *Meteoridium* (Müll. Hal.) Broth. in Engler & Prantl, Nat. Pflanzenfam. 1(3): 825. 1906.

Primary stem creeping, secondary stems elongate, irregularly branched, pendent. Stem and branch leaves spreading from insertion, ovate to ovate-lanceolate, apex acuminate, contorted, margin sharply serrate, costa ending above midleaf; laminal cells smooth, linear, wider and porose at base, alar cells little differentiated. Dioicous. Perichaetial leaves lanceolate, long-acuminate, ecostate. Seta 2-3 mm long, capsule ovoid, operculum conic-rostrate, exostome teeth lanceolate, striate in basal part, papillose in upper part, endostome segments papillose, cilia lacking. Calyptra cucullate.

Distribution: Neotropical; in the Guianas 1 species.

Note: *Meteoriopsis* sect. *Meteoridium* was raised to generic status by Manuel (1977a), because of the cucullate calyptra and transversely striate exostome teeth.

1. **Meteoridium remotifolium** (Müll. Hal.) Manuel, Lindbergia 4: 49. 1977. – *Leskea remotifolia* Müll. Hal., Linnaea 19: 216 1847. – *Meteoriopsis remotifolia* (Müll. Hal.) Broth. in Engler & Prantl, Nat. Pflanzenfam. 1(3): 825. 1906. Lectotype (Manuel 1977): Mexico, Deppe & Schiede 1088 (H).

Description see Musci I in Fl. Suriname 6: 262. 1964, as *Meteoriopsis remotifolia*.

Distribution: Mexico, West Indies, C and tropical S America.

Ecology: Epiphytic on branches and leaves in the understory of tropical rainforest at higher altitudes; often collected in French Guiana, seldom collected in Suriname and Guyana.

4. **ORTHOSTICHELLA** Müll. Hal., Bull. Herb. Boissier 5:204. 1897. Lectotype (Buck 1994b): O. filamentosula Müll. Hal.

Neckera sect. *Pseudopilotrichum* subsect. *Orthostichella* Müll. Hal., Syn. Musc. Frond. 2: 123. 1850. – *Pseudopilotrichum* (Müll. Hal.) W.R. Buck & B.H. Allen in W.R. Buck, J. Hattori Bot. Lab. 75: 69.1994. Lectotype (Buck 1994a): N. hexasticha (Schwägr.) Müll. Hal.

Slender plants with elongate, creeping stems, sparingly foliate; secondary stems stipitate or pendent, irregularly branched; leaves often in conspicuous rows, concave, oblong-obovate, often panduriform, apex cuspidate to short-acuminate, margins broadly inflexed above, serrulate, costa very short or absent. Cells long-hexagonal to linear, smooth, alar cells few, quadrate. Dioicous. Perichaetial leaves oblong-lanceolate, gradually acuminate. Seta short, capsule erect, cylindric, annulus present, operculum rostrate, peristome double, exostome teeth narrowly lanceolate, papillose on outer surface, endostome with low basal membrane, cilia rudimentary or absent. Calyptra cucullate with sparse erect hairs.

Distribution: Pantropical; in the Guianas 1 species.

Notes: Buck (1994a) divided *Pilotrichella* in 2 genera; he recognized the small-statured species as a separate genus *Pseudopilotrichum*. This name proved to be antedated by *Orthostichella* (Buck 1994b).
Allen & Magill (2007) evaluated the different characters of *Orthostichella*: the secondary stems are essentially stipitate (in *Pilotrichella* never), the leaves are often spirally ranked and have sparsely developed alar cells (in *Pilotrichella* not spirally ranked and well-developed, often bulging alar cells).

1. **Orthostichella versicolor** (Müll. Hal.) B.H. Allen & W.R. Buck, Mem. New York Bot. Gard. 76: 140. 2003. – *Neckera versicolor* Müll. Hal., Syn. Musc. Frond. 2: 127. 1850. – *Pilotrichella versicolor* (Müll. Hal.) A. Jaeger. Lectotype (Allen & Magill 2007): Brazil, Serra d'Estrella, Beyrich s.n. (BM).

Description see Musci I in Fl. Suriname 6: 253. 1964, as *Pilotrichella versicolor*.

Distribution : Tropical S America.

Ecology : Epiphytic in light forest at higher altitudes.

Note : Robinson (1967) synonymized *Pilotrichella versicolor* and several other species with *P. pentasticha* (Brid.) Wijk & Margad., the oldest name. But Allen & Magill (2007) saw the type from Mauritius and concluded that this was *Orthostichidium*. The correct name is *Orthostichella versicolor*.

5. **PAPILLARIA** (Müll. Hal.) Lorentz, Moosstudien 165. 1864, nom. cons.
Type: P. nigrescens (Sw. ex Hedw.) A. Jaeger (Hypnum nigrescens Sw. ex Hedw.), typ. cons.

Neckera subsect. *Papillaria* Müll. Hal., Syn. Musc. Frond. 2: 134. 1850.

Description see Musci I in Fl. Suriname 6: 254. 1964.

Distribution : Pantropical; in the Guianas 1 species.

Note : Buck (1998) transferred the *Papillaria* specimens of the West Indies to the genus *Meteorium*, a genus with mostly unipapillose laminal cells, but occasionally 2-3 papillae per cell. In the *Papillaria* specimens from the Guianas only cells with one papilla could be observed.

1. **Papillaria nigrescens** (Sw. ex Hedw.) A. Jaeger, Ber. St. Gallischen Naturwiss. Ges. 1875-76: 256. 1877. – *Hypnum nigrescens* Sw. ex Hedw., Sp. Musc. Frond. 250. 1801. Type: Jamaica, Swartz s.n. (S).

Description see Musci I in Fl. Suriname 6: 255. 1964.

Distribution : Pantropical.

Ecology: Epiphytic on solitary trees and in the canopy of lowland rainforest; common in the coastal region of Suriname, uncommon in Guyana and French Guiana.

Note: In Musci I in Fl. Suriname 6: 257. 1964 is *P. deppei* (Hornsch.) A. Jaeger recorded for Suriname with one specimen (Anakam s.n.). This specimen proved to be *P. nigrescens*; no other collections are known from the Guianas.

6. **SQUAMIDIUM** (Müll. Hal.) Broth. in Engler & Prantl, Nat. Pflanzenfam. 1(3): 807. 1906.
Type: S. lorentzii (Müll. Hal.) Broth. (Meteorium lorentzii Müll. Hal.)

Description see Musci I in Fl. Suriname 6: 257. 1964.

Distribution: Neotropical and a single species disjunct in eastern Africa; in the Guianas 2 species.

Note: Determining the sections of *Squamidium* Allen & Crosby (1986) accentuated the importance of stolon leaves. Stolons are described as stoloniferous shoots at the end of stems in regions of rapid elongation. Two sections are distinguished:
sect. *Squamidium*: immersed capsules, stolon leaves entire or irregularly serrate at apex (*S. leucotrichum*).
sect. *Macrosquamidium*: exserted capsules, stolon leaves with recurved marginal teeth (*S. macrocarpum*).

KEY TO THE SPECIES

1 Branch leaves abruptly long-acuminate, hairpointed; stolon leaves entire or weakly serrulate. Capsules immersed . . . . *1. S. leucotrichum*
Branch leaves acute or short-acuminate; stolon leaves with large, recurved teeth. Capsules exserted . . . . *2. S. macrocarpum*

1. **Squamidium leucotrichum** (Taylor) Broth. in Engler & Prantl, Nat. Pflanzenfam. 1(3): 809. 1906. – *Hypnum leucotrichum* Taylor, London J. Bot. 7: 196. 1848. Type: Ecuador, Pinchincha, Jameson s.n. (FH, NY).

Description see Musci I in Fl. Suriname 6: 258. 1964 (gametophyte).

Description sporophyte (Allen & Crosby 1986): Dioicous. Perichaetia on short lateral branches, inner perichaetial leaves lanceolate, gradually acuminate, sheathing at base, to 6 mm long, costa indistinct or absent. Seta ca. 1 mm long, capsule immersed, ca. 4 mm long and 1 mm wide, operculum short-rostrate, exostome teeth slender, ca. 1 mm long, trabeculate, slightly papillose at outer surface, endostome segments from a high basal membrane, papillose, cilia rudimentary or absent.

Distribution: West Indies, Mexico, C and tropical S America.

Ecology: Epiphytic on trunks and branches of trees at higher altitudes.

Note: Collections without sporophytes can be confused with *Orthostichopsis praetermissa* W.R. Buck, a representative of the PTEROBRYACEAE. The capsules with well-developed double peristome and the absence of pseudoparaphyllia are characteristic of METEORIACEAE (see also note under PTEROBRYACEAE). Other characters to recognize *Squamidium leucotrichum* without capsules are the long, pendent stems with many short, broad branches, the branch leaves not arranged in rows and the cordate leaf base with a round, bulging group of alar cells.

2. **Squamidium macrocarpum** (Mitt.) Broth. in Engler & Prantl, Nat. Pflanzenfam. 1(3): 809. 1906. – *Meteorium macrocarpum* Mitt., J. Linn. Soc., Bot. 12: 437. 1869. Types: Peru, Andes Peruvianum, in monte Guayrapurima, Spruce 1191 & 1194 (NY). – Fig. 94

Medium sized plants with long, creeping or pending stems, often stoloniferous, to 20 cm long, appressed-foliate; branches erect, stout, to 1 cm long, seriate-foliate. Stem leaves ovate-lanceolate, to 3.5 mm long and 0.8 mm wide, long-acuminate, entire to serrulate, clasping and decurrent at base, costa slender, extending to base of acumen; stolon leaves similar but abruptly acuminate with a long and flexuose acumen, dentate with strong, recurved teeth; branch leaves imbricate, to 2 mm long and 0.8 mm wide, concave with inflexed margins in upper part, apex flat, acute, costa ending below apex; laminal cells smooth, firm-walled, linear, at midleaf 90-140 μm long and 5-7 μm wide, shorter in apex and towards base; alar cells rounded-quadrate to short-rectangular, forming a distinct coloured group, slightly bulging. Perichaetial leaves to 2.5 mm long, inner leaves strongly sheathing at base, ovate-acuminate. Seta 2-3 mm long, smooth, capsule exserted, erect, 2-2.7 mm long, constricted at base, operculum 1-1.5 mm long, short-rostrate. Calyptra 3 mm long, densely hairy. (Sporophyte description after Allen & Crosby 1986).

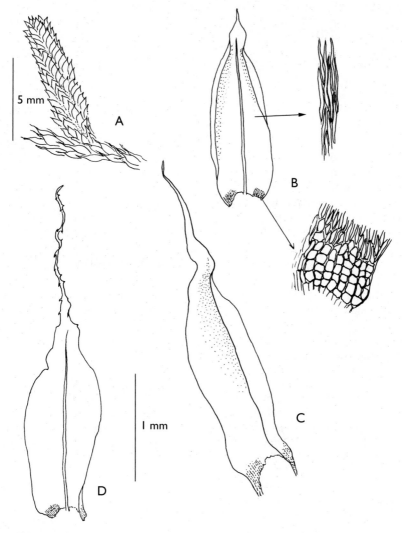

Fig. 94. *Squamidium macrocarpum* (Mitt.) Broth.: A, part of stem with branch; B, branch leaf; C, stem leaf; D, stolon leaf (A-D, Allen 25464).

Distribution: C and tropical S America.

Ecology: Epiphytic on trunks and branches; in the Guianas rare.

Specimen examined: Suriname: Eilerts De Haan Mts., alt. 300 m, Allen 25464 (L, MO).

7. **ZELOMETEORIUM** Manuel, J. Hattori Bot. Lab. 43: 110. 1977. Lectotype (Manuel 1977): Z. patulum (Hedw.) Manuel (Hypnum patulum Hedw.)

Meteoriopsis sect. *Squarridium* Broth. p.p., in Engler & Prantl, Nat. Pflanzenfam. 1(3): 825. 1906.

Stem leaves erect to squarrose from a sheathing base, ovate to ovate-lanceolate with cordate base, apex short- to long-acuminate, costa ending above midleaf, margin entire or serrulate; branch leaves erect to squarrose-recurved, not sheathing, ovate , apex obtuse, acute or long-acuminate, margin entire or denticulate; laminal cells linear, smooth, alar cells occasionally differentiated, porose. Dioicous. Perichaetial leaves ovate-lanceolate, ecostate. Seta short, capsule oblong-ovoid, operculum conic-rostrate; peristome double, exostome teeth lanceolate, papillose, endostome segments papillose, cilia lacking. Calyptra mitrate, pilose.

D i s t r i b u t i o n : Neotropical, tropical Africa; in the Guianas 1 species.

N o t e : Manuel (1977b) segregated from *Meteoriopsis* sect. *Squarridium* the species with smooth laminal cells and papillose exostome teeth as a new genus *Zelometeorium*; those with unipapillose laminal cells and exostome teeth with striate basal part remained in *Meteoriopsis* s.s. (no representatives in the Guianas).

1. **Zelometeorium patulum** (Hedw.) Manuel, J. Hattori Bot. Lab. 43: 118. 1977. – *Hypnum patulum* Hedw., Sp. Musc. Frond. 279. 1801. – *Meteoriopsis patula* (Hedw.) Broth. in Engler & Prantl, Nat. Pflanzenfam. 1(3): 825. 1906. Lectotype (Manuel 1977): Jamaica, Swartz s.n. (G, H, NY).

Description and synonymy see Musci I in Fl. Suriname 6: 259. 1964, as *Meteoriopsis patula*.

D i s t r i b u t i o n : Florida, Mexico, West Indies, C and tropical S America.

E c o l o g y : Epiphytic on branches and stems of small trees, also on leaves and terrestrial on detritus; very common in the canopy and in open areas in the understory of lowland rainforest.

N o t e : In most collections this species is easily recognized by the short branches with squarrose-recurved leaves. In some collections with long elongate stems the short branches are less conspicuous and here the long-acuminate stem leaves with clasping bases are critical for determination.

20. NECKERACEAE

Primary stems creeping, leaves scalelike; secondary stems ascending or pendent, irregularly to pinnately branched. Leaves complanate, ovate to oblong-lanceolate or lingulate, asymmetric to cultriform, often undulate, apex acute to obtuse or truncate, costa single, reaching beyond midleaf or short and forked or absent; laminal cells smooth, rhomboidal to linear. Autoicous or dioicous. Perichaetial leaves differentiated, usually sheathing. Seta usually short, capsule erect (except in Isodrepanium), operculum rostrate, peristome double, exostome teeth 16, endostome low with slender segments, cilia lacking. Calyptra cucullate or mitrate.

Distribution: Worldwide; in the Guianas 2 genera.

Note: In the description of the NECKERACEAE in Musci II in Fl. Suriname 6: 273. 1986 the family is treated in a broader concept, including *Pinnatella* and *Porotrichum*. These genera are now transferred to the THAMNOBRYACEAE (Sastre-de Jesús 1987).

LITERATURE

Enroth, J. 1994. On the evolution and circumscription of the Neckeraceae. J. Hattori Bot. Lab. 76: 13-20.
Sastre-De Jesús, I. 1987. A revision of the Neckeraceae Schimp. and the Thamnobryaceae Marg. & Dur. in the Neotropics. PhD-dissertation, City Univ. New York.

KEY TO THE GENERA

1 Costa short and indistinct; apex acute or short-acuminate . . . .
. . . . *1. Isodrepanium*
Costa single, extending beyond midleaf; apex broad-obtuse or truncate . .
. . . . *2. Neckeropsis*

1. **ISODREPANIUM** (Mitt.) E. Britton, Torreya 14: 28. 1914. – *Lepidopilum* sect. *Isodrepanium* Mitt., J. Linn. Soc., Bot. 12: 369. 1869.
Type: I. lentulum (Wilson) E. Britton (Homalia lentula Wilson)

Distribution: Neotropical; 1 species.

1. **Isodrepanium lentulum** (Wilson) E. Britton, Torreya 14: 28. 1914. – *Homalia lentula* Wilson, Ann. Mag. Nat. Hist. 20: 379. 1847. Type: Jamaica, Mc Nab s.n. (BM).

Description and synonymy see Musci II in Fl. Suriname 6: 278. 1986.

D i s t r i b u t i o n : Mexico, West Indies, C and tropical S America.

E c o l o g y : Epiphytic on branches in the canopy of forest at higher altitudes.

2. **NECKEROPSIS** Reichardt, Verh. Zool.-Bot. Ges. Wien 18: 192. 1868. Type: N. undulata (Hedw.) Reichardt (Neckera undulata Hedw.)

Description see Musci II in Fl. Suriname 6: 273. 1986.

D i s t r i b u t i o n : Pantropical; in the Guianas 2 species.

KEY TO THE SPECIES

1   Leaves plane when moist, at least on young branches. Calyptra naked . . . . .
    . . . . *1. N. disticha*
    Leaves transversely undulate when moist. Calyptra with erect, leaflike
    appendages . . . . *2. N. undulata*

1. **Neckeropsis disticha** (Hedw.) Kindb., Canad. Rec. Sci. 6: 21. 1894. – *Neckera disticha* Hedw., Sp. Musc. Frond. 201. 1801. Type: Jamaica, Swartz s.n. (G).

Description and synonymy see Musci II in Fl. Suriname 6: 274. 1986.

D i s t r i b u t i o n : Southern U.S.A., Mexico, West Indies, C and tropical S America, tropical Africa.

E c o l o g y : Epiphytic on branches and trunks of trees in lowland rainforest, occasionally on decaying wood.

2. **Neckeropsis undulata** (Hedw.) Reichardt, Verh. Zool.-Bot. Ges. Wien 18: 192. 1868. – *Neckera undulata* Hedw., Sp. Musc. Frond. 201. 1801. Type: Jamaica, Swartz s.n. (G).

Description see Musci II in Fl. Suriname 6: 276. 1986.

Distribution: Southern U.S.A., Mexico, West Indies, C and tropical S America.

Ecology: Epiphytic on branches and stems of trees and on decaying wood in lowland rainforest.

21. PHYLLOGONIACEAE

Distribution: Neotropical; 1 genus.

Note: Shan-Hsiung Lin (1983) re-evaluated the status of the genera in the PHYLLOGONIACEAE and as a result only *Phyllogonium* stayed in this family.

LITERATURE

Lin, S.-H. 1983. A taxonomic revision of Phyllogoniaceae. Part 1. J. Taiwan Mus. 36: 37-86.

1. **PHYLLOGONIUM** Brid., Bryol. Univ. 2: 671. 1827.
   Type: P. fulgens (Hedw.) Brid. (Pterigynandrum fulgens Hedw.)

Medium sized to robust plants. Primary stems stoloniferous with small, scale-like leaves. Secondary stems erect to pendent, simple to pinnately branched. Leaves distichous, densely imbricate, strongly concave to conduplicate, oval-oblong to oblong-lanceolate with cucullate apex, margins entire, costa absent or short and double; laminal cells linear, incrassate and porose, short and coloured at leaf base. Pseudoparaphyllia foliose, semicircular. Dioicous. Seta short. Capsule exserted to immersed, erect, operculum conic, rostellate; peristome single or double, exostome teeth 16, linear-lanceolate, smooth, endostome membranous or absent. Calyptra cucullate to mitrate, smooth or hairy.

Distribution: Neotropical; in the Guianas 2 species.

KEY TO THE SPECIES

1 Stem and branches 1.5-3.5 mm wide with leaves. Leaf apex pungent; alar cells not conspicuous. Capsules exserted. . . . . *1. P. fulgens*
Stem and branches to 6 mm wide with leaves. Leaf apex mucronate, recurved; alar cells in a conspicuous round group. Capsules immersed . . .
. . . . *2. P. viride*

1. **Phyllogonium fulgens** (Hedw.) Brid., Bryol. Univ. 2: 671. 1827. – *Pterigynandrum fulgens* Hedw., Sp. Musc. Frond. 86. 1801. Type: Jamaica, Swartz s.n. (G). – Fig. 95 C

Description see Musci I in Fl. Suriname 6: 264. 1964.

D i s t r i b u t i o n : Mexico, West Indies, C and tropical S America.

E c o l o g y : Epiphytic on tree trunks and branches in humid forests, mostly at higher altitudes.

N o t e : The distinct alar group mentioned in the description in Musci I is characteristic for *P. viride*. In *P. fulgens* the basal cells are coloured, but usually not arranged in a dark brown, round group.

2. **Phyllogonium viride** Brid., Bryol. Univ. 2: 673. 1872. Type: Brazil, "in brasiliae arboribus pendulum habitat", sine coll. (B). – Fig. 95 A-B

Yellow-green to golden-brown, glossy plants. Primary stems creeping, leaves appressed, broad-ovate, ca. 1 mm long. Secondary stems pendent, to 35 cm long, but usually much shorter, irregularly to pinnately branched; leaves distichous, more or less complanate, shorter and more distant than branch leaves. Branches to 9 cm long, strongly complanate, 3-6(-7) mm wide with leaves, narrower distally; leaves closely imbricate, oblong-lanceolate, conduplicate, to 3.5 mm long and 1.5 mm wide, auriculate at base, at apex abruptly narrowed to a short, recurved mucro, margins crenulate at apex, costa absent; laminal cells strongly incrassate and porose, linear, 50-90 µm long, shorter at apex; alar cells rounded-quadrate, dark brown in a conspicuous round group. Perichaetial leaves ovate, inner leaves acuminate, to 1.5 mm long, denticulate at apex. Perigonia not seen. Seta very short, to 0.8 mm long. Capsule immersed, ovoid, to 2.5 mm long, operculum conic with inclined rostrum; peristome single, exostome teeth ca. 300 µm long. Calyptra small, cucullate or mitrate. (Sporophyte description after Lin 1983).

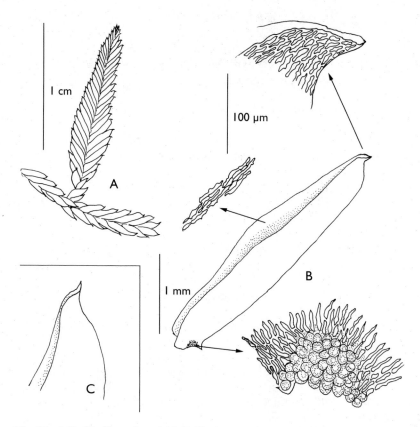

Fig. 95. A-B. *Phyllogonium viride* Brid.: A, part of secondary stem with branch; B, leaf (A-B, Lindeman & Stoffers *et al*. 220). C. *Phyllogonium fulgens* (Hedw.) Brid.: C, leaf apex (C, Gradstein 5413a).

Distribution: Mexico, West Indies, C and tropical S America.

Ecology: Epiphytic on branches and upper trunks in humid forests, mostly at higher altitudes.

Selected specimens: Guyana: Upper Mazaruni R., Karowtipu Mt., alt. 920 m, Boom *et al*. 7390 (NY). Suriname: Lely Mts., SW plateau, alt. 550-710 m, Lindeman & Stoffers *et al*. 588 (L). French Guiana: Mt. de l'Inini, alt. 800 m, Cremers 9038 (CAY, L).

Note: The width of stem and branches is not reliable as distinguishing character. In some collections the largest stems are not over 3 mm wide. The presence of dark-coloured, round groups of alar cells is a good additional character.

22. PTEROBRYACEAE

Description see Musci I in Fl. Suriname 6: 246. 1964.
Amplification: Secondary stems erect to pendent, pseudoparaphyllia present, nearly always filamentous.

Distribution: Pantropical; in the Guianas 8 genera.

Note: The distinction between PTEROBRYACEAE and METEORIACEAE is not obvious. The differences are mainly determined by the sporophytes: double peristomes with well-developed endostomes (METEORIACEAE) or peristomes appearing single, endostomes rudimentary (PTEROBRYACEAE). However, sporophytes are not common in these families. Allen (1987) added a gametophytic character to distinguish the families adequately: pseudoparaphyllia are present in PTEROBRYACEAE and absent in METEORIACEAE.

LITERATURE

Allen, B. 1987. On distinguishing Pterobryaceae and Meteoriaceae by means of pseudoparaphyllia. Bryol. Times 42: 1-3.
Arzeni, C.B. 1954. The Pterobryaceae of the southern United States, Mexico, Central America and West Indies. Amer. Midl. Naturalist 52: 1-67.
Buck, W.R. 1989. Henicodium replaces Leucodontopsis. Bryologist 92: 534.
Buck, W.R. 1991. Notes on neotropical Pterobryaceae. Brittonia 43: 96-101.

KEY TO THE GENERA

1 Leaves ecostate 2
  Leaves with single or double costa 3

2 Leaves in distinct spiral rows . . . . *3. Hildebrandtiella*
  Leaves not in spiral rows . . . .*7. Renauldia*

3 Leaves at least 5 mm long, with long piliform acumen. . . . *8. Spiridentopsis*
  Leaves shorter, acute or acuminate 4

4 Costa extending to leaf apex . . . . *6. Pireella*
  Costa extending 3/4 of leaf-length or less 5

5   Secondary stems pendent, leaves arranged in spiral rows, leaf apex abruptly acuminate or mucronate . . . . *5. Orthostichopsis*
Secondary stems erect or ascending, if pendent leaves not arranged in spiral rows, leaf apex acute or acuminate 6

6   Laminal cells papillose, alar cells quadrate, incrassate forming a conspicuous group. . . . . *2. Henicodium*
Laminal cells smooth, alar cells scarcely differentiated 7

7   Leaves complanate-spreading. . . . . *1. Calyptothecium*
Leaves squarrose-spreading . . . . *4. Jaegerina*

1. **CALYPTOTHECIUM** Mitt., J. Linn. Soc., Bot. 10: 190. 1868. Type: C. praelongum Mitt.

Medium sized plants. Primary stem creeping, secondary stem erect to pendent, simple, irregularly pinnate to frondose from a stipitate base. Leaves complanate-spreading, broad-ovate to short-lanceolate, concave, apex acute to short-acuminate, base auriculate, margin entire or weakly serrulate at apex, costa single, slender; gemmae frequently present in leaf axils; laminal cells smooth, elongate, often porose, alar cells variously differentiated. Dioicous. Perichaetial leaves oblong-lanceolate, convolute. Capsule immersed to short-exserted, ovoid-cylindric, operculum conic-rostrate. Peristome double, often rudimentary or lacking. Calyptra mitrate.

D i s t r i b u t i o n : Mexico, West Indies, C and tropical S America, tropical Asia and Africa; in the Guianas 1 species.

1.   **Calyptothecium planifrons** (Renauld & Paris) Argent, J. Bryol. 7: 565. 1973. – *Garovaglia planifrons* Renauld & Paris, Rev. Bryol. Lichénol. 29: 7. 1902. Type: Madagascar, Antoagazi, 10 March 1901, sin coll. (REN).   – Fig. 96

Creeping primary stem with scalelike leaves. Secondary stems erect to pendent, simple or with one or two short branches, to 2 cm long. Leaves complanate-spreading, concave to conduplicate, ovate-lanceolate, apex short-acuminate, often with reflexed apiculus, 2-2.6 mm long and 0.6-0.8 mm wide, margin entire, slightly denticulate near apex, costa extending 1/2-2/3 of leaf-length; laminal cells porose, vermicular, linear, 60-120 μm long and ca. 7 μm wide, shorter in apex; basal cells inflated and coloured in two or three rows at insertion, but not in the alar region. Sporophyte not seen.

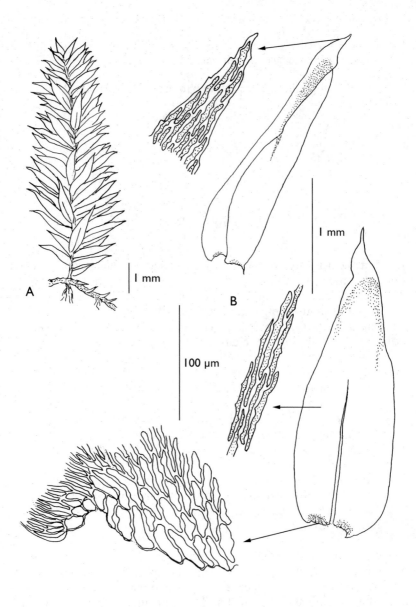

Fig. 96. *Calyptothecium planifrons* (Renauld & Paris) Argent: A, secondary stem; B, leaves (A-B, Allen 25498).

Distribution: Amazon basin from Colombia to Brazil, tropical Africa.

Ecology: Epiphytic in moist lowland to lower montane forests.

Specimens examined: Suriname: Eilerts de Haan Mt., trail around the summit, on tree trunk, alt. 739 m, Allen 25483 and 25498 (MO).

2. **HENICODIUM** (Müll. Hal.) Kindb., Enum. Bryin. Exot. 16. 1888. – *Hypnum* sect. *Henicodium* Müll. Hal., Linnaea 39: 470. 1875. Type: H. niam-niamiae (Müll. Hal.) Kindb. (Hypnum niam-niamiae Müll. Hal.)

Leucodontopsis Renauld & Cardot, Bull. Soc. Roy. Bot. Belgique 32: 177. 1893. Type: L. plicata Renauld & Cardot

Distribution: Pantropical; 1 species.

Note: This genus is moved from the LEUCODONTACEAE to the PTEROBRYACEAE on account of the presence of filiform pseudoparaphyllia, a single costa, elongate lamina cells, well-defined alar cells and a rudimentary endostome (Buck 1989).

1. **Henicodium geniculatum** (Mitt.) W.R. Buck, Bryologist. 92: 534. 1989. – *Leucodon geniculatus* Mitt., J. Linn. Soc., Bot. 12. 409. 1869. – *Leucodontopsis geniculata* (Mitt.) H.A. Crum & Steere, Sci. Surv. Porto Rico & Virgin Islands 7: 511. 1957. Type: Peru, Guayarapurina Spruce s.n. (NY).

Description and synonymy see Musci I in Fl. Suriname 6: 242. 1964, as *Leucodontopsis geniculata*.

Ecology: Epiphytic; rather common in the canopy of mesophytic rainforest and on solitary trees.

3. **HILDEBRANDTIELLA** Müll. Hal., Linnaea 40: 257. 1876. Type: H. endotrichelloides Müll. Hal.

Orthostichidium Müll. Hal. ex Dusén, Bih. Kongl. Svenska Vetensk.-Akad. Handl. 28(2): 19. 1895. Type: O. perpinnatum (Broth.) Dusén (Hildebrandtiella perpinnata Broth.)

Description see Musci I in Fl. Suriname 6: 246. 1964, as *Orthostichidium*.

D i s t r i b u t i o n : Pantropical; in the Guianas 1 species.

N o t e : In comparing Neotropical and African species of *Orthostichidium* and *Hildebrandtiella*, Buck (1991) concluded that the differences are not convincing; many intermediate forms make the distinction impossible and so the *Orthostichidium* species are transferred to *Hildebrandtiella*.

1. **Hildebrandtiella guyanensis** (Mont.) W.R. Buck, Brittonia 43: 96. 1991. – *Neckera guyanensis* Mont., Syll. Gen. Sp. Crypt. 24. 1856. – *Orthostichidium guyanensis* (Mont.) Broth. in Engler & Prantl, Nat. Pflanzenfam. 1(3): 795. 1906. Type: French Guiana, Leprieur 325 (PC). – Fig. 97

Description see Musci I in Fl. Suriname 6: 247. 1964, as *Orthostichidium guyanensis*.
In Musci I this species was reported for French Guiana. A complete description and illustration are given here:
Primary stems creeping with small, sometimes scalelike leaves. Secondary stems loosely erect to pendent, to 5 cm long, irregularly to pinnately branched; branches densely foliate. Leaves arranged in 4-5 spiral rows, 1-1.8(-2.5) mm long and 0.5-0.7(-1.5) mm wide, oblong, concave, more or less cordate at base, abruptly short-acuminate, costa absent, upper margins broadly inflexed, weakly serrulate. Midleaf cells linear, flexuose, incrassate, 50-80 μm long, shorter and more incrassate towards base, in alar region often coloured in basal rows. Dioicous. Perichaetial leaves longer than normal leaves, gradually acuminate. Capsule immersed, ovoid; peristome teeth slender, smooth. Calyptra small, mitrate.

D i s t r i b u t i o n : West Indies, Mexico, Guyana, French Guiana.

E c o l o g y : Epiphytic on tree trunks and branches; in the Guianas rare, apparently confined to higher altitudes.

S p e c i m e n s   e x a m i n e d : Guyana: Para Mt., 3 km SW of Paramakatoi, alt. 1050 m, Newton *et al.* 3552 (L, US). French Guiana: Mt. de l'Inini, alt. 800 m, Cremers 8981, 9103 (CAY, L).

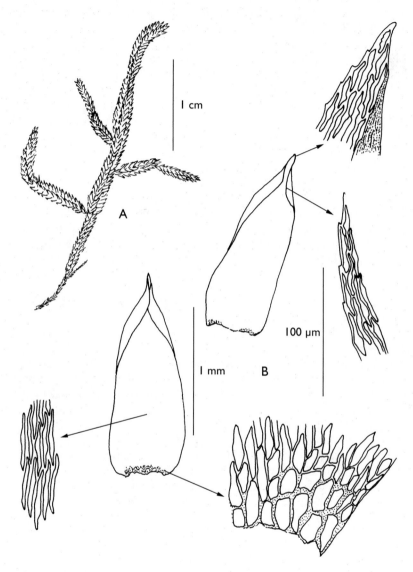

Fig. 97. *Hildebrandtiella guyanensis* (Mont.) W.R. Buck: A, secondary stem with branches; B, leaves (A-B, Newton *et al.* 3552).

4. **JAEGERINA** Müll. Hal., Linnaea 40: 274. 1876.
Type: J. stolonifera (Müll. Hal.) Müll. Hal. (Pilotrichum stoloniferum Müll. Hal.)

Medium-sized plants, secondary stems erect, simple or sparingly branched. Leaves squarrose-spreading, ovate to lanceolate, margins plane, costa single or short and double, or lacking. Leaf cells linear, incrassate and porose, smooth or minutely prorulose, alar group scarcely differentiated. Dioicous. Seta short, smooth. Capsule oval, operculum long-rostrate, peristome single.

Distribution: Pantropical and subtropical; in the Guianas 1 species.

1. **Jaegerina scariosa** (Lorentz) Arzeni, Amer. Midl. Naturalist 52: 12. 1954. – *Meteorium scariosum* Lorentz, Moosstud. 165. 1864. – *Jaegerinopsis scariosa* (Lorentz) Broth. in Engler & Prantl, Nat. Pflanzenfam. 1(3): 791. 1906. Type: Panama, Chiriqui, Wagner s.n. (BM). – Fig. 98

Jaegerinopsis squarrosa E. Britton, Bryol. 21: 48. 1918. Type: Cuba, Sierra de las Yeguas, Bros., Léon & Clement s.n.

Light green plants growing in loose tufts. Primary stems creeping, naked or with small, scalelike leaves; secondary stems ascending from a prostrate base, to 3 cm long, sparingly branched, densely foliate. Leaves squarrose-spreading, concave when moist, broad-ovate to ovate-lanceolate with a cordate base and acute apex, to ca. 2 mm long and 1 mm wide, smaller along the basal part of the stem, margin minutely serrulate in upper part, sometimes nearly to base; costa variable, usually single and extending 3/4 of leaf-length, occasionally forked or short and double or indistinct; laminal cells linear, smooth, incrassate and porose, often sigmoid, 35-55 µm long and 5-7 µm wide, shorter towards apex, wider and coloured at leaf base, alar group indistinct, consisting of a few subquadrate cells. Perichaetial leaves longer than normal leaves, ecostate, acuminate. Seta ca. 4 mm long, reddish. Capsule oblong-oval, 2 mm long, operculum long-rostrate, peristome teeth short, blunt, fragile. (Sporophyte description after Arzeni 1954).

Distribution: Southern U.S.A., Mexico, West Indies, C and northern S America.

Ecology: Epiphytic on upper trunks and branches of the canopy. Not common.

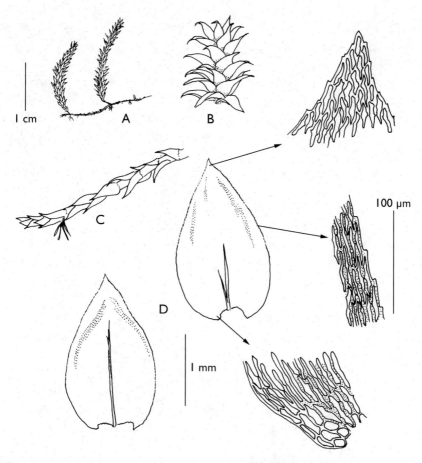

Fig. 98. *Jaegerina scariosa* (Lorentz) Arzeni: A, habit; B, end of branch (dry); C, end of primary stem; D, leaves (A-D, Florschütz-de Waard & Zielman 5736).

Selected specimens: Guyana: Kanuku Mts., alt. ca. 100 m, Jansen-Jacobs *et al*. 220 (L). Suriname: Corantijn R., Frederik Willem IV Falls, Florschütz-de Waard & Zielman 5736 (L). French Guiana: Saül, 'Sentier Limonade', Montfoort & Ek 631 (L).

Note: Based on the variability of the costa 2 species of *Jaegerina* have been described: *J. scariosa* with short, double costa and *J. squarrosa* with elongate, single costa. Arzeni (1954) concluded that the 2 species are not clearly separated. In collections from the Guianas the costa-variation may occur within one plant.

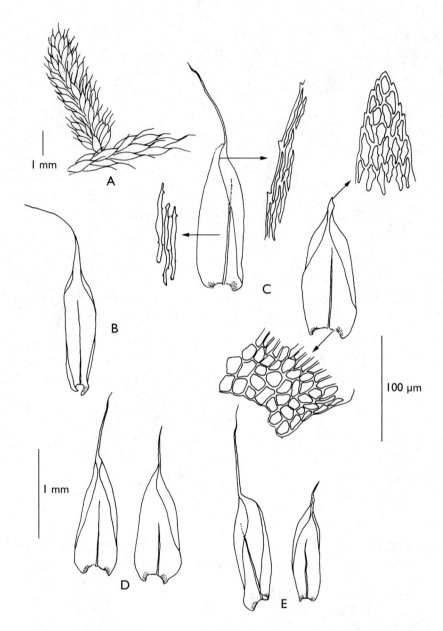

Fig. 99. *Orthostichopsis praetermissa* W.R. Buck: A, part of stem with branch; B, stem leaf; C-E, branch leaves (A-C, Florschütz 4835; D, Maas *et al.* 2667; E, Smith 3604).

5. **ORTHOSTICHOPSIS** Broth. in Engler & Prantl, Nat. Pflanzenfam.
1(3): 804. 1906.
Type: O. tetragona (Hedw.) Broth. (Hypnum tetragonum Hedw.)

Description see Musci I in Fl. Suriname 6: 247. 1964.

Distribution: Pantropical; in the Guianas 2 species.

KEY TO THE SPECIES

1   Branch leaves not plicate, leaf apex long-cuspidate to hairpointed; alar cells
    fewer, in a roundish group at leaf base . . . . *1. O. praetermissa*
    Branch leaves plicate, leaf apex mucronate; alar cells numerous in a large,
    triangular group, extending up the margins . . . . *2. O. tetragona*

1. **Orthostichopsis praetermissa** W.R. Buck, Brittonia 43: 98. 1991.
Type: Peru, Mt. Guayrapurina, Spruce 1207 (NY).   – Fig. 99

Bright green to olive-green, often orange-tinged plants. Secondary
stems long, usually pendent, sparingly branched; stem leaves erect-
appressed, oblong-lanceolate, concave, at apex gradually narrowed to
a long, flexuose hairpoint, lamina 1.5-2 mm long, hairpoint to 1 mm
long. Branches simple to (occasionally) pinnately branched, to 3 cm
long; branch leaves imbricate in 5 spiral rows, concave, oblong-ovate,
1.5-2 mm long, at apex abruptly contracted to a short, cuspidate point or
to a hairpoint of various length, to 1 mm long, margins broadly inflexed
below apex, minutely serrulate nearly to base, costa slender, 1/2-3/4
of leaf length; laminal cells linear, incrassate, porose, shorter in the
mucronate apex, alar cells rounded quadrate, thickwalled, in an irregular
roundish group. Dioicous. Perichaetial leaves convolute, lanceolate, to
7 mm long, gradually acuminate. Seta short, ca. 0.5 mm long, smooth.
Capsule immersed, broad-cylindric, to 3 mm long and 1.5 mm wide,
operculum conic-rostrate, ca. 1 mm long, peristome double, exostome
teeth narrowly lanceolate, smooth, endostome rudimentary. Calyptra
cucullate, pilose with long, crispate hairs. (Sporophyte description after
Buck 1991).

Distribution: West Indies, C and tropical S America.

Ecology: Epiphytic on trunks and branches in mesophytic rainforest,
also at higher altitudes.

Selected specimens: Guyana: Mt. Latipu, alt. 600 m, Gradstein 5558 (L); Kaieteur Falls Nat. Park, alt. 450 m, Newton *et al.* 3411 (L, US). Suriname: Lely Mts., alt. 650 m, Florschütz 4815 (L). French Guiana: Mt. de l'Inini, alt. 700 m, Cremers 8999 (CAY, L).

N o t e : In Musci I in Fl. Suriname 6: 249. 1964 this species was reported as *Orthostichopsis crinita* (Sull.) Broth. for Guyana. Buck (1991) concluded that this name was used for 2 different species. He distinguished *O. praetermissa* from *O. crinita* by its considerably coarser habit, shorter hairpoints and more darkly coloured alar cells. In the collections from the Guianas the length of the hairpoint may vary from cuspidate to long-acuminate with an acumen as long as the lamina. The 2 species here possibly overlap.

2. **Orthostichopsis tetragona** (Hedw.) Broth. in Engler & Prantl, Nat. Pflanzenfam. 1(3): 805. 1906. – *Hypnum tetragonum* Hedw., Sp. Musc. Frond. 246. 1801. Type: Jamaica, Swartz s.n. (G, NY).

Description and synonymy see Musci I in Fl. Suriname 6: 248. 1964.

D i s t r i b u t i o n : Mexico, West Indies, C and tropical S America, tropical Asia.

E c o l o g y : Epiphytic; very common on trees in a sunny habitat as well as in dark rainforest.

6. **PIREELLA** Cardot, Rev. Bryol. 40: 17.1913.
Type: P. mariae (Cardot) Cardot (Pirea mariae Cardot)

Description see Musci I in Fl. Suriname 6: 249. 1964.

D i s t r i b u t i o n : Pantropical and subtropical; in the Guianas 2 species.

KEY TO THE SPECIES

1　Secondary stems irregularly branched, not dendroid. Alar cells numerous, arranged in longitudinal rows forming a distinct group, extending up the margins . . . . *1. P. cymbifolia*
　Secondary stems pinnately branched, often dendroid. Alar cells few, in a small, indistinct group . . . . *2. P. pohlii*

1. **Pireella cymbifolia** (Sull.) Cardot, Rev. Bryol. 40: 17 1913. –
Pilotrichum cymbifolium Sull. in A. Gray, Manual ed. 2: 681. 1856.
Type: Florida, ex. hb. Gray (NY).   – Fig. 100

Bright green to yellowish plants growing in loose tufts. Primary stems
creeping, scarcely foliate, leaves minute. Secondary stems erect, 0.5-2 cm
long, irregularly branched, sometimes subpinnate, short-stipitate at base.
Leaves arranged in rows, erect-spreading when moist, ovate-oblong, to
1.5 mm long and 0.7 mm wide, deeply concave, apex flat, short-acuminate,
margins minutely serrulate nearly to base, costa percurrent; laminal cells
linear, flexuose, minutely prorulose at both ends, 25-45 μm long and ca.
5 μm wide, at leaf base shorter and incrassate, alar cells numerous, small,
subquadrate in a distinct group, extending up the margin at least 10 cells.
Dioicous. Inner perichaetial leaves to 2 mm long, lanceolate, abruptly
long-acuminate, coarsely dentate. Seta reddish, smooth, 4-6(-10) mm
long. Capsule cylindric, ca. 3 mm long including the long-rostrate
operculum; exostome teeth irregular, perforate, endostome rudimentary.
Calyptra cucullate.

Distribution: Southern U.S.A., Mexico, West Indies, C and northern
S America.

Ecology: Epiphytic; in the Guianas rare, only a single collection from
Guyana and Suriname at higher altitudes.

Specimens examined: Guyana: Potaro-Siparuni region, alt. 580 m,
Newton *et al.* 3626 (L, US). Suriname: Brownsberg, alt. 400 m, Gradstein
4644, 4649 (L).

2. **Pireella pohlii** (Schwägr.) Cardot, Bryologist 40: 18. 1913. –
Leucodon pohlii Schwägr., Sp. Musc. Frond. Suppl. 3(1): 232. 1828.
Type: Brazil, Pohl s.n. (G).

Description and synonymy see Musci I in Fl. Suriname 6: 250. 1964.

Distribution: Southern U.S.A., Mexico, West Indies, C and northern
S America.

Ecology: Epiphytic on trunks and branches in dense forest, not
uncommon in the canopy.

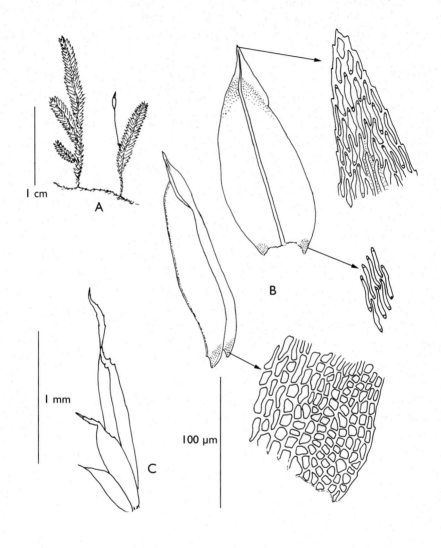

Fig. 100. *Pireella cymbifolia* (Sull.) Cardot: A, primary stem with secondary stems; B, leaves; C, perichaetial leaves (A-C, Gradstein 4649).

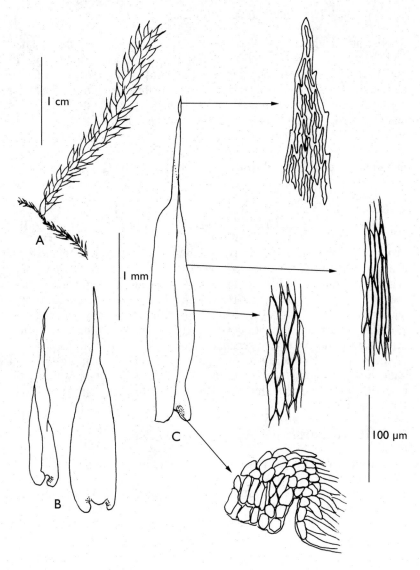

Fig. 101. *Renauldia paradoxica* B.H. Allen: A, primary stem with young secondary stem; B, secondary stem leaves; C, branch leaf (A-C, Cremers 12711).

7. **RENAULDIA** Müll. Hal. in Renauld, Rev. Bot. Bull. Mens. 9:
221. 1891.
Type: R. hildebrandtielloides Müll. Hal.

Robust plants growing in loose tufts. Primary stems creeping, slender,
with scale-like leaves. Secondary stems erect to pendent, simple to
pinnately branched, pseudoparaphyllia filiform; leaves concave, ecostate
or indistinctly double-costate, apex acute to long-acuminate, leaf base
cordate or auriculate; laminal cells linear, usually incrassate and porose,
alar cells strongly incrassate and coloured in a small, rounded group. Seta
short, capsule immersed, peristome double. Calyptra mitrate, naked.

Distribution: Tropical America and Africa; in the Guianas 1 species.

1. **Renauldia paradoxica** B.H. Allen, J. Bryol. 14: 677. 1987. Type:
Panama, Veraguas, Crosby 10154 (MO, PMA).   – Fig. 101

Primary stems creeping, leaves to 3 mm long, ovate-lanceolate, with
acuminate apex; secondary stems erect to pendent to ca. 10 cm long,
sparingly branched, leaves loosely appressed, slender, acuminate,
gradually larger and merging to the size and shape of branch leaves; branch
leaves patent-spreading, ovate-oblong, 3-4 mm long, deeply concave with
auriculate, clasping base, at apex abruptly narrowed to a flexuose, long-
acuminate point, minutely serrulate; laminal cells elongate-hexagonal,
flexuose, variously incrassate and porose, 35-60 μm long and 5-8 μm
wide, narrower along the margin, shorter in the acumen and towards base,
alar cells strongly incrassate, orange-red, in a roundish group. Sporophyte
not known.

Distribution: Panama, Guyana, French Guiana.

Ecology: Epiphytic on trees in light forest; rare.

Selected specimens: Guyana, north slope of Mt. Roraima, alt. 700-
1000 m, Gradstein 5187 (L). French Guiana: Mts. de la Trinité, Bassin
de la Mana, alt. 400 m, Cremers 12711 (L); Mataroni R., Florschütz-de
Waard 6162 (L).

8. **SPIRIDENTOPSIS** Broth. in Engler & Prantl., Nat. Pflanzenfam.
1(3): 805. 1906.
Type: S. longissima (Raddi) Broth. (Hypnum longissimum Raddi)

Distribution: C America, southeastern Brazil; 1 species.

1. **Spiridentopsis longissima** (Raddi) Broth. in Engler & Prantl, Nat. Pflanzenfam. 1(3): 806. 1906. – *Hypnum longissimum* Raddi, Critt. Bras. 9. 1822. Type: Brazil, Raddi s.n.

Large plants, primary stems creeping, secondary stems pendent, irregularly branched. Leaves ovate-lanceolate, at least 5 mm long, auriculate at base, abruptly narrowed to a long capillary hairpoint, costa single, extending 2/3 of lamina length; laminal cells linear, smooth, alar cells little differentiated. Seta to 4 mm long, capsule erect, operculum long-rostrate, peristome reduced.

N o t e : A single record of this species is known from the Guianas: a collection of hb. Kaulfuss (L) labelled 'Rio Essequibo, Mart. ded.' Since no Martius collections are known from Guyana the occurrence of this species in the Guianas is doubtful.

23. RACOPILACEAE

Description see Musci I in Fl. Suriname 6: 236. 1964.

D i s t r i b u t i o n : Pantropical; in the Guianas 1 genus.

1. **RACOPILUM** P. Beauv., Prodr. Aethéogam. 36. 1805. Type: R. tomentosum (Hedw.) Brid. (Hypnum tomentosum Hedw.)

Description see Musci I in Fl. Suriname 6: 236. 1964.

D i s t r i b u t i o n : Pantropical; in the Guianas 1 species.

1. **Racopilum tomentosum** (Hedw.) Brid., Bryol. Univ. 2: 719. 1827. – *Hypnum tomentosum* Hedw., Sp. Musc. Frond. 240. 1801. Type: Hispaniola, Swartz s.n. (G).

Description see Musci I in Fl. Suriname 6: 236. 1964.

D i s t r i b u t i o n : Southern U.S.A., Mexico, West Indies, C and tropical S America, tropical Africa, SE Asia.

E c o l o g y : Epiphytic on tree trunks, tree bases and logs, also terrestrial on detritus; common at higher altitudes.

24. RHACOCARPACEAE

N o t e : *Rhacocarpus*, traditionally belonging to the family HEDWIGIACEAE, has been placed in a separate family based on its distinctive characters, such as the fiddle-shaped leaves with well-marked auricles, the remarkable cell ornamentation and immersed stomata (Crum 1994).

D i s t r i b u t i o n : Pantropical; in the Guianas 1 genus.

LITERATURE

Crum, H. 1994. Rhacocarpaceae. In Sharp, Crum and Eckel, Moss Fl. Mexico 2. Mem. New York Bot. Gard. 69: 667.
Frahm, J.-P. 1996. Revision der Gattung Rhacocarpus Lindb. Cryptog. Bryol. Lichénol. 17: 39-65.

1. **RHACOCARPUS** Lindb., Öfvers. Kongl. Vetensk.-Akad. Förh. 19: 607. 1863.
Type: R. humboldtii (Hook.) Lindb. (Anictangium humboldtii Hook.)

Harrisonia Spreng., Syst. Veg. 4: 145. 1827, nom. illeg.

Medium-sized plants. Stems prostrate, pinnately branched. Leaves oval, oblong or panduriform, auriculate, apex apiculate, lanceolate or hairpointed, costa absent; laminal cells oblong-linear, obscure, appearing finely pluripapillose, marginal cells usually smooth and shiny, forming a distinct border, alar cells differentiated. Dioicous. Perichaetial leaves lanceolate, alar cells not differentiated. Seta elongate, capsule erect, obovoid, with a short neck and a wide mouth, operculum long-subulate, peristome absent. Calyptra large, cucullate, smooth.

D i s t r i b u t i o n : Pantropical; in the Guianas 1 species.

1. **Rhacocarpus purpurascens** (Brid.) Paris, Index Bryol. Suppl. 292. 1900. – *Hypnum purpurascens* Brid., Muscol. Recent. Suppl. 2: 121. 1812. Type: Ins. Bourbon, Bory St. Vincent s.n. – Fig. 102

Rhacocarpus humboldtii (Hook.) Lindb., Öfvers. Kongl. Vetensk. Akad.-Förh. 19: 603. 1863. – *Anictangium humboldtii* Hook., Pl. Crypt. 1a. 1816. Type: Nova Granata, Humboldt & Bonpland s.n. (BM).

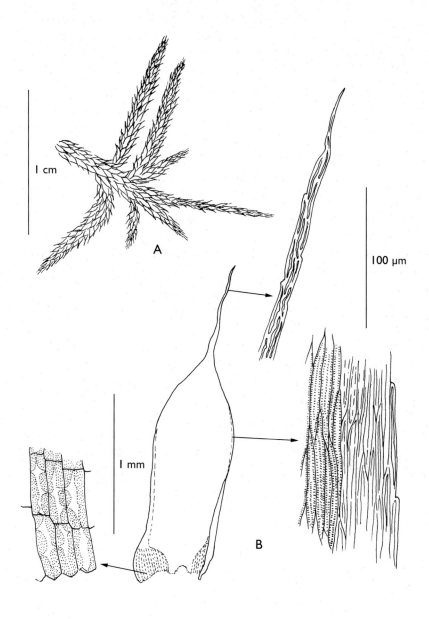

Fig. 102. *Rhacocarpus purpurascens* (Brid.) Paris: A, end of stem with branches; B, leaf (A-B, Gradstein 5348).

Dull green, yellowish brown or reddish plants in flat mats. Stems elongate, prostrate, pinnately branched; branches ascending, to 2 cm long. Leaves ca. 2 mm long, oblong-panduriform, narrowed above the large, dark-brown auricles, rounded at apex and abruptly contracted to a long, flexuose hairpoint, bordered with a broad yellowish band of elongate cells, subentire or denticulate in upper part; laminal cells at midleaf linear, incrassate, 60-90 μm long, densely covered with a fine reticulum, appearing finely papillose, border cells smooth, coloured, alar cells reddish-brown, irregularly quadrate or rectangular, strongly incrassate and porose forming conspicuous auricles, unequal in size. Perichaetial leaves hairpointed, the innermost more or less notched at the base of the awn, seta 6-20 mm long, capsule 2mm long. (Sporophyte description after Crum 1994).

D i s t r i b u t i o n : Pantropical.

E c o l o g y : On exposed, wet rocks at high altitudes.

S p e c i m e n s e x a m i n e d : Guyana: North slope of Mt. Roraima, alt. 2000-2300 m, Gradstein 5348, 5405. (GOET, L); Upper slopes of Mt. Wokomung, alt. 1540-1600 m, Boom *et al*. 9166, 9177, 9179 (NY).

N o t e : In Musci I in Fl. Suriname 6: 241. 1964 this species was reported as *R. humboldtii* (Hook.) Lindb..

Order **Hookeriales**

Plants pleurocarpous. Stems generally creeping, leaves complanate, lateral leaves asymmetric, costa often double, sometimes single or absent, alar cells not differentiated. Sporophytes lateral on stem and branches, setae elongate, capsules erect or inclined, peristome double, exostome teeth usually long and slender, endostome of the same length with high basal membrane and keeled segments.

KEY TO THE FAMILIES

1   Leaves ecostate, laminal cells lax, pellucid . . . . *26. LEUCOMIACEAE*
     Costa single or double, laminal cells of various shapes 2

2   Costa single, leaves with a broad border of elongate cells, margin entire . . .
     . . . . *25. DALTONIACEAE*
     Costa double, leaves generally not bordered, margin often serrate . . . .
     . . . . *27. PILOTRICHACEAE*

25. DALTONIACEAE

Small plants. Stems simple or sparingly branched, erect or prostrate, complanate or evenly foliate. Leaves lanceolate to ovate, margin entire or serrulate, mostly bordered by narrow, elongated cells, costa single, sometimes short or absent. Laminal cells short, smooth, firm-walled, alar cells not differentiated. Perichaetial leaves small. Seta elongate, capsule erect or inclined, exothecial cells collenchymatous, operculum conic-rostrate; peristome double, exostome teeth furrowed and cross-striolate or unfurrowed and papillose, slightly trabeculate at back, endostome with a low basal membrane and keeled segments. Calyptra mitrate, fringed at base.

Distribution: Pantropical, usually at higher altitudes; in the Guianas 2 genera.

Note: Crosby (1974) considered the 'daltoniaceous' peristome, characterized by the non-thickened exostome teeth, as the principal character of the DALTONIACEAE. Buck (1987) argued that the 'daltoniaceous' peristome could be the result of repeated modifications of the 'hookeriaceous' peristome under ecological pressures and advised to evaluate both gametophytic and sporophytic characters. Whittemore & Allen (1989) suggested that the non-thickened teeth of the 'daltoniaceous' peristome are result of reduction of the thickened 'hookeriaceous' peristome teeth. As a result of the thickenings the teeth in the latter are reflexed when dry and incurved when moist. In 'daltoniaceous' peristomes, the teeth are incurved when dry and reflexed when moist. This relationship between peristome morphology and function occurs also in other families, and indicates that these characters may be understood as adaptations to climatic conditions. Whittemore & Allen have also found additional support for recognition of the DALTONIACEAE by gametophytic characters such as rhizoids and axillary hairs.

LITERATURE

Buck, W.R. 1987. Taxonomic and nomenclatural rearrangment in the Hookeriales with notes on the West Indian taxa. Brittonia 39: 210-224.

Crosby, M.R. 1974. Toward a revised classification of the Hookeriaceae. J. Hattori Bot. Lab. 38: 129-141.

Whittemore, A. & B. Allen. 1989. The systematic position of Adelothecium Mitt. and the familial classification of the Hookeriales. Bryologist 92: 261-272.

KEY TO THE GENERA

1   Stems erect, evenly foliate, leaves lanceolate . . . . *1. Daltonia*
    Stems prostrate, complanate-foliate, lateral leaves larger than median leaves,
    ovate to spathulate . . . . *2. Leskeodon*

1. **DALTONIA** Hook. & Taylor, Muscol. Brit. 80. 1818.
Type: D. splachnoides (Sm.) Hook & Taylor (Neckera splachnoides
Sm.)

Stems erect, evenly foliate. Leaves lanceolate to linear-lanceolate, acuminate, bordered by several rows of elongate cells, broadest at base; costa single, extending beyond midleaf, ending below apex; laminal cells usually thick-walled, oval to long-hexagonal, marginal cells linear, basal cells incrassate and porose. Perichaetial leaves small, costa indistinct. Seta elongate, rough in upper part; capsule erect or inclined, ovoid-cylindric, exothecial cells quadrate, collenchymatous, exostome teeth papillose, endostome segments of the same length as the exostome, papillose. Calyptra mitrate, ciliate at base, small, just covering the operculum.

Distribution: Pantropical; in the Guianas 1 species.

1. **Daltonia longifolia** Taylor, London J. Bot. 7: 284. 1848. Type:
Ecuador, Pinchincha, Jameson s.n. (FH).   – Fig. 103

Small plants growing scattered or in small tufts. Stems erect, 1-3 cm long, little divided. Leaves contorted when dry, erect-spreading when moist, 2-3.5 mm long, lanceolate, short-acuminate; costa firm, extending 3/4-5/6 of leaf length, margin plane or recurved at apex, entire or distantly serrulate near apex, bordered with linear, incrassate cells, 6-10 cells wide at base, 1-3 cells wide in apex, confluent in acumen; laminal cells firm-walled, oval to oblong, 20-30 µm long and 9-10 µm wide, basal cells broader and porose, coloured at insertion. Autoicous or synoicous. Perichaetial leaves ovate, acute to short-acuminate, ca. 1 mm long, indistinctly bordered, cells elongate, thin-walled. Seta reddish, 5-12 mm long, scabrous in upper part. Capsule erect, ovoid, operculum conic-rostrate, exostome teeth recurved when dry, finely papillose, trabeculate at back, endostome segments slender from a low basal membrane, papillose. Calyptra densely ciliate at base.

Distribution: Mexico, West Indies, C and tropical S America; at high altitudes.

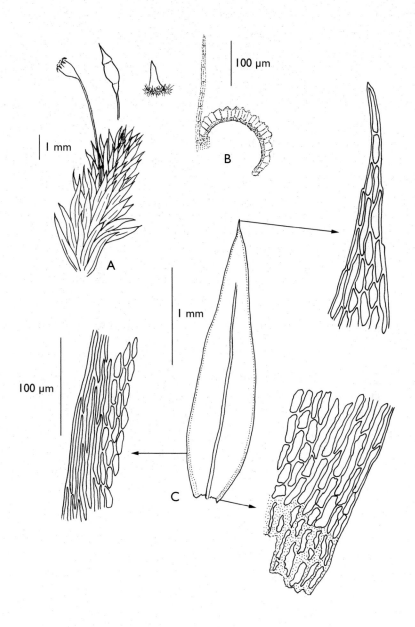

Fig. 103. *Daltonia longifolia* Taylor: A, plant with capsule and calyptra; B, reflexed exostome tooth with erect endostome segment; C, leaf (A-C, Gradstein 5469).

Ecology: Epiphytic on tree trunks and branches in the canopy; rare, reported from Guyana and French Guiana.

Specimen examined: Guyana: N slope of Mt. Roraima, alt. 1200-1600 m, Gradstein 5469 (L).

2. **LESKEODON** Broth. in Engler & Prantl, Nat. Pflanzenfam. 1(3): 925. 1907.
Type: L. auratus (Müll. Hal.) Broth. (Mniadelphus auratus Müll. Hal.)

Stems prostrate to suberect, usually undivided. Leaves more or less complanate, ovate, obovate, oblong-lanceolate or spathulate, narrow at insertion, apex acute to abruptly acuminate, bordered with one to several rows of narrow, elongate cells, costa single, extending 1/2-3/4 of leaf length; laminal cells firm-walled to collenchymatous, isodiametric to rounded-hexagonal. Perichaetia small, usually bordered. Seta smooth, capsule erect or inclined, exostome teeth papillose with a zig-zag median line, endostome segments from a low basal membrane, papillose. Calyptra mitrate, ciliate at base.

Distribution: Mexico, West Indies, C and tropical S America, tropical Asia; in the Guianas 1 species.

1. **Leskeodon cubensis** (Mitt.) Thér., Mem. Soc. Cub. Hist. Nat. "Felipe Poey" 14: 364. 1940. – *Distichophyllum cubense* Mitt., J. Linn. Soc., Bot. 12: 395. 1869. Type: Cuba, Wright 87 (NY).   – Fig. 104

Stems short, ca. 1 cm long, prostrate or ascending, complanate foliate; leaves crispate when dry, spreading when moist, 1-2.2 mm long and 0.5-0.8 mm wide, median leaves ovate or obovate, lateral leaves longer, oblong-ovate or spathulate, apex rounded, abruptly apiculate, margin entire, plane, border of 2(-3) rows of incrassate, linear cells merging in apex, apiculus to 0.2 mm long; costa firm, extending 1/2-3/4 of leaf length, occasionally subpercurrent in lateral leaves; laminal cells incrassate, isodiametric, rounded quadrate to hexagonal, 10-16 μm long. Autoicous or synoicous. Perichaetial leaves small, ca. 0.5 mm long, ovate, ecostate, bordered. Seta reddish, smooth, 3-4 mm long; capsule erect or slightly inclined, short-cylindric with a distinct neck, exothecial cells elongate, collenchymatous, operculum conic-rostrate; exostome teeth recurved when dry, finely papillose, with a zig-zag median line, trabeculate at back, endostome segments lanceolate from a low basal membrane, papillose. Calyptra covering only the operculum, mitrate, fimbriate at base, smooth.

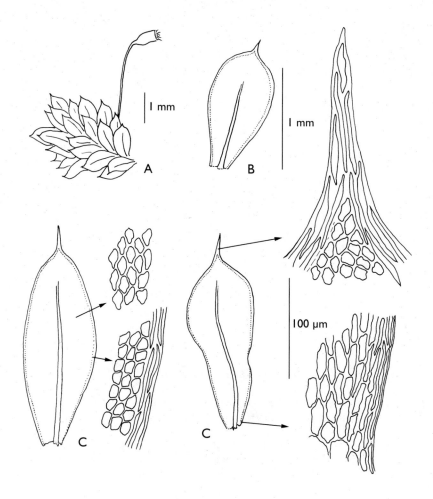

Fig. 104. *Leskeodon cubensis* (Mitt.) Thér.: A, part of stem with capsule; B, median leaf; C, lateral leaves (A-C, Gradstein 5268).

Distribution: Mexico, West Indies, C and tropical S America.

Ecology: Epiphytic on tree trunks and logs; collected in Guyana at high altitudes.

Specimen examined: Guyana: North slope of Mt. Roraima, alt. 1200-1600 m, Gradstein 5268 (L).

26. LEUCOMIACEAE

Description see Musci III in Fl. Guianas, ser. C, 1: 365. 1996.

D i s t r i b u t i o n : Pantropical; in the Guianas 1 genus.

LITERATURE

Allen, B.H. 1987. A revision of the genus Leucomium. Mem. New York Bot. Gard. 45: 661-677.
Florschütz-de Waard, J. & K. Veling. 1996. Leucomiaceae. In: A.R.A. Görts-van Rijn (ed.), Flora of the Guianas. Ser. C, 1: 365-370.

1. **LEUCOMIUM** Mitt., J. Linn. Soc., Bot. 10: 181. 1868.
Type: L. debile (Sull.) Mitt. (Hookeria debilis Sull.)

Description see Musci III in Fl. Guianas, ser. C, 1: 366. 1996.

D i s t r i b u t i o n :Pantropical; in the Guianas 2 species.

KEY TO THE SPECIES

1   Seta 25-30 mm long, capsule (without operculum) 1-2 mm long. Leaves long-acuminate with piliferous, hyaline acumen, 1.5-3 mm long . . . .
. . . . 1. L. steerei
Seta 8-20 mm long, capsule (without operculum) to 1 mm long. Leaves acute or short-acuminate, 1.3-2 mm long . . . . 2. L. strumosum

1. **Leucomium steerei** B. Allen & Veling, Mem. New York Bot. Gard. 45: 674. 1987. Type: Venezuela, Chimantá Massif, Steyermark 75964A (FH).

Description see Musci III in Fl. Guianas, ser. C, 1: 366. 1996.

D i s t r i b u t i o n : Guayana Highlands.

E c o l o g y : Epiphytic on twigs, roots and logs; only collected in Guyana at high altitudes, rare.

2.   **Leucomium strumosum** (Hornsch.) Mitt., J. Linn. Soc., Bot. 12: 502.
1869. – *Hookeria strumosum* Hornsch. in Mart., Fl. Bras. 1(2): 69.
1840. – *Hypnum strumosum* (Hornsch.) Müll. Hal., Syn. Musc. Frond.
2: 238. 1851. Lectotype (Allen 1987): Brazil, Tijucam, Olfers s.n. (BM).

Vesicularia eligianum W.R. Buck, Mem. New York Bot. Gard. 76: 161.
2003. Type: French Guiana, Commune de Saül, Cr. Saint-Eloi, Buck 33113
(holotype NY).

Description and synonymy see Musci III in Fl. Guianas, ser. C, 1: 368. 1996.

Distribution: Pantropical.

Ecology: Epiphytic on treebases and logs in lowland rainforest,
occasionally terrestrial or epilithic in moist areas; common.

Note: For variability of this species see remarks under *L. strumosum* in
Musci III: 370.

27. PILOTRICHACEAE

Plants with creeping stems, irregularly branched; branches erect or
prostrate, simple or divided, sometimes bipinnate (*Pilotrichum*). Leaves
generally in 8 rows (4 in *Crossomitrium*) usually complanate with the
median and lateral leaves different in shape, broad-ovate to lanceolate,
costae strong and double, sometimes absent (*Crossomitrium*); laminal
cells smooth, prorulose or papillose, often narrowed towards margins, alar
cells not differentiated. Autoicous or dioicous. Seta elongate, smooth or
roughened, capsule erect or inclined, operculum conic-rostrate; peristome
double, exostome teeth on outer side papillose and thin-walled or cross-
striolate and furrowed, endostome with basal membrane and keeled
segments, cilia lacking. Calyptra mitrate.

Distribution: Worldwide; in the Guianas 12 genera.

Notes: In the last decade the family classification within the *Hookeriales*
has been discussed by several authors (e.g., Miller 1971, Crosby
1974, Buck 1987, Whittemore & Allen 1989). Moreover several new
gametophytic characters have been introduced such as stem anatomy,
axillary hairs, pseudoparaphyllia, and rhizoids. Whittemore & Allen (1989)
using sporophytic and gametophytic characters distinguished 2 groups:
HOOKERIACEAE in the traditional sense (including LEUCOMIACEAE)
and DALTONIACEAE (including *Adelothecium* Mitt.).

Buck (1987) emphasized the differentiation of the stem cortex as an important character and recognized 5 families:

— HOOKERIACEAE with costa absent, leaves unbordered, cells short, lax, stem cortex undifferentiated.
— PILOTRICHACEAE with costa double, leaves unbordered, cells often elongate, inner and outer cortex well-differentiated.
— DALTONIACEAE with costa single and strong, leaves bordered, cells short, stem cortex little differentiated.
— ADELOTHECIACEAE with costa single and strong, leaves unbordered, cells small and thickwalled, cortex well-differentiated.
— LEUCOMIACEAE without costa, leaves unbordered, cells elongate, stem cortex undifferentiated.

In a recent phylogenetic study of the relationships within the *Hookeriales* (Buck *et al.* 2005), 7 families are distinguished adding the HYPOPTERYGIACEAE, SAULOMATACEAE fam. nov. and SCHIMPEROBRYACEAE fam. nov. The ADELOTHECIACEAE are comprised in the DALTONIACEAE.

Most genera with well-developed double costa, traditionally belonging in the HOOKERIACEAE, are now placed in the PILOTRICHACEAE. In a proposed classification *Crossomitrium* was tentatively placed in the HOOKERIACEAE as was also suggested by Allen (1990). I have here followed the classification used in Gradstein *et al.* (2001).

LITERATURE

Allen, B. 1990. A revision of the genus Crossomitrium, Trop. Bryol. 2: 3-34.

Bartram, E.B. 1928. Costa Rican Mosses collected by Paul C. Standley in 1924-26. Contr. U.S. Natl. Herb. 26: 51-114.

Buck, W.R. 1987. Taxonomic and nomenclatural rearrangement in the Hookeriales with notes on West Indian taxa. Brittonia 39: 210-224.

Buck, W.R. 1998. Pleurocarpous mosses of the West Indies. Pilotrichaceae. Mem. New York Bot. Gard. 82: 20-91.

Buck, W.R. & R.R. Ireland. 1985. A reclassification of the Plagiotheciaceae. Nova Hedwigia 41: 89-125.

Buck, W.R., C.J. Cox, A.J. Shaw & B. Goffinet. 2005. Ordinal relationships of pleurocarpous mosses, with special emphasis on the Hookeriales. Systematics and Biodiversity 2: 121-145.

Crosby, M.R. 1974. Toward a revised classification of the Hookeriaceae. J. Hattori Bot. Lab. 38: 129-141.

Crosby, M.R. 1975. Lectotypification of Schizomitrium B.S.G.. Taxon 24: 353-355.

305

Crosby, M.R., B.H. Allen & R.E Magill. 1985: A review of the moss genus Hypnella. Bryologist 88: 121-129.

Florschütz-de Waard, J. 1986. Hookeriaceae. In Flora of Suriname 6, 1: 289-350.

Gradstein, S.R. & J. Florschütz-de Waard. 1989. Results of a botanical expedition to Mount Roraima, Guyana. 1. Bryophytes. Trop. Bryol. 1: 25-54.

Kuc, M. 2000. The distribution of Hemiragis aurea (Brid.) Ren. & Card. and related notes of interest. Trop. Bryol.18: 55-64.

Miller, H.A. 1971. An overview of the Hookeriales. Phytologia 21: 243-252.

Welch, W.H. 1971. The Hookeriaceae of Jamaica, Hispaniola, and Puerto Rico. Bryologist 74: 77-130.

Welch, W.H. 1974. The Hookeriaceae of Central America. Bryologist 77: 328-404.

Whittemore, A. & B.H. Allen. 1989. The systematic position of Adelothecium Mitt. and the familial classification of the Hookeriales. Bryologist 92: 261-272.

KEY TO THE GENERA

1 Laminal cells pluri-papillose, papillae in rows over the lumina. .7. *Hypnella*
Laminal cells unipapillose or smooth 2

2 Plants with frondlike secondary stems, pinnately to bipinnately branched . .
. . . . *10. Pilotrichum*
Plants irregularly branched or undivided 3

3 Leaves linear-lanceolate, longitudinally plicate . . . . *6. Hemiragis*
Leaves ovate, oblong or lanceolate, not plicate .4

4 Leaf margins narrowly revolute from base to apex, entire or slightly denticulate in upper part . . . . *1. Actinodontium*
Leaf margins flat or partly recurved, usually serrate-dentate in upper part . 5

5 Costae absent. Stem and branches tightly attached to the substrate; leaves in 4 rows at upper side of stem only . . . . *4. Crossomitrium*
Costae distinct. Branches prostrate or ascending; leaves in 6 to 8 rows . . . 6

6 Costae extending 1/4-1/2 of leaf length 7
Costae longer, extending beyond 1/2 of leaf length 8

7 Seta smooth, exostome teeth striate and furrowed . . . . . . *8. Lepidopilidium*
Seta scabrous or papillose, exostome teeth papillose and lacking furrow . . .
. . . . *9. Lepidopilum p.p.*

8   Delicate plants, laminal cells wide, leaves bordered by 1 row of narrow
     cells. . . . . 5. *Cyclodictyon*
     More robust plants; laminal cells, if wide then leaves bordered by 2 or more
     rows of narrow cells 9

9   Leaves distinctly bordered by narrow cells . . . . 9. *Lepidopilum p.p.*
     Leaves bordered by quadrate cells or not bordered 10

10   Leaves oval-ovate, apex obtuse or apiculate (see also *Thamniopsis incurva*)
     . . . . .3. *Callicostella*
     Leaves oblong-lanceolate, apex acute-acuminate 11

11   Robust plants, leaves undulate in upper part 12
     Medium sized plants, leaves not undulate 13

12   Leaves oblong-lingulate, apex acute. . . . . 2. *Brymela*
     Leaves lanceolate, apex acuminate. . . . . 11. *Thamniopsis p.p.*

13   Leaves not falcate, acute or acuminate; marginal teeth if bifid not swollen. .
     . . . . 11. *Thamniopsis p.p.*
     Leaves falcate, long-acuminate; marginal teeth often bifid and swollen . . . .
     . . . . .12. *Trachyxiphium*

1. **ACTINODONTIUM** Schwägr., Sp. Musc. Frond. Suppl. 2(2):
    75. 1826.
    Type: A. adscendens Schwägr.

Small plants with erect or ascending, usually undivided stems, evenly
foliate to slightly complanate. Leaves lanceolate, acuminate, costa double;
laminal cells smooth, elongate-rhomboidal to oblong, narrower along the
margins, forming indistinct borders. Autoicous. Seta elongate, smooth,
capsule erect, operculum conic-rostrate, peristome papillose. Calyptra
mitrate, laciniate at base.

Distribution: Pantropical; in the Guianas 1 species.

1. **Actinodontium sprucei** (Mitt.) A. Jaeger, Ber. St. Gallischen
    Naturwiss. Ges. 1875-76: 325. 1877. – *Lepidopilum sprucei* Mitt., J.
    Linn. Soc., Bot. 12: 307. 1869. Type: Ecuador, Canelos, Spruce s.n.
    (NY).   – Fig. 105

Actinodontium portoricense H.A. Crum & Steere, Bryologist 59: 251. 1956.
Type: Puerto Rico, Sierra de Luquillo, Steere 6458 (NY).

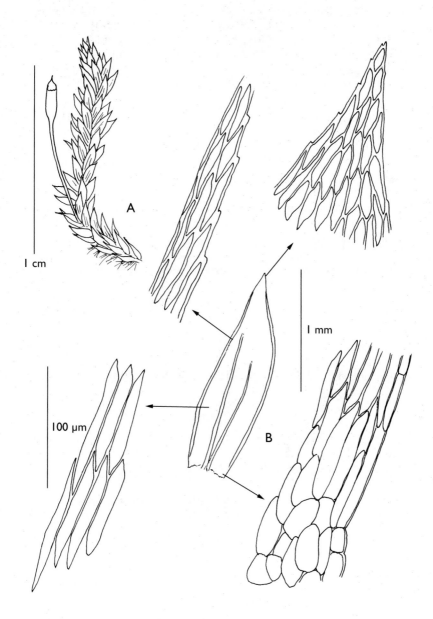

Fig. 105. *Actinodontium sprucei* (Mitt.) A. Jaeger: A, stem with capsule; B, leaf (A-B, Allen 20867).

Glossy plants growing in small tufts. Stems 1-2 cm long, ascending, usually simple. Leaves uniform, erect-spreading to slightly complanate, oblong-lanceolate, subsymmetric, 1.5-2.5 mm long and 0.3-0.5 mm wide, acute or short-acuminate, margins denticulate in upper half, often narrowly revolute from base nearly to apex, costae extending 1/2-3/4 of leaf-length, unequal; laminal cells in apex elongate-rhomboidal, more elongate towards midleaf, oblong-linear, to 120 μm long and 12 μm wide, narrower in a few rows along the margins, shorter and lax at insertion. Monoicous or synoicous. Perichaetia and perigonia similar, ca. 1 mm long, leaves ovate, ecostate, margins plane. Seta smooth, 6-12 mm long, capsule erect, cylindric, 1.5-2 mm long, exothecial cells rectangular with thickened vertical walls, operculum rostrate; exostome teeth whitish, erect, linear-lanceolate, papillose with median zigzagline, endostome segments linear, from a low basal membrane, as long as exostome teeth, densely papillose. Calyptra mitrate, deeply laciniate, smooth.

Distribution: West Indies, Venezuela, Suriname, French Guiana.

Ecology: Epiphytic on tree trunks and branches in moist forests at higher altitudes; not seen from Guyana.

Specimens examined: Suriname: Tafelberg, alt. 435 m, Allen 20867 (MO). French Guiana: Mt. Galbao, alt. 590 m, Buck 33258 (NY); Approuague-Kaw, Pic Matécho, alt. 590 m, Buck 37982 (NY).

2. **BRYMELA** Crosby & B.H. Allen, Monogr. Syst. Bot. Missouri Bot. Gard. 11: 211. 1985.
   Type: B. tutezona Crosby & B.H. Allen

Robust plants, stems creeping or ascending, leaves erect to complanate-spreading, oblong-lanceolate, usually undulate in upper part, apex acute to subobtuse, costae reaching 1/2-3/4 leaf length; laminal cells homogeneous, elongate, incrassate and porose. Dioicous. Seta elongate, capsule inclined to horizontal, operculum rostrate, peristome double, exostome teeth furrowed, densely striate, endostome segments papillose and keeled on high basal membrane.

Distribution: Neotropical; in the Guianas 1 species.

1. **Brymela parkeriana** (Hook. & Grev.) W.R. Buck, Brittonia 39: 210. 1987. – *Hookeria parkeriana* Hook. & Grev., Edinburgh J. Sci. 2: 229. 1825. – *Hookeriopsis parkeriana* (Hook. & Grev.) A. Jaeger, Ber. St. Gallischen Naturwiss. Ges. 1875-76: 360. 1877. Type: Guyana, Demerara R., Parker s.n. (BM).

Description and synonymy see Musci II in Fl. Suriname 6: 314. 1986, as *Hookeriopsis parkeriana*.

D i s t r i b u t i o n : Barbados, tropical S America.

E c o l o g y : Epiphytic on twigs in understory of light forest, also at higher altitudes; rather common in Guyana, rare in Suriname, not collected in French Guiana.

N o t e : Buck (1987) transferred *Hookeriopsis parkeriana* to *Brymela*, a genus characterized by large size, undulate leaves and homogeneous, elongate, incrassate laminal cells.

3. **CALLICOSTELLA** (Müll. Hal.) Mitt., J. Linn. Soc., Bot. Suppl.1: 136. 1859, nom. cons. – *Hookeria* sect. *Callicostella* Müll. Hal., Syn. Musc. Frond. 2: 216. 1851.
Type: C. papillata (Mont.) Mitt. (Hookeria papillata Mont.)

Schizomitrium Schimp. in Bruch *et al.*, Bryol. Europ. 5: 59. 1851. Lectotype (Crosby 1975): S. martinianum (Hornsch.) Crosby (Hookeria martiana Hornsch.)

Description see Musci II in Fl. Suriname 6: 317. 1986, as *Schizomitrium*, nom. rej..

D i s t r i b u t i o n : Pantropical; in the Guianas 6 species.

N o t e : The name *Callicostella* was replaced by *Schizomitrium* (Crosby 1975), but was later conserved.

KEY TO THE SPECIES

1 Leaf apex obtuse (sometimes apiculate), costae reaching apex and usually converging in upper part. 2
Leaf apex acute or short-acuminate, costae ending below apex, diverging throughout 3

2   Laminal cells smooth, apical margins crenulate. Seta smooth . . . .
. . . . *3. Callicostella merkelii*
Laminal cells papillose, apical margins regularly serrate. Seta scabrous at
least in upper part . . . . *4. Callicostella pallida*

3   Upper laminal cells lax, over 10 μm wide (except in ventral leaves) . . . .
. . . . *1. Callicostella colombica*
Upper laminal cells narrower 4

4   Costae in older leaves usually reddish; midleaf cells oblong, incrassate . . . .
. . . . 6. *Callicostella rufescens*
Costae not reddish; midleaf cells elongate-hexagonal, thin-walled 5

5   Upper margins toothed, teeth often conic. Seta smooth . . . .
. . . .*2. Callicostella guatemalensis*
Upper margins serrulate. Seta papillose . . . . *5. Callicostella rivularis*

1. **Callicostella colombica** R.S. Williams, Bryologist 28: 61. 1925. –
Schizomitrium colombicum (R.S. Williams) W.R. Buck & Steere,
Moscosoa 2:47. 1983. Type: Colombia, Cordoba, Killip 5044
(BM, NY).

Callicostella grossiretis E.B. Bartram, Bryologist 42: 156. 1939. –
Schizomitrium grossiretis (E.B. Bartram) H.A. Crum, Bryologist 87: 208.
1984. Type: Panama, Barro Colorado Island, Willis 41 (FH).

Description see Musci II in Fl. Suriname 6: 324. 1986, as *Schizomitrium
grossiretis*.

Distribution: West Indies, Panama, Colombia, Suriname, French
Guiana.

Ecology: Epiphytic in undergrowth of moist forest, occasionally on rock.

Note: *Callicostella grossiretis* is synonymized with *C. colombica* (Buck
1998). The only difference between the 2 species seems to be in the setae:
papillose in *C. grossiretis* and smooth in *C. colombica*. This can also be
noted in the types, but no gametophytic differences can be observed.

2. **Callicostella guatemalensis** (E.B. Bartram) J. Florsch., Fl. Suriname
6: 324. 1986. – *Hookeriopsis guatemalensis* E.B. Bartram, Bryologist
49: 120. 1946. – *Schizomitrium guatemalensis* (E.B. Bartram)
J. Florsch., Fl. Suriname 6: 326. 1986. Type: Guatemala, Izabal,
Steyermark 38243 (FH).

Description see Musci II in Fl. Suriname 6: 326. 1986, as *Schizomitrium guatemalensis*.

Distribution: Guatemala, Suriname.

Ecology: Epiphytic on log, rare.

3. **Callicostella merkelii** (Hornsch.) A. Jaeger, Ber. St. Gallischen Naturwiss. Ges. 1875-76: 351. 1877. – *Hookeria merkelii* Hornsch. in Mart., Fl. Bras. 1(2): 62. 1840. – *Schizomitrium merkelii* (Hornsch.) J. Florsch., Fl. Suriname 6: 322. 1986. Type: Brazil, Sebastianopolin, Merkel s.n. (not seen).

Description see Musci II in Fl. Suriname 6: 322. 1986, as *Schizomitrium merkelii*.

Distribution: Guatemala, Brazil, Suriname, French Guiana.

Ecology: Epiphytic and epilitic in creeks, not common.

4. **Callicostella pallida** (Hornsch.) Ångstr., Öfvers. Kongl. Vetensk.-Akad. Förh. 33(4): 27. 1876. – *Hookeria pallida* Hornsch. in Mart., Fl. Bras. 1(2): 64. 1840. – *Schizomitrium pallidum* (Hornsch.) H.A. Crum & L.E. Anderson, Mosses E.N. America: 822. 1981. Type: Brazil, Minas Geraes, Martius s.n. (M).

Description and synonymy see Musci II in Fl. Suriname 6: 319. 1986, as *Schizomitrium pallidum*.

Distribution: Southern U.S.A., Mexico, West Indies, C and tropical S America.

Ecology: Epiphytic on bark of trees or decaying wood, occasionally on rocks or terrestrical; very common.

5. **Callicostella rivularis** (Mitt.) A. Jaeger, Ber. St. Gallischen Naturwiss. Ges. 1875-76: 355. 1877. – *Hookeria rivularis* Mitt., J. Linn. Soc., Bot. 12: 353. 1869. – *Schizomitrium rivulare* (Mitt.) H.A. Crum, Bryologist 87: 208. 1984. Lectotype (Welch 1971): Ecuador, near Bombonasa R., Spruce 668 (BM). – Fig. 106

312

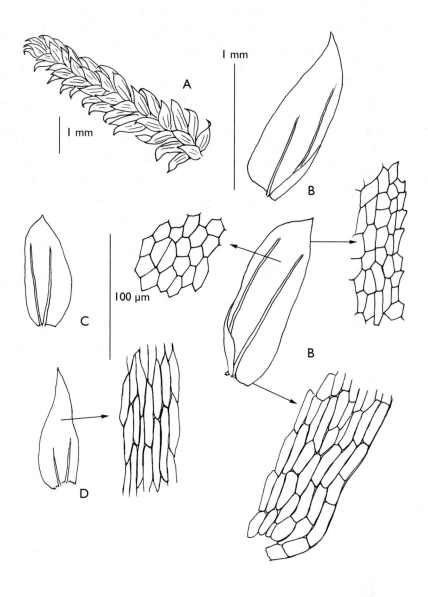

Fig. 106. *Callicostella rivularis* (Mitt.) A. Jaeger: A, end of stem; B, lateral leaves; C, dorsal leaf; D, ventral leaf (A-D, Newton *et al.* 3591).

Slender, dull-green plants growing in flat mats. Stems creeping, to ca. 3 cm long, irregularly branched; branches short, prostrate. Leaves slightly contorted when dry, complanate spreading when dry, usually leaf tips curved to the substrate; median leaves subsymmetric, ovate-oblong, to 1 mm long and 0.5 mm wide, lateral leaves asymmetric, oblong-lanceolate, to 1.5 mm long and 0.5 mm wide; apex broad-acute to short-acuminate (in ventral leaves sharp-acute), margins irregularly serrulate in upper half, smooth below, costae variable in length, extending 1/2-4/5 of leaf-length, diverging below and parallel above or diverging throughout, at back smooth or with a few blunt teeth at end; laminal cells smooth, thin-walled, in upper part rhomboidal to elongate-hexagonal, 15-24 µm long and 8-10 µm wide, basal cells more elongate, oblong or rectangular; in ventral leaves all cells elongate. Dioicous. Perichaetial leaves lanceolate, to 1.5 mm long, gradually acuminate, margins sharply serrate, costae toothed at back. Seta to 1 cm long, rough with low, rounded papillae, capsule inclined, ovoid with distinct neck, operculum conic-rostrate; peristome with brown, incurved exostome teeth, 250-300 µm long, furrowed at back, transversely striate below, papillose above, endostome segments of the same length from a high basal membrane, keeled, papillose, cilia rudimentary. Calyptra mitrate, scabrous at apex, lobed at base.

D i s t r i b u t i o n : West Indies, Guatemala, Guyana, Ecuador, Bolivia, Peru.

E c o l o g y : At higher altitudes, on wet stones in streams and epiphytic on roots and logs in periodically submerged forest.

S p e c i m e n s   e x a m i n e d : Guyana: Potaro-Siparuni region NW of Paramakatoi, Paramaka R., alt. 700 m, Newton *et al.* 3567, 3591 (L, US); Kamarang R., alt. 500-600 m, Robinson 85-0036 (L, US).

N o t e : This species is characterized by acute leaves and short upper laminal cells. In ventral leaves, however, the apex is acuminate, the costae short and the laminal cells thin-walled and elongate.

6.   **Callicostella rufescens** (Mitt.) A. Jaeger, Ber. St. Gallischen Naturwiss. Ges. 1875-76: 355. 1877. – *Hookeria rufescens* Mitt., J. Linn. Soc., Bot. 12: 352. 1869. – *Schizomitrium rufescens* (Mitt.) J. Florsch., Fl. Suriname 6: 328. 1986. Type: Brazil, Pará, Spruce 629 (BM).

Description and synonymy see Musci II in Fl. Suriname 6: 328. 1986, as *Schizomitrium rufescens*.

314

Distribution: Trinidad?, Guianas, Brazil, Ecuador.

Ecology: Epiphytic on bark and decaying wood, occasionally on rock; very common on rotten logs in moist lowland rainforests.

4. **CROSSOMITRIUM** Müll. Hal., Linnaea 38: 611. 1874.
Type: C. patrisiae (Brid.) Müll. Hal. (Hypnum patrisiae Brid.)

Small to medium-sized, glossy plants with long, creeping stems, irregularly branched, often firmly attached to the substrate. Leaves in 4 rows on upper side of the stem, ecostate, lanceolate, oblong, (ob) ovate or suborbicular, apex acuminate or rounded, short-acuminate to umbonate; margins serrulate to serrate, teeth often bifid; laminal cells elongate, thinwalled, smooth or finely papillose in upper leaf. Propagules filiform, septate, in clusters at leaf base or on specialized brood-branches. Dioicous. Seta scabrous in upper part or densely papillose throughout. Capsule suberect to inclined, ovoid to elliptic with faintly differentiated neck, operculum conic-rostrate; peristome double, exostome teeth pale, papillose with median zig-zag line, endostome segments from a basal membrane, hyaline. Calyptra mitrate, fimbriate at base.

Distribution: Neotropical; in the Guianas 3 species..

KEY TO THE SPECIES

1   Lateral leaves oblong-lanceolate with acuminate apex; leaves erect at base, apices recurved to the substrate, only flat and overlapping in propagules-forming parts of the stem . . . . 2. *C. patrisiae*
     Lateral leaves oval-oblong or oval-obovate with rounded apex; leaves complanate, flat to the substrate, overlapping 2

2   Lateral leaves to 1(-1.2) mm long, laminal cells pluri-papillose in upper leaf (mainly in marginal cells). . . . . *1. C. epiphyllum*
     Lateral leaves to 1.8 mm long; laminal cells smooth . . . . . . *3. C. sintenisii*

1. **Crossomitrium epiphyllum** (Mitt.) Müll. Hal., Linnaea 38: 613. 1874. – *Lepidopilum epiphyllum* Mitt., J. Linn. Soc., Bot. 12: 370. 1869. Lectotype (Allen 1990): Ecuador, Chimborazo, Spruce 801 (NY).   – Fig. 107

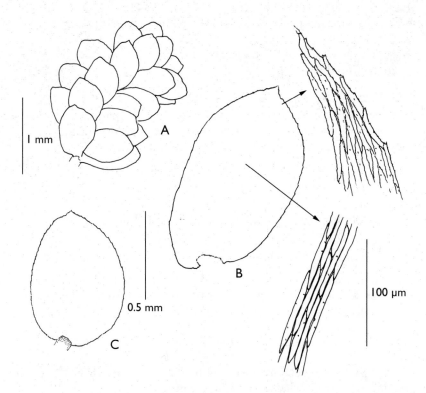

Fig. 107. *Crossomitrium epiphyllum* (Mitt.) Müll. Hal.: A, end of stem (moist); B, lateral leaf; C, median leaf (A-C, Cremers 6778).

Tiny plants growing flat to the substrate. Stems creeping, irregularly branched. Leaves complanate, lateral laeves oval or obovate, rounded at insertion, to 1(-1.2) mm long and ca. 0.6 mm wide, apex rounded broad-acute, umbonate to very short acuminate, margins serrulate, minutely papillose in upper part; median leaves shorter, to 0.5 mm long, suborbicular, overlapping; laminal cells linear, 70-120 µm long and 5-7 µm wide, wider near insertion, in apical part irregularly papillose, especially along the margins. Propagules on specialized, erect broodbranches with spreading leaves. Seta 4-7 mm long, papillose above; capsule inclined to suberect, ovoid to elliptic, 0.7-1 mm long, operculum 1 mm long. (Sporophyte description after Allen 1990).

Distribution: West Indies, C and tropical S America.

Ecology: Epiphyllous, also epiphytic on bark of trees; in the Guianas extremely rare, but probably overlooked by the small size.

Specimen examined: French Guiana: Sommet Tabulaire, alt. 750 m, Cremers 6778 (CAY, L).

Note: This very small species with flattened, round leaves, appressed to the substrate can be mistaken for a species of hepatics. It has much in common with *C. sintenisii* but size and papillose upper laminal cells make it easy to distinguish. Also the laminal cells are narrower. If present the erect brood-branches with propagules on all sides of the stem are characteristic.

2. **Crossomitrium patrisiae** (Brid.) Müll. Hal., Linnaea 38: 612. 1874.
   – *Hypnum patrisiae* Brid., Bryol. Univ. 2: 539. 1827. Type: hb. Bridel nr. 905.

Description and synonymy see Musci II in Fl. Suriname 6: 291. 1986. For complete synonymy and description of several expressions of this species see Allen 1990.

Distribution: Mexico, West Indies, C and tropical S America.

Ecology: Mainly epiphyllous and on twigs and stems of small trees in the understory of lowland rainforest; common.

3. **Crossomitrium sintenisii** Müll. Hal., Hedwigia 37: 244. 1898.
   Type: Puerto Rico, Sierra de Luquillo, Sintenis s.n. (BM, FH, G).
   – Fig. 108

Crossomitrium rotundifolium Herzog, Bibl. Bot. 87: 134. 1916. Lectotype (Allen 1990): Bolivia, San Miquelito, Herzog 2726 (H, JE, S).

Medium-sized plants growing in dense mats. Stems creeping, sparingly branched, branches prostrate. Leaves complanate, contorted when dry, flat to the substrate when moist, overlapping in basal part; lateral leaves oval-oblong, to 1.8 mm long and 1 mm wide, apex broad-acute, rounded and short-acuminate, sometimes carinate at the tip, margins serrate in upper half, serrulate towards base; median leaves not different, but shorter, ovate to suborbicular; laminal cells smooth, at midleaf elongate-hexagonal or linear, 90-130 μm long and ca. 9 μm wide, shorter in extreme apex, wider at insertion. Propagules borne in clusters at leaf base. Sporophyte unknown.

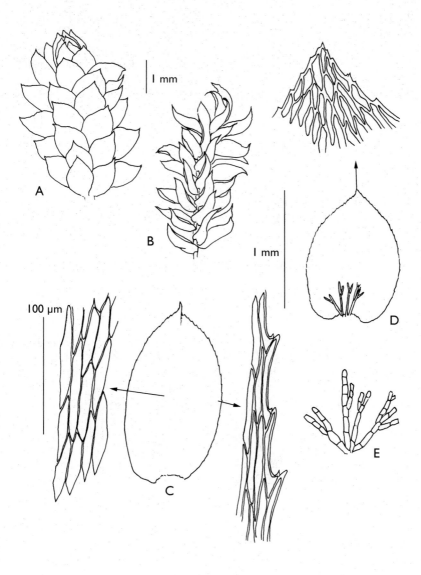

Fig. 108. *Crossomitrium sintenisii* Müll. Hal.: A, end of stem (moist); B, end of stem (dry); C, lateral leaf; D, median leaf; E, propagula (A-E, Gradstein 4951).

Distribution: West Indies, C and tropical S America.

Ecology: Epiphytic on branches and tree trunks, occasionally epiphyllous (on palmleaf roof); rare in the Guianas.

Specimens examined: Guyana: Upper Mazaruni Distr., Jawalla, alt. 500 m, Gradstein 4951 (L). Suriname: Upper Saramacca R. upstreams Paka-paka, Florschütz 1631 (L).

Note: Collections from Central French Guiana are reported as *C. rotundifolium* (Buck 2003).

5. **CYCLODYCTION** Mitt., J. Linn. Soc., Bot. 7: 163. 1864.
   Type: C. laetevirens (Hook. & Taylor) Mitt. (Hookeria laetevirens Hook. & Taylor)

Description see Musci II in Fl. Suriname 6: 306. 1986.

Distribution: Pantropical; in the Guianas 1 species.

1. **Cyclodictyon varians** (Sull.) O. Kuntze, Rev. Gen. Pl. 2: 835. 1891.
   – *Hookeria varians* Sull., Proc. Amer. Acad. Arts 5: 285. 1861. Type: Cuba, Wright, Pl. Cubens. 88 (FH).

Description see Musci II in Fl. Suriname 6: 307. 1986.

Distribution: Southern U.S.A., West Indies, tropical S America.

Ecology: Terrestrial and epiphytic in wet forest; rare in the Guianas, only one collection known from Suriname (Emma Mts., alt. 775 m).

6. **HEMIRAGIS** (Brid.) Besch., Ann. Sci. Nat., Bot. sér. 6,3: 242. 1876. – *Leskea* subgenus *Hemiragis* Brid., Bryol. Univ. 2: 334. 1827. – *Hookeria* subgenus *Hemiragis* (Brid.) Mitt., J. Linn. Soc., Bot. 12: 335. 1869.
   Type: H. striata (Schwaegr.) Besch. (Leskea striata Schwaegr.)

Distribution: West Indies, southern C America and northern S America.; 1 species.

1. **Hemiragis aurea** (Brid.) Kindb., Enum. Bryin. Exot. 16: 1888.
– *Hypnum aureum* Lam. ex Brid., Muscol. Recent. 2: 146. 1801. –
Harpophyllum aureum (Brid.) Spruce, Cat. Musc. Amaz. And.: 11.
1867. Lectotype (Welch 1971): Guadeloupe, Richard s.n. (PC).

Description and synonymy see Musci II in Fl. Suriname 6: 305. 1986.

Distribution: West Indies, southern C America and northern S
America. For detailed data see Kuc 2000.

Ecology: Epiphytic on lower trunks and branches, sometimes terrestrial
on detritus; in Guyana at higher altitudes (Mt. Roraima, alt. 1200-1600m).

7. **HYPNELLA** (Müll. Hal.) A. Jaeger, Ber. St. Gallischen Naturwiss.
Ges. 1875-76: 365. 1877. – *Hookeria* sect. *Hypnella* Müll. Hal., Syn.
Musc. Frond. 2: 208. 1851.
Lectotype (Crosby *et al.* 1985): H. leptorrhyncha (Hook. & Grev.) A.
Jaeger (Hookeria leptorrhyncha Hook. & Grev.)

Neohypnella E.B. Bartram, Contr. U.S. Natl. Herb. 26: 104. 1928.
Type: N. mucronifolia E.B. Bartram

Description see Musci II in Fl. Suriname 6: 295. 1986.
Additional: Autoicous, synoicous or dioicous; peristome double, exostome
teeth densely striate and furrowed or papillose with zig-zag median line,
endostome segments weakly papillose, basal membrane high or low.

Distribution: Neotropical; in the Guianas 2 species.

Note: The genus *Neohypnella* as treated in Musci II, was created
by Bartram (1928) for *Neohypnella diversifolia*, a species with a
different peristome type: the exostome teeth are not furrowed at back
as in *Hypnella* but have a zig-zag median line. Crosby, Allen & Magill
(1985) concluded that the peristomes are not essentially different. The
exostome teeth of *Hypnella* also have a zig-zag median line, but by
thickenings on the dorsal plates this median line becomes invisible at
the bottom of the furrow. However, Buck *et al.* (2005), using multigene
molecular data, considered *Neohypnella* to be a good genus, distinct
from *Hypnella*.

KEY TO THE SPECIES

1 Leaves broad-acute or obtuse, apiculate; costae extending beyond midleaf;
   papillae simple . . . . *1. H. diversifolia*
   Leaves obtuse, often canaliculate at apex; costae extending to midleaf or
   shorter; papillae multifid, often stalked . . . . *2. H. pallescens*

1. **Hypnella diversifolia** (Mitt.) A. Jaeger, St. Gallischen Naturwiss.
   Ges. 1875-76: 366. 1877. – *Hookeria diversifolia* Mitt., J. Linn. Soc.,
   Bot. 12: 364.1869. – *Neohypnella diversifolia* (Mitt.) W.H. Welch
   & H.A. Crum, Bryologist 62: 174. 1959. Type: Ecuador, Andes
   Quitenses, Spruce 678 (NY).

   Description see Musci II in Fl. Suriname 6: 297. 1986, as *Neohypnella
   diversifolia*.

   D i s t r i b u t i o n : West Indies, C and tropical S America.

   E c o l o g y : Epiphytic on exposed branches and stems, often pendent;
   only collected in Guyana at higher altitudes.

   S p e c i m e n s   e x a m i n e d : Guyana: N slope of Mt. Roraima, alt. 700 m,
   Aptroot 17092 (L); ibid. alt. 1200-1600 m, Gradstein 5299, 5311, 5408
   (L).

2. **Hypnella pallescens** (Hook.) A. Jaeger, Ber. St. Gallischen Naturwiss.
   Ges. 1875-76: 365. 1877. – *Hookeria pallescens* Hook., Musci Exot.
   1: 38. 1818. Type: Venezuela, Orinoco R. near Esmeralda, Humboldt
   & Bonpland s.n. (BM, NY).

   Hypnella cymbifolia (Hampe) A. Jaeger, Ber. St. Gallischen Naturwiss. Ges.
   7: 365. 1877. – *Hookeria cymbifolia* Hampe, Linnaea 25: 362. 1852. Type:
   Porto Rico, Schwanecke 35 (BM).
   Pseudohypnella guianensis P.W. Richards, Bull. Misc. Inform. Kew 1934:
   336. 1934. Type: Guyana, Moraballi Cr., Richards 407 (BM).
   Hypnella guayanensis B.H. Allen & W.R. Buck, in Buck, Mem. New York
   Bot. Gard. 64: 1990. Type: Guyana, Mt. Wokomung, Boom & Samuels
   9131 (NY).

   Description see Musci II in Fl. Suriname 6: 295. 1986, as *Hypnella
   cymbifolia*.

   D i s t r i b u t i o n : West Indies, tropical S America.

Ecology: Epiphytic on branches and tree trunks, also on logs, occasionally on wet rocks; usually at higher altitudes, sometimes in lowland savanna forest.

Notes: This species is extremely variable: the plants vary from small, pinnately branched with leaves to 0.4 mm long, to more elongate with leaves to 1.4 mm long. Also a great variation in shape of the papillae can be observed, from highly stalked multifid to low and simple or sometimes absent in most leaves. Also the length of the costae is variable: extending half the leaf length to very short and indistinct. No correlation between these characters could be observed.

Allen & Buck (in Buck 1990) described a new species: *Hypnella guayanensis* for the ecostate specimens. In my opinion these collections all belong to *H. pallescens*. In the many collections from Mt. Roraima (Gradstein 1989) all variations in costa length are present, also very short and indistinct. But always traces of costae could be observed, even in the paratypes cited for *H. guayanensis*.

8. **LEPIDOPILIDIUM** (Müll. Hal.) Broth. in Engler & Prantl, Nat. Pflanzenfam. 1(3): 942. 1907. – *Hookeria* sect. *Lepidopilidium* Müll. Hal., Hedwigia 39: 273. 1900.
Type: L. tenuisetum (Müll. Hal.) Broth. (Hookeria tenuiseta Müll. Hal.)

Description see Musci II in Fl. Suriname 6: 330. 1986.

Distribution: Mexico, West Indies, C and tropical S America and tropical Africa; in the Guianas 1 species.

1. **Lepidopilidium portoricense** (Müll. Hal.) H.A. Crum & Steere, Bryologist 59: 253. 1956. – *Crossomitrium portoricense* Müll. Hal., Hedwigia 37: 244. 1898. Type: Porto Rico, Mt. Cuyón, Sintenis s.n. (BM, NY).

Description see Musci II in Fl. Suriname 6: 331. 1986.

Distribution: Mexico, West Indies, C and tropical S America.

Ecology: Epiphytic on bark of trees and logs; rather common on branchlets and leaves in forests at higher altitudes.

9. **LEPIDOPILUM** (Brid.) Brid., Bryol. Univ. 2: 267. 1827, nom. cons. – *Pilotrichum* subgenus *Lepidopilum* Brid., Muscol. Recent. Suppl. 4: 141. 1819. Type: L. subenerve Brid.

Description see Musci II in Fl. Suriname 6: 333. 1986.

Distribution: Pantropical; in the Guianas 9 species.

KEY TO THE SPECIES

1 Costae extending to midleaf or beyond 2
   Costae shorter, not reaching midleaf 6

2 Leaf apex distinctly bordered by 2 or more rows of elongate cells 3
   Leaf apex bordered by a single row of elongate cells or not bordered 5

3 Leaves to 6 mm long, costae extending to midleaf. . . . . *4. L. polytrichoides*
   Leaves not over 3 mm long; costae extending beyond midleaf 4

4 Leaves ovate-lanceolate, apex acuminate. . . . . *3. L. denticulatum*
   Leaves obovate-oblong, apex obtuse, apiculate or abruptly short-acuminate
   . . . . *9. L. tortifolium*

5 Branches prostrate, leaves complanate. . . . . *2. L. cubense*
   Branches usually erect, circumfoliate. . . . . *8. L. surinamense*

6 Leaves to 3.5 mm long, purplish red. Seta ca. 3 mm long . . . .
   . . . . *5. L. purpurascens*
   Leaves shorter, greenish. Seta elongate 7

7 Leaves oblong, often transversely undulate, apex obtuse, apiculate . . . .
   . . . . *1. L. affine*
   Leaves ovate-lanceolate, apex acute or acuminate. 8

8 Lateral leaves scalpelliform, subentire . . . . *6. L. radicale*
   Lateral leaves strongly falcate, coarsely serrate-dentate . . . . *7. L. scabrisetum*

1. **Lepidopilum affine** Müll. Hal., Linnaea 21: 192. 1848. Type: Suriname, Kegel 741 (GOET, L).

Description see Musci II in Fl. Suriname 6: 340. 1986.

Distribution: C and tropical S America.

Ecology: Epiphytic in marshforest; not common, not collected in Guyana.

323

2. **Lepidopilum cubense** (Sull.) Mitt., J. Linn. Soc., Bot. 12: 384. 1869.
– *Hookeria cubensis* Sull., Proc. Amer. Acad. Arts 5: 285. 1861. Type:
Cuba, Wright, Musci Cubenses 85 (FH, NY).

Description and synonymy see Musci II in Fl. Suriname 6: 342. 1986.

Distribution: Mexico, Guatemala, West Indies, Suriname, French
Guiana.

Ecology: Epiphytic in dense rainforest; in the Guianas rare, not known
from Guyana.

Note: Buck (1998) treated *L. cubense* as a synonym of *L. amplirete*
(Sull.) Mitt.. However, *L. cubense* is distinct in the elongate costae
extending far beyond midleaf (in *L. amplirete* 1/3-1/2 leaf length). The
collections reported for French Guiana (Buck 2003) are representatives of
L. cubense, including the typical propagulum-forming at the leaf apices.

3. **Lepidopilum denticulatum** (Thér.) F.D. Bowers, Mem. New York
Bot. Gard. 69: 801. 1994. – *Amblytropis denticulata* Thér., Mem.
Soc. Cub. Hist. Nat. "Felipe Poey" 14: 367. 1940. Type: Cuba, Pico
Turquino, Acuña Gale 391 (NY, not seen). – Fig. 109

Rather robust plants growing in loose tufts. Stems creeping, densely
tomentose; branches elongate, to 5 cm long, prostrate or ascending,
sparingly divided, distantly foliate. Leaves strongly contorted when dry,
twisted at apex when moist, ovate-oblong, abruptly acuminate, 1.5-2.5 mm
long, margin dentate in upper half, bordered, costae strong, often reddish,
extending ca. 3/4 of leaf length; laminal cells elongate-hexagonal at
midleaf, 50-65 µm long and ca. 20 µm wide, slightly incrassate, smaller
in apex, border cells linear, in 3-5 rows. Dioicous. Perichaetial leaves
ovate, abruptly long-acuminate, ca. 1 mm long. Seta 5-7 mm long,
densely papillose; capsule ovoid, ca. 2 mm long, operculum conic-
rostrate, peristome teeth ca. 700 µm long. Calyptra smooth.

Distribution: Mexico, West Indies, C and tropical S America.

Ecology: At higher altitudes in moist localities, on rock and trunk
base, rare.

Specimen examined: Guyana: N slope of Mt. Roraima, alt. 1200-
1600 m, Gradstein 5242 (L).

324

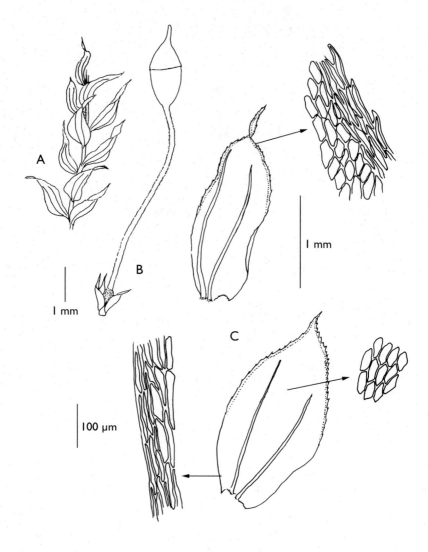

Fig. 109. *Lepidopilum denticulatum* (Thér.) F.D. Bowers: A, part of branch; B, perichaetium with sporophyte; C, leaves (A-C, Gradstein 5242).

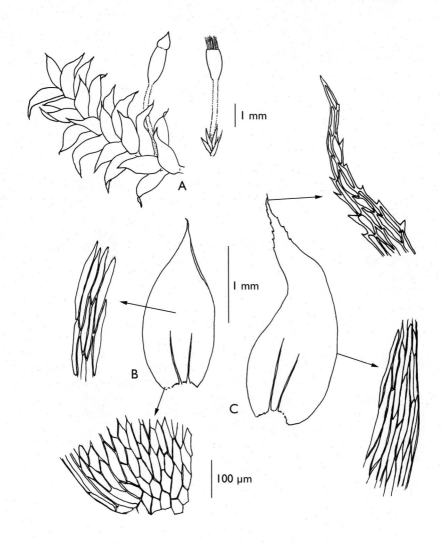

Fig. 110. *Lepidopilum purpurascens* Schimp. ex Besch.: A, part of branch with sporophytes; B, median leaf; C, lateral leaf (A-C, Aptroot 17101).

4. **Lepidopilum polytrichoides** (Hedw.) Brid., Bryol. Univ. 2: 269. 1827. – *Hypnum polytrichoides* Hedw., Sp. Musc. Frond.: 244. 1801. Type: Jamaica, Swartz s.n.

Desciption see Musci II in Fl. Suriname 6: 337. 1986.

Distribution: Mexico, West Indies, C and tropical S America.

Ecology: Epiphytic on tree bases and trunks in the undergrowth of rainforest, also in the lower canopy.

5. **Lepidopilum purpurascens** Schimp. ex Besch., Ann. Sci. Nat., Bot. sér. 6, 3: 229. 1876. Lectotype (Welch 1971): Guadeloupe, l'Herminier s.n. (BM, NY). – Fig. 110

Rather robust plants with glossy purplish-red leaves, pale-green when young. Branches prostrate or ascending, to 4(-6) cm long, little divided, complanate-foliate; median leaves ovate, subsymmetric, abruptly acuminate, often twisted at the tip, lateral leaves wide-spreading (also dry), lanceolate, gradually acuminate, strongly falcate, 2.5-3.5 mm long; margins serrulate at apex to sharp-serrate in lateral leaves, costae slender extending 1/3-1/2 of leaf length; laminal cells at midleaf firm-walled, oblong-linear, to 250 µm long and ca. 18 µm wide, apical cells shorter, basal cells wider and shorter, oval-hexagonal. Propagules rare. Dioicous? Perigonia not seen. Perichaetial leaves small, lanceolate, to 1 mm long. Seta short, ca. 3 mm long, densely papillose, papillae round, ca. 50 µm high; capsule erect, ovoid, to 2 mm long, operculum conic-rostrate, exostome teeth slender, 600-900 µm long, papillose, endostome segments of the same length, keeled, finely papillose. Calyptra mitriform, with some erect hairs, laciniate at base.

Distribution: West Indies, Colombia, Guyana.

Ecology: Epiphytic on branches in mossy forest, occasionally on mossy rock face; only at higher altitudes, rare.

Specimens examined: Guyana: N slope of Mt. Roraima, alt. 700 m, Aptroot 17101 (L); ibid, alt. 1200-1600 m, Gradstein 5221, 5383, 5454 (L).

Note: Buck (1998) treated *L. purpurascens* as a form of *L. scabrisetum*, adapted to high light conditions. However, apart from the larger size and the characteristic glossy purplish-red colour, *L. purpurascens* is distinct in the longer costae and larger cells; moreover, the sporophyte is quite different with a firm and short seta and capsules with conspicuous, elongate peristome teeth.

6. **Lepidopilum radicale** Mitt., J. Linn. Soc., Bot. 12: 378. 1869. Type: Ecuador, Andes Quitenses, Spruce 765 (NY).

Description see Musci II in Fl. Suriname 6: 348. 1986.

Distribution: Mexico, West Indies, C and tropical S America.

Ecology: Terrestrial on clay and wet rocks, occasionally epiphytic in the undergrowth of humid forest; rare.

Note: Buck (1998) synonymized this species with *L. scabrisetum*, but the entire leaf margins and slenderly acuminate apex of *L. radicale* are clear differences with *L. scabrisetum*.

7. **Lepidopilum scabrisetum** (Schwägr.) Steere, Bryologist 51: 140. 1948. – *Neckera scabriseta* Schwägr., Sp. Musc. Frond. Suppl. 1(2): 153. 1816. Type: French Guiana, Richard s.n. (BM, G, PC).

Lepidopilum stolonaceum Müll. Hal., Hedwigia 37: 245. 1898. Type: Puerto Rico, near Manati, Sintenis 127 (NY).
Lepidopilum brevipes Mitt., J. Linn. Soc., Bot. 12: 376. 1869. Type: Peru, Tavalosos, Spruce s.n. (NY).

Description and synonymy see Musci II in Fl. Suriname 6: 344. 1986. Additional: branches often curved and partly defoliated, stoloniferous; lateral leaves extremely variable in shape as described for *L. stolonaceum,* see Musci II in Fl. Suriname 6: 346. 1986.

Distribution: Mexico, West Indies, C and tropical S America.

Ecology: Epiphytic on branches and young stems in the undergrowth of lowland rainforest; rather common.

Notes: *L. stolonaceum* as distinguished in Musci II with rather short and broad, often caducous leaves is considered as a form of *L. scabrisetum*, adapted to poor growth conditions (Buck 1998). This corresponds with the rare development of sporophytes in these specimens.

L. brevipes was reported for the Guianas by Buck (2003), being distinguished from *L. scabrisetum* by the shorter seta and the naked calyptra. However, seta length in the 2 species overlaps and additional gametophytic characters to distinguish *L. brevipes* have not been observed. Therefore *L. brevipes* is here treated as a synonym of *L. stolonaceum*.

8. **Lepidopilum surinamense** Müll. Hal., Linnaea 21: 193. 1848. – *Hookeria surinamensis* (Müll. Hal.) Müll. Hal., Syn. Musc. Frond. 2: 207. 1851, nom. illeg. Type: Suriname, Mariepaston, Kegel 1406 (GOET).

Description and synonymy see Musci II in Fl. Suriname 6: 334. 1986.

Distribution: Tropical S America.

Ecology: Epiphytic on branches, lianas, palm leaves, tree bases and logs, in undergrowth of lowland rainforest rather common, also at higher altitudes.

9. **Lepidopilum tortifolium** Mitt., J. Linn. Soc., Bot. 12: 374. 1869. Type: Colombia, Andes Bogotenses, Bucaramanga, Weir 147 (NY).
– Fig. 111

Mediumsized plants growing in loose mats. Stems creeping, irregularly branched, branches prostrate, elongate. Leaves complanate-spreading when moist, strongly crispate when dry, median leaves orbicular, lateral leaves obovate-oblong, 1.5-2.5 mm long and 1-1.5 mm wide, apex obtuse, apiculate or abruptly short-acuminate, serrate; costae firm, 1/2-3/4 of leaf length; laminal cells at midleaf hexagonal, 35-50 µm long and ca. 40 µm wide, gradually more elongate towards base, marginal cells narrower in 3-4 rows, forming a distinct border. Dioicous. Seta 12-15 mm long, papillose throughout, capsule 1.5-2 mm long. Calyptra slightly roughened at apex, laciniate at base. (Capsule description after Buck, 2003).

Distribution: Mexico, West Indies, C and tropical S America.

Ecology: On wet rock, in streams and waterfalls in moist forest; in the Guianas rare.

Specimen examined: French Guiana: Mt. Galbao, source of La Mana R., alt. 500 m, Buck 33333 (NY).

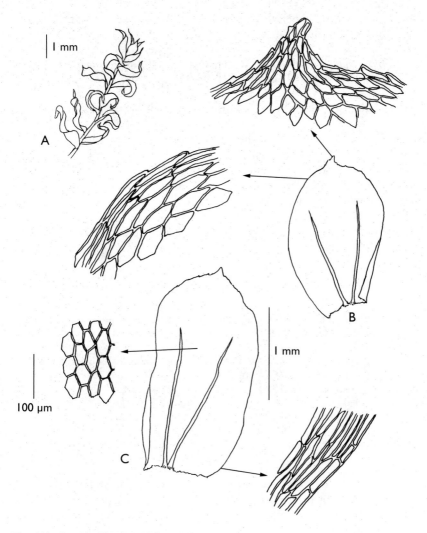

Fig. 111. *Lepidopilum tortifolium* Mitt.: A, end of stem (dry); B, median leaf; C, lateral leaf (A-C, Buck 33333).

N o t e : This species resembles *L. surinamense* in leaf shape and cell size. It is different in the complanate habit when moist. Also the longer costae, the distinct border and the elongate, papillose setae are good decisive characters; also the habitat in which the species occur are very different.

10. **PILOTRICHUM** P. Beauv., Mag. Encycl. 5: 327. 1804.
Lectotype (Crosby 1969): P. bipinnatum (Schwägr.) Brid. (Neckera
bipinnata Schwägr.)

Pilotrichum Brid., Muscol. Recent. Suppl. 4: 140. 1819, nom. illeg.
Callicosta Müll. Hal., Linnaea 21: 188. 1848.

Description see Musci II in Fl. Suriname 6: 298. 1986, as *Callicosta*.

N o t e : Crosby (1969) used the name *Pilotrichum* in his revision of this
genus. When he realized that this was an illegitimate name, he proposed
to conserve it. After rejection of this proposal he replaced it by *Callicosta*
(Crosby 1978). In 1988 *Pilotrichum* proved not to be illegitimate and the
name was used again.

D i s t r i b u t i o n : Mexico, West Indies, C and tropical S America; in the
Guianas 3 species.

KEY TO THE SPECIES

1   Leaf apex acute to short-acuminate; costae diverging throughout, extending
3/4 of leaf length or less, ending bluntly or in a low tooth at back . . . .
. . . . .2. *P. evanescens*
Leaf apex rounded and apiculate; costae becoming parallel in upper leaf,
extending more than 3/4 of leaf length, ending in a conspicuous tooth
composed of several cells. 2

2   Costae smooth or with a few scattered teeth, no propagules. Plants
synoicous . . . .*1. P. bipinnatum*
Costae with a dorsal crest, often bearing propagules. Plants dioicous . . . . . .
. . . .*3. P. fendleri*

1.   **Pilotrichum bipinnatum** (Schwägr.) Brid., Muscol. Recent. Suppl.
4: 140. 1819. – *Neckera bipinnata* Schwägr., Sp. Musc. Frond.
Suppl. 1(2): 156. 1816. – *Callicosta bipinnata* (Schwägr.) Müll. Hal.,
Linnaea 21: 189. 1848. Type: French Guiana, Richard s.n. (BM, PC).

Description see Musci II in Fl. Suriname 6: 301. 1986, as *Callicosta
bipinnata*.

D i s t r i b u t i o n : Mexico, West Indies, C and tropical S America.

E c o l o g y : Epiphytic on tree trunks, branches and lianas; rather common
in the understory of lowland rainforest, seldom in the canopy.

331

2. **Pilotrichum evanescens** (Müll. Hal.) Crosby, Bryologist 72: 326.
    1969. – *Callicosta evanescens* Müll. Hal., Linnaea 21: 189. 1848.
    – *Pilotrichum bipinnatum* (Schwägr.) Brid. var. *evanescens* (Müll.
    Hal.) Müll. Hal., Syn. Musc. Frond. 2: 179. 1851. Type: Suriname,
    Paramaribo, Kegel 742 (GOET, L).

Description and synonymy see Musci II in Fl. Suriname 6: 299. 1986, as
Callicosta evanescens.

Distribution: Mexico, West Indies, C and tropical S America.

Ecology: Epiphytic on lianas and tree branches in the understory of
savanna forest, also in the canopy of dense forest; not uncommon.

3. **Pilotrichum fendleri** Müll. Hal., Linnaea 42: 492. 1879. – *Callicosta
    fendleri* (Müll. Hal.) Crosby, Bryologist 81: 436. 1978. Type:
    Venezuela, near Tovar, Fendler 93 (FH).

Description see Musci II in Fl. Suriname 6: 303. 1986, as *Callicosta
fendleri*.

Distribution: Mexico, West Indies, C and northern S America.

Ecology: Epiphytic on tree bases and branches in the undergrowth of
lowland rainforest; rare; unknown from Guyana.

11. **THAMNIOPSIS** (Mitt.) M. Fleisch., Musc. Buitenzorg 3: 952.
    1908. – *Hookeria* sect. *Callicostella* subsect. *Thamniopsis* Mitt., J.
    Linn. Soc., Bot. 12: 338. 1869.
    Type: T. pendula (Hook.) M. Fleisch. (Hookeria pendula Hook.)

Distribution: Southern U.S.A., West Indies, C and tropical S America;
in the Guianas 4 species.

Notes: The genus *Hookeriopsis* as treated in Musci II (Fl. Suriname 6:
311. 1986) has been subject to an extensive study in which it is split in
4 different genera (Buck 1987). In this concept the name *Hookeriopsis*
is preserved for a small group of species with short and double costa
and a stem without hyalodermis. This group has no representatives in
the Guianas.
Most species, traditionally belonging in the genus *Hookeriopsis*, are
transferred to the genus *Thamniopsis*. This genus is characterized by a

thin-walled, hyaline stem epidermis and leaves with a strong, double costa, heterogeneous areolation, an (often obscure) border and (often) strongly toothed margins.

A group of slender plants with bifid, swollen marginal teeth is segregated in the genus *Trachyxiphium* (*T. guadalupensis*) with undifferentiated stem epidermis and leaves with homogeneous areolation (laminal cells often prorulose).

A group of large plants with simple marginal teeth is segregated in the genus *Brymela* (*B. parkeriana*) with undifferentiated stem epidermis and leaves with homogeneous areolation (cells thick-walled and porose).

The character that is most emphasized in this classification is the structure of the stem epidermis, which is difficult to examine. Moreover, other characters are not always correlated. The classification of Buck is therefore accepted here with some hesitation.

KEY TO THE SPECIES

1   Leaves with thickened margins at midleaf; costae diverging at base, following the margins in upper part . . . . *1. T. gradsteinii*
    Leaf margins not thickened; costae diverging throughout or parallel in upper part 2

2   Leaves ovate-oblong with obtuse or broad-acute apex. 3
    Leaves lanceolate with acute-acuminate apex 4

3   Leaf apex broad-acute, sharply toothed, stem and branch leaves not different . . . . *2. T. incurva*
    Leaves polymorphous; stem leaves ovate with obtuse, entire apex, at end of stem and branches leaves lanceolate with serrate apex . . . . *3. T. killipii*

4   Only upper leaves sometimes undulate and sharp-dentate, lower leaves shorter and apex gradually more rounded and entire . . . . *3. T. killipii*
    All leaves strongly undulate and sharp-dentate in upper part . . . . *4. T. undata*

1. **Thamniopsis gradsteinii** J. Florsch., nov. sp. Type: Guyana: North slope of Mt. Roraima, alt. 1200-1600 m, Gradstein 5244 (holotype L, isotypes GOET, MO, NY).   – Fig. 112

Caules et rami prostrati. Folia uniformia, erecto-patentia, ovato-lanceolata, supra medium folii abrupte attenuata, apice acuto, margine integro in apice obtuse serrulato, costa duplici e basi divergente in

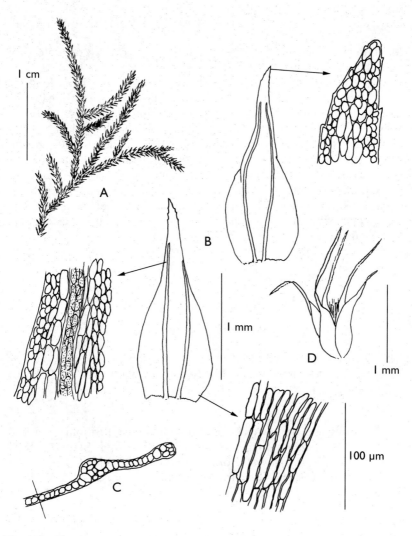

Fig. 112. *Thamniopsis gradsteinii* J. Florsch.: A, part of plant; B, leaves; C, cross section at midleaf showing one costa and thickened margin; D, perichaetium (A-D, Gradstein 5244).

dimidio distali pseudomarginali, circa 3/4 folii attingente; cellulis laminaribus magnitudine variabilibus, marginalibus pluristratosis parvis, juxtacostalibus magnis pellucidisque. Folia perichaetalia ecostata, ovata, longe acuminata. Seta rubella, circa 3 cm longa, laevis. Capsula ignota.

Medium-sized plants growing in loose mats. Stems elongate, prostrate, irregularly complanate-branched, branches simple or repeatedly divided. Leaves erect-spreading, uniform, ovate-lanceolate, 1.5-2 mm long and ca. 0.7 mm wide in basal part, above midleaf rather abruptly narrowed to a short acumen; apex acute, margins at midleaf thickened, entire, at apex bluntly serrate; costae firm, diverging from base to midleaf, in upper part following the margins, extending 3/4-7/8 of leaf-length, smooth at back; laminal cells incrassate, porose, variable in size, midleaf cells quadrate to oval, ca. 15 µm wide, along the margin in more than one layer, along the costae and the marginal cells a few rows of large pellucid cells, to 70 µm long and 15 µm wide, basal cells also elongate and wide, rectangular. Dioicous? Perigonia not seen. Perichaetial leaves ecostate, to 1.5 mm long, ovate, abruptly narrowed to a linear acumen, serrulate at apex, cells elongate, thinwalled. Seta reddish, ca. 3 cm long, smooth. Capsule not seen.

Distribution: Guyana, only known from the type locality.

Ecology: On logs in humid mossy forest at high altitudes.

Specimens examined: Guyana: North slope of Mt. Roraima, alt. 1200-1600 m, Gradstein 5244 (holotype L, isotypes GOET, MO, NY), 5271 (GOET, L).

2. **Thamniopsis incurva** (Hornsch.) W.R. Buck, Brittonia 39: 218. 1987. – *Chaetophora incurva* Hornsch., Horae Phys. Berol. 65. 1820. – *Hookeriopsis incurva* (Hornsch.) Broth. in Engler & Prantl, Nat. Pflanzenfam. 1(3): 942. 1907. Type: Chile, Chamisso 1487 (BM).

– Fig. 113

Medium-sized plants growing in loose mats. Stems creeping, irregularly branched, branches prostrate. Leaves contorted when dry, complanate-spreading when moist, median leaves round to ovate or obovate, subsymmetric, to 1.2 mm long and 1 mm wide, lateral leaves obovate-oblong, asymmetric, to 2 mm long, apex obtuse or broadly rounded-acute, margins sharply toothed in upper leaf, teeth conic-inflated and often bifid, lower margin serrate or entire; costae diverging from base, extending 1/2-2/3 of leaf length, smooth at back with sometimes one sharp tooth at end; laminal cells lax, thin-walled or slightly incrassate,

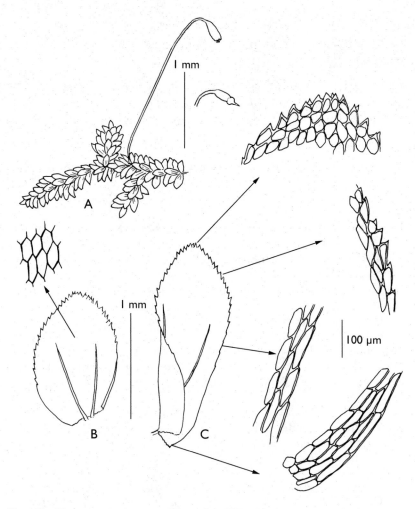

Fig. 113. *Thamniopsis incurva* (Hornsch.) W.R. Buck: A, part of plant with sporophyte; B, median leaf; C, lateral leaf. (A-C, Newton *et al*. 3549)

midleaf cells elongate-hexagonal or rhomboidal, 25-30 μm wide and ca. 70 μm long, shorter to isodiametric in apex, more elongate towards base, along the basal margins narrower in a few rows. Autoicous or synoicous. Perichaetial leaves small, ca. 1 mm long, ecostate, with acuminate, serrate apex. Seta 2-2.5 cm long, reddish, smooth or roughened just below the capsule, capsule inclined, cylindric, ca. 2.5 mm long, operculum conic-rostrate, peristome with the characters of the genus. Calyptra mitrate, lobed at base, smooth.

Distribution: Mexico, West Indies, C and tropical S America.

Ecology: Epiphytic on rotten logs in forest at higher altitudes; rare, only collected in Guyana.

Specimens examined: Guyana: Para Mt., 3 km S of Paramakatoi, alt. 850-1050 m, Newton *et al.* 3528B, 3549 (L, US); Trail Paramakatoi to Kato, alt. 700-800 m, Newton *et al.* 3651 (L, US).

3. **Thamniopsis killipii** (R.S. Williams) E.B. Bartram, Contr. U.S Natl. Herb. 26: 103. 1928. – *Hookeriopsis killipii* R.S. Williams, Bryologist 28: 62. 1925. Type: Colombia, Cordoba, Killip 5246 (BM, NY)

Description see Musci II in Fl. Suriname 6: 308. 1986.

Distribution: Colombia, Venezuela, Peru, Guyana, Suriname.

Ecology: On decaying wood in forest at higher altitudes.

Note: Williams described this species with only ovate leaves with blunt, entire apex. The lanceolate leaves with acute, sharp-dentate apex are also present in the type collection, but apparently were overlooked. In most collections these lanceolate leaves are present and sometimes even dominating over the short, ovate leaves.

4. **Thamniopsis undata** (Hedw.) W.R. Buck, Brittonia, 39: 219. 1987. – *Leskea undata* Hedw., Sp. Musc. Frond. 214. 1801. Type: Jamaica, Swartz s.n. (NY).

Hookeriopsis crispa (Müll. Hal.) A. Jaeger, Ber. St. Gallischen Naturwiss. Ges. 1875-76: 358. 1877. – *Hookeria crispa* Müll. Hal., Bot. Zeitung (Berlin) 13: 768. 1855. Type: Venezuela, Moritz s.n. (BM).

Description see Musci II in Fl. Suriname 6: 316. 1986, as *Hookeriopsis crispa*.

Distribution: West Indies, C and tropical S America; only collected in Guyana at higher altitudes.

Ecology: On rock, terrestrial or epiphytic on roots and logs in moist areas at higher altitudes.

12. TRACHYXIPHIUM W.R. Buck, Brittonia 39: 219. 1987.
Type: T. guadalupense (Spreng.) W.R. Buck (Hypnum guadalupense
Spreng.)

Small to medium-sized plants. Stems creeping or ascending, irregularly branched, hyalodermis absent. Leaves falcate, often homomalous, lanceolate to oblong-lanceolate, apex acuminate, margins usually strongly serrate, teeth bifid and often swollen; costae double, parallel, extending ca. 3/4 leaf length; laminal cells linear, incrassate, often prorulose. Autoicous. Seta elongate, capsule erect to inclined, operculum conic-rostrate; peristome double, exostome teeth strongly furrowed, endostome with high basal membrane, segments papillose. Calyptra mitrate, naked.

Distribution: Mexico, West Indies, C and tropical S America; in the Guianas 1 species.

1. **Trachyxiphium guadalupense** (Spreng.) W.R. Buck, Brittonia 39: 220. 1987. – *Hypnum guadalupense* Spreng., Spec. Musc. 2: 96. 1812. – *Hookeriopsis guadalupense* (Spreng.) A. Jaeger, Ber. St. Gallischen Naturwiss. Ges. 1875-76: 362. 1877. Type: Guadeloupe, ex. hb. Candolle, sine col. (B).

Hookeriopsis falcata (Hook.) A. Jaeger, Ber. St. Gallischen Naturwiss. Ges. 1875-76: 363. 1877. – *Hookeria falcata* Hook., Musci Exot. 1: 54. 1818. Type: Colombia, Humboldt & Bonpland s.n.(BM).

Description see Musci II in Fl. Suriname 6: 312. 1986, as *Hookeriopsis falcata*.

Distribution: Mexico, West Indies, C and tropical S America.

Ecology: Epiphytic on bark of trees and rotten logs in forest, at higher altitudes.

Note: This variable moss, widespread in S America has been described under various names. Buck (1987) synonymized these with *Hypnum guadalupense* and placed it in a new genus *Trachyxiphium*.

Order **Hypnales**

Plants pleurocarpous. Stems usually prostrate, secondary stems, if present variously branched; leaves usually ovate-lanceolate, acute or acuminate, costa single or short and double, sometimes absent, laminal cells mostly elongate, alar cells often differentiated. Sporophytes lateral, setae usually elongate, capsules erect or inclined and asymmetric, peristome double, exostome teeth abruptly tapering in upper part, endostome with high basal membrane and keeled segments alternating with cilia.

KEY TO THE FAMILIES

1   Costa single, well-developed 2
     Costa short and double or lacking. 5

2   Stems pinnately branched 3
     Stems irregularly branched 4

3   Secondary stems erect, frondose or dendroid, often stipitate. Laminal cells
     rhomboid to linear, smooth or prorulate . . . . *33. THAMNOBRYACEAE*
     Stems prostrate, 1- to 3-pinnately branched, usually covered with paraphyllia.
     Laminal cells short, oval or hexagonal, often pluripapillose . . . .
     . . . . *34. THUIDIACEAE*

4   Alar cells, if differentiated, in equal groups at both sides of the leaf base . . .
     . . . . *28. FABRONIACEAE*
     Leaves with differentiated alar cells in a large triangular group, often only on
     one side of the leaf base . . . . *32. STEREOPHYLLACEAE* p.p.

5   Plants from aquatic habitats; stems floating. Alar cells seldom differentiated.
     Capsules immersed, setae short . . . . *29. HYDROPOGONACEAE*
     Plants not exclusively aquatic, generally growing in mats. Alar cells usually
     differentiated. Setae elongate 6

6   Stems complanate-foliate. Ventral leaves lacking, lateral leaves asymmetric
     with a conspicuous group of alar cells only on one side of the leaf base . .
     . . . . *32. STEREOPHYLLACEAE* p.p.
     Stems not complanate-foliate. Alar cells if differentiated, in equal groups at
     both sides of the leaf base. 7

7   Leaves often asymmetric and homomallous, alar cells if differentiated
     quadrate and small (partly inflated in *Ectropothecium*) . .*30. HYPNACEAE*
     Leaves subsymmetric, often concave, basal cells usually conspicuously
     inflated, oval to oblong . . . . *31. SEMATOPHYLLACEAE*

28. FABRONIACEAE

Small plants with creeping stems, irregularly branched; stem and branch leaves slightly differentiated, ovate to narrowly lanceolate, acute to long-acuminate, margin entire, serrate or ciliate, costa single, extending beyond midleaf; laminal cells smooth, rhombic to oblong-rhomboidal, alar cells usually differentiated, quadrate. Autoicous. Seta elongate; capsule erect, cylindric, operculum conic to conic-rostrate, peristome double, single or absent, exostome teeth (when present) lanceolate, often in pairs, reflexed when dry, endostome (when present) well developed with basal membrane or reduced to single segments. Calyptra cucullate, smooth.

Distribution: Worldwide; in the Guianas 1 genus.

LITERATURE

Buck, W.R. & H. Crum. 1978. A re-interpretation of the Fabroniaceae with notes on selected genera. J. Hattori Bot. Lab. 44: 347-369.
Buck, W.R. 1980. A re-interpretation of the Fabroniaceae: additions and corrections. J. Hattori Bot. Lab. 47: 45-55.
Buck, W.R. 1981. A re-interpretation of the Fabroniaceae III: Anacamptodon and Fabronidium revisited, Mamillariella, Helicodontiadelphus and Bryobartlettia gen. nov. Brittonia 33: 473-481.
Fleischer, M. 1915. Fabroniaceae. In: Musc. Buitenzorg 4: 1115-1134.

Note: Fleischer's (1915) broad concept of the FABRONIACEAE was revised by Buck & Crum (1978). Mainly based on peristome characters the family was divided in FABRONIACEAE s.s. and MYRINIACEAE. The family FABRONIACEAE in this concept is characterized as slender pleurocarps with erect capsules. The peristomes are diplolepideous but are often reduced (Buck 1980, 1981).

1. **ANACAMPTODON** Brid., Muscol. Recent. Suppl. 4: 136. 1819.
   Type: A. splachnoides (Brid.) Brid. (Orthotrichum splachnoides Brid.)

Small plants growing in flat mats. Stems creeping, freely branching. Leaves ovate-lanceolate, acute, costa ending near midleaf or extending to apex; laminal cells rhombic to elongate-rhomboidal, alar cells gradually differentiated, subquadrate. Seta erect, smooth, capsule oblong-cylindric with a distinct neck and strongly contracted below the mouth when dry, operculum conic, blunt or short-rostrate, peristome double, exostome teeth strongly hygroscopic, broad-lanceolate, at base fused in pairs, densely papillose at outer surface, endostome segments linear, free, not arising from a basal membrane.

Distribution: Worldwide; in the Guianas 1 species.

1. **Anacamptodon cubensis** (Sull.) Mitt., J. Linn. Soc., Bot. 12: 540. 1869. – *Fabronia cubensis* Sull., Proc. Amer. Acad. Arts 5: 283. 1861. Lectotype (Buck & Crum 1978): Cuba, Wright, Musci Cubensis 66 (NY). – Fig. 114

Tiny plants growing in low to dense, green mats. Stems creeping, irregularly to pinnately branched, branches horizontal or ascending, to 1 cm long. Stem leaves ovate-lanceolate, sometimes slightly falcate, acuminate, 0.5-0.7 mm long, branch leaves usually smaller, distant, acute; costa firm, not reaching apex; margin entire or crenulate; laminal cells thin- to firm-walled, elongate rhomboidal to oblong, 20-35 μm long and 8-15 μm wide, alar cells irregularly quadrate in an inconspicuous group. Perichaetial leaves lanceolate, acuminate, ca. 1 mm long. Seta 3-4 mm long, capsule cylindric, ca. 1 mm long, constricted at the neck and below the mouth when dry, operculum conic rostrate, peristome with the characters of the genus, exostome teeth strongly reflexed with erect endostome segments when dry.

Distribution: Cuba, Puerto Rico, Bolivia, Peru, Venezuela, Guyana, Suriname.

Ecology: Epiphytic on tree trunks or terrestrial between rocks; rare.

Specimens examined: Guyana: Rupununi Savanna, Shea Rock, alt. 200 m, Jansen-Jacobs 5604 (L); Rupununi Savanna, 5 km N of Dadanawa Ranch, Florschütz-de Waard 6025 (L). Suriname: Tafelberg, Lisa Cr., alt. 615 m, Allen 20676A (MO).

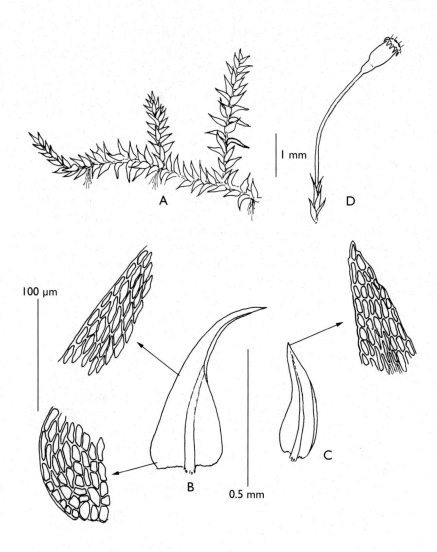

Fig. 114. *Anacamptodon cubensis* (Sull.) Mitt.: A, end of stem with branches; B, stem leaf; C, branch leaf; D, perichaetium with capsule (A-D, Allen 20676A).

29. HYDROPOGONACEAE

Description see Musci I in Fl. Suriname 6: 238. 1964.

Distribution: Neotropical; in the Guianas 2 genera.

LITERATURE

Churchill, S.P. 1991. Bryologia Novo Granatensis 5. Additional records for Colombia and Antioquia, with review to the distribution of Hydropogon fontinaloides in South America. Bryologist 94: 44-48.
Mägdefrau, K. von. 1973. Hydropogon fontinaloides (Hook.) Brid., ein periodisch hydro-aerophytisch Laubmoos des Orinoco und Amazonas. Herzogia 3: 141-149.
Welch, W.H. 1943. The systematic position of the genera Wardia, Hydropogon and Hydropogonella. Bryologist 46: 25-46.

KEY TO THE GENERA

1 Rather rigid plants with erect, short often club-shaped branches. Leaves imbricate, apex acute-acuminate, margins revolute. Peristome simple . . . .
. . . . 1. *Hydropogon*
Flaccid plants, irregularly branched. Leaves distant, spreading, apex rounded, margins plane. Peristome absent . . . . 2. *Hydropogonella*

1. **HYDROPOGON** Brid., Bryol. Univ. 1: 769. 1826.
Type: H. fontinaloides (Hook.) Brid. (Grimmia fontinaloides Hook.)

Distribution: Colombia, Ecuador, Venezuela, Suriname, French Guiana; 1 species.

1. **Hydropogon fontinaloides** (Hook.) Brid., Bryol. Univ. 1: 770. 1826. – *Grimmia fontinaloides* Hook., Musci Exot. 1: 2. 1818. Type: Venezuela, Orinoco R., Humboldt & Bonpland s.n. (BM, NY).

Description see Musci I in Fl. Suriname 6: 238. 1964.

Distribution: Colombia, Ecuador, Venezuela, Suriname, French Guiana.

Ecology: Epiphytic on shrubs in rapids, not common in the Guianas. Probably this special habitat is the reason why this moss is confined to the Upper Amazon Basin as reported by Churchill (1991).

Note: Mägdefrau (1973) noticed that the short, erect branches developped in the dry season when the substrate was not submerged. On these club-shaped branches perichaetia and perigonia are produced.

2. **HYDROPOGONELLA** Cardot, Rev. Bryol. 22: 18. 1895.
Type: H. gymnostoma (Bruch. & Schimp.) Cardot (Fontinalis gymnostoma Bruch. & Schimp.)

Distribution: Venezuela, Guyana, Suriname, Bolivia, Brazil; 1 species.

1. **Hydropogonella gymnostoma** (Bruch. & Schimp.) Cardot, Rev. Bryol. 22: 18. 1895. – *Fontinalis gymnostoma* Bruch. & Schimp., Flora 22: 451. 1839. – *Cryptangium schomburgkii* Müll. Hal., Linnaea 17: 599. 1843 (nom. illeg.). Lectotype (Welch 1943): Guyana, R.H. Schomburgk s.n. (BM, NY). – Fig. 115

Hydropogonella gymnostoma (Bruch. & Schimp.) Cardot fo. *obtusifolia* P.W. Richards, Kew Bull. 1934: 327. Type: Guyana, Richards 845 (BM, NY).

Slender, dullgreen plants growing submerged in thin mats. Stems elongate, creeping, irregularly branched; branches short, distantly foliate. Leaves spreading, flaccid, obovate, oval or oblong, to 2.3 mm long, apex rounded acute or obtuse and short-acuminate, margins entire, plane, costa absent, laminal cells thin-walled, rhomboidal or hexagonal, to 40 μm long and 15 μm wide, marginal cells rectangular, alar cells quadrate to rectangular, thinwalled, not forming a distinct group. Monoicous. Sporophyte not seen, description after Welch (1943): perigonia small, budlike; perichaetial leaves imbricate, acuminate, sheathing, apices spreading, capsule sessile, immersed to emergent, subcylindric, 0.8-1 mm long, operculum plane, rostellate, peristome absent. Calyptra conic, slightly laciniate at base, covering upper part of capsule.

Distribution: Venezuela, Guyana, Suriname, Bolivia, Brazil.

Ecology: On submerged rocks, shrubs, overhanging branches along rivers; not common.

344

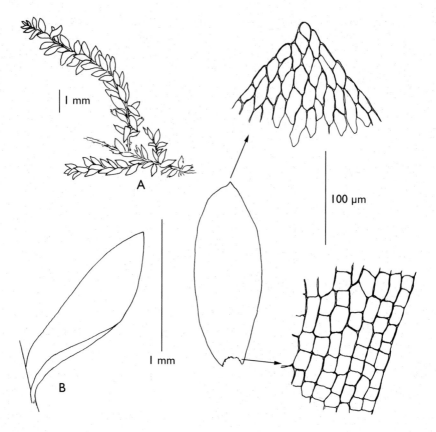

Fig. 115. *Hydropogonella gymnostoma* (Bruch. & Schimp.) Cardot: A, part of stem with branches; B, leaves (A-B, Florschütz-de Waard & Zielman 5272).

Specimens examined: Suriname: Kabalebo R., Davis Falls, Florschütz-de Waard & Zielman 5182, 5272 (L); Coppename R., 20 km downstream of Kabalebo airstrip, Florschütz & Maas 2649c (L).

30. HYPNACEAE

Description see Musci III in Fl. Guianas, ser. C, 1: 439. 1996.

Distribution: Worldwide; in the Guianas 7 genera.

LITERATURE

Buck, W.R. 1984. Taxonomical and nomenclatural notes on West Indian Hypnaceae. Brittonia 36: 178-183.

Buck, W.R. 1987. Notes on Asian Hypnaceae and associated taxa. Mem. New York Bot. Gard. 45: 519-527.

Ireland, R.R. 1992. The moss genus Isopterygium in Latin America. Trop. Bryol. 6: 111-132.

Robinson, H. 1965. Notes on Oreoweisia and Hypnella from Latin America. Bryologist 68: 331-334.

KEY TO THE GENERA

1   Midleaf cells oval to elongate-hexagonal, lax, 10-40 μm wide 2
      Midleaf cells linear, narrow, less than 10 μm wide 3

2   Midleaf cells 30-100 μm long, 10-18 μm wide. Capsule with conic, short-rostrate or apiculate operculum, exostome teeth not furrowed . . . .
      . . . . *7. Vesicularia*
      Midleaf cells 100-200 μm long, 10-40 μm wide. Capsule with slenderly rostrate operculum, exostome teeth with median furrow . . . .
      . . . . see *26. LEUCOMIACEAE*

3   Leaf cells smooth (sometimes with a luminous spot at cell ends, but in side-view not prorulose). 4
      Leaf cells with papillae or prorulae (in side-view visible as projections) 6

4   Branch leaves dimorphous, dorsal and lateral leaves ovate, asymmetric to falcate, complanate-spreading, ventral leaves narrower, subsymmetric, erect . . . . *6. Rhacopilopsis*
      Branch leaves not dimorphous 5

5   Plants subpinnately branched, leaves usually strongly falcate and homomallous, serrate at apex. Perichaetia conspicuous, to 3.5 mm high . . . .
      . . . . *2. Ectropothecium*
      Plants irregularly branched, leaves subsymmetric, complanate or erect-spreading, bluntly serrulate at apex. Perichaetia small, to 1.5 mm high . . .
      . . . . *3. Isopterygium*

6   Leaf apex blunt or truncate, leaf cells with 1-4 papillae. . . . . . . *5. Phyllodon*
      Leaf apex acute or acuminate, leaf cells prorulose 7

7   Plants with prostrate secondary stems, not stipitate. Leaf cells prorulose at both cell ends . . . . *1. Chryso-hypnum*
      Plants with ascending, curved secondary stems with stipitate base. Leaf cells prorulose at distal cell ends only . . . . *4. Mittenothamnium*

1. **CHRYSO-HYPNUM** Hampe, Bot. Zeitung (Berlin) 28: 35. 1870.
Type: C. patens Hampe

Description see Musci III in Fl. Guianas, ser. C, 1: 441. 1996.

Distribution: Pantropical; in the Guianas 1 species.

1. **Chryso-hypnum diminutivum** (Hampe) W.R. Buck, Brittonia 36: 182. 1984. – *Hypnum diminutivum* Hampe, Linnaea 20: 86. 1847. – *Mittenothamnium diminutivum* (Hampe) E. Britton, Bryologist 17: 9. 1914. Type: Venezuela, Caracas, Moritz 20 (BM).

Description see Musci III in Fl. Guianas, ser. C, 1: 442. 1996.

Distribution: Southern U.S.A., Mexico, West Indies, C and tropical S America.

Ecology: Epiphytic on bark of trees, sometimes on rocks or terrestrial, preferably in light habitats.

2. **ECTROPOTHECIUM** Mitt., J. Linn. Soc., Bot. 10:180. 1868.
Type: E. tutuilum (Sull.) Mitt. (Hypnum tutuilum Sull.)

Descripion see Musci III in Fl. Guianas, ser. C, 1: 444. 1996.

Distribution: Pantropical; in the Guianas 1 species.

1. **Ectropothecium leptochaeton** (Schwägr.) W.R. Buck, Brittonia 35: 311. 1983. – *Hypnum leptochaeton* Schwägr., Sp. Musc. Frond. Suppl. 1(2): 296. 1816. Type: French Guiana, Cayenne, Richard s.n. (PC).

Description and synonymy see Musci III in Fl. Guianas, ser. C, 1: 444. 1996.

Distribution: Mexico, West Indies, C and tropical S America.

Ecology: Epiphytic, occasionally on stones; rather common in moist rainforest.

3. **ISOPTERYGIUM** Mitt., J. Linn. Soc., Bot. 12: 21. 1869.
Type: I. tenerum (Sw.) Mitt. (Hypnum tenerum Sw.).

Description see Musci III in Fl. Guianas, ser. C, 1: 446. 1996.

Distribution: Worldwide; in the Guianas 2 species.

KEY TO THE SPECIES

1 Leaves slenderly lanceolate, 3-4 times as long as broad, usually symmetric;
   apex acute to gradually acuminate. Mature capsules erect or inclined . . . .
   . . . . .1. I. subbrevisetum
   Leaves ovate-lanceolate, 2-3 times as long as broad, usually asymmetric to
   subfalcate; apex acuminate. Mature capsules horizontal to pendent . . . . . .
   . . . . .2. I. tenerum

1. **Isopterygium subbrevisetum** (Hampe) Broth. in Engler & Prantl,
   Nat. Pflanzenfam. 1(3): 1081. 1908. – *Hypnum subbrevisetum* Hampe,
   Vidensk. Meddel. Naturhist. Foren. Kjøbenhavn ser. 3, 6: 165. 1875.
   Type: Brazil, vicinity of Rio de Janeiro, Glaziou 6356 (BM).

Description see Musci III in Fl. Guianas, ser. C, 1: 447. 1996.

Distribution: West Indies, C and tropical S America.

Ecology: Epiphytic on bark of trees and logs in moist forest; in the
Guianas rare.

2. **Isopterygium tenerum** (Sw.) Mitt., J. Linn. Soc., Bot. 12: 499. 1869.
   – *Hypnum tenerum* Sw., Fl. Ind. Occ. 3: 1817. 1806. Type: Jamaica,
   Swartz 2719 (BM).

Description and synonymy see Musci III in Fl. Guianas, ser. C, 1: 449. 1996.

Distribution: Southern U.S.A., Mexico, West Indies, C and tropical
S America.

Ecology: Epiphytic, sometimes terrestrial; common in open areas in
moist forest, seldom collected in French Guiana.

Note: *Isopterygium tenerifolium* Mitt. reported from the Guianas (Ireland
1992, Buck 2003) is mainly different from *I. tenerum* in size. The Buck
collection from French Guiana fits largely within the variation of *I. tenerum*.

4. **MITTENOTHAMNIUM** Henn., Hedwigia 41 (Beibl.): 225. 1902, nom. cons.
Type: M. reptans (Hedw.) Cardot (Hypnum reptans Hedw.)

Microthamnium Mitt., J. Linn. Soc., Bot. 12: 503. 1869, nom. illeg.

Description see Musci III in Fl. Guianas, ser. C, 1: 451. 1996.

D i s t r i b u t i o n : Neotropical, tropical Africa and India; in the Guianas 1 species.

1. **Mittenothamnium reptans** (Hedw.) Cardot, Rev. Bryol. 40: 21. 1913. – *Hypnum reptans* Hedw., Sp. Musc. Frond.: 264. 1801. Type: Jamaica, 'et Insulae meridionales', Swartz s.n. (G).

Description see Musci III in Fl. Guianas, ser. C, 1: 451. 1996.

D i s t r i b u t i o n : Mexico, West Indies, C and tropical S America.

E c o l o g y : Epiphytic in open vegetations at higher altitudes.

5. **PHYLLODON** Schimp., Bryol. Eur. 5: 60. 1851.
Type: P. truncatulus (Müll. Hal.) W.R. Buck (Hypnum truncatulum Müll. Hal.)

Glossadelphus M. Fleisch., Musc. Buitenzorg 4: 1351. 1923.

Description see Musci III in Fl. Guianas, ser. C, 1: 453. 1996.

D i s t r i b u t i o n : Pantropical; in the Guianas 1 species.

1. **Phyllodon truncatulus** (Müll. Hal.) W.R. Buck, Mem. New York Bot. Gard. 45: 519. 1987. – *Hypnum truncatulum* Müll. Hal., Syn. Musc. Frond. 2: 263. 1851. – *Glossadelphus truncatulus* (Müll. Hal.) M. Fleisch., Musc. Buitenzorg 4: 1352. 1923. Type: Peru, Pöppig s.n. (BM).

Description and synonymy see Musci III in Fl. Guianas, ser. C, 1: 454. 1996.

D i s t r i b u t i o n : West Indies, C and tropical S America.

Ecology: On stones and tree bases in moist forest; rare, only known from French Guiana.

Note: This species was previously placed in the genus *Glossadelphus* (SEMATOPHYLLACEAE). It is also described as *Hypnella* (Robinson 1965) and by some authors it still belongs in the SEMATOPHYLLACEAE. Buck (1987) transferred *Glossadelphus truncatulus* with some closely related species to *Phyllodon* and placed it in the HYPNACEAE.

6. **RHACOPILOPSIS** Renauld & Cardot, Rev. Bryol. 27: 47. 1900.
   Type: R. dupuisii (Renauld & Cardot) Renauld & Cardot (Cyathophorum dupuisii Renauld & Cardot) [= R. trinitensis (Müll. Hal.) E. Britton & Dixon]

Description and synonymy see Musci III in Fl. Guianas, ser. C, 1: 456. 1996.

Distribution: Neotropics, tropical Africa; in the Guianas 1 species.

1. **Rhacopilopsis trinitensis** (Müll. Hal.) E. Britton & Dixon, J. Bot. 60: 86. 1922. – *Hypnum trinitense* Müll. Hal., Syn. Musc. Frond. 2: 284. 1851. – *Ectropothecium trinitense* (Müll. Hal.) Mitt., J. Linn. Soc., Bot. 12: 86. 1869. Type: Trinidad, Mt. Tocuche, Crüger s.n. (BM).

Description and synonymy see Musci III in Fl. Guianas, ser. C, 1: 456. 1996.

Distribution: Mexico, West Indies, C and tropical S America.

Ecology: Epiphytic on trunks and branches, rather common in forest at higher altitudes.

7. **VESICULARIA** (Müll. Hal.) Müll. Hal., Bot. Jahrb. Syst. 23: 330. 1896 (non P.A. Micheli ex G. Targioni Tozetti 1826), nom. cons. prop. – *Hypnum* subsect. *Vesicularia* Müll. Hal., Syn. Musc. Frond. 2: 233. 1851.
   Type: V. vesicularis (Schwägr.) Broth. (Hypnum vesiculare Schwägr.)

Distribution: Pantropical; in the Guianas 1 species.

1. **Vesicularia vesicularis** (Schwägr.) Broth. in Engler & Prantl, Nat. Pflanzenfam. 1(3): 1094. 1908. – *Hypnum vesiculare* Schwägr., Sp. Musc. Frond. Suppl. 2(2): 167. 1827. Type: Jamaica, Richmond, Reider s.n. (L).

Description and synonymy see Musci III in Fl. Guianas, ser. C, 1: 459. 1996.

Distribution: Neotropical.

Ecology: On tree bases and logs, on stones and terrestrial; very common in the understory of moist forest.

Note: *Vesicularia eligiana* W.R. Buck, described from French Guiana (Buck 2003) belongs in *Leucomium* considering the leaf size, cell size and capsule with furrowed exostome teeth. Microscopically it is identical with *Leucomium strumosum* (expression mosenii). The type (Buck 33113) has ovate-oblong, short-acuminate leaves with firm-walled cells and a capsule with distinct furrowed peristome teeth. See also note under *Leucomium strumosum* (see Musci III in Fl. Guianas, ser. C, 1: 370. 1996).

31. SEMATOPHYLLACEAE

Description see Musci III in Fl. Guianas, ser. C, 1: 384. 1996.

Distribution: Pantropical; in the Guianas 9 genera.

LITERATURE

Buck, W.R. 1982. On Meiothecium. Contr. Univ. Michigan Herb. 15: 137-140.

Buck, W.R. 1986. Wijkia in the New World. Hikobia 9: 297-303.

Buck, W.R. 1988. Donnellia resurrected and refound in Florida after 110 years. Bryologist 91: 134-135.

Buck, W.R. 1990. Contributions to the moss flora of Guyana. Mem. New York Bot. Gard. 64: 184-196.

Buck, W.R. 1998. Sematophyllaceae. In Pleurocarpous mosses of the West Indies. Mem. New York Bot. Gard.82: 339-379.

Churchill, S.P. 2001. Mosses. In Gradstein, S.R. *et al.* Guide to the Bryophytes of tropical America. Mem. New York Bot. Gard. 86: 240-577.

Churchill, S.P. & E. Linares C. 1995. Sematophyllaceae. In Prodromus Bryologiae. Novo Granatensis. Biblioth. José Jerónimo Triana 12: 751-785.

Crum, H. 1977. Meiothecium, a new record for North America. Bryologist 80: 188-193.

Florschütz-de Waard, J. 1992. A revision of the genus Potamium. Trop. Bryol. 5: 109-121.

KEY TO THE GENERA

1   Alar cells conspicuously inflated and curved to the insertion, 70-170 μm long. Leaf apex tubulose by incurved margins . . . . *1. Acroporium*
     Alar cells, if inflated, not over 100 μm long. Leaf apex not tubulose 2

2   Leaf cells papillose 3
     Leaf cells smooth 4

3   Leaf cells pluripapillose . . . . *7. Taxithelium*
     Leaf cells unipapillose . . . . *8. Trichosteleum*

4   Alar cells quadrate, in basal row sometimes oval, seldom inflated 5
     Alar cells in basal row distinctly inflated 6

5   Median leaf cells elongate-rhomboidal, regularly incrassate with fusiform lumen . . . . *3. Donnellia*
     Median leaf cells linear, thin-walled. . . . . *5. Pterogonidium*

6   Plants bipinnately branched; branch leaves considerably smaller than stem leaves . . . . *9. Wijkia*
     Plants irregularly to subpinnately branched; branch and stem leaves little differentiated in size 7

7   Leaf margins narrowly reflexed nearly to apex. Peristome single, exostome teeth remote . . . . *4. Meiothecium*
     Leaf margins plane or only loosely reflexed. Peristome double, exostome teeth closely spaced 8

8   Plants with capsules 9
     Plants without capsules. 10

9   Capsules suberect; exostome teeth pale, slender, at inner side without traverse lamellae, endostome with filiform, often fragile segments on a low basal membrane . . . . *2. Colobodontium*
     Capsules inclined to pendent; exostome teeth brown, firm, at inner side thickened with high transverse lamellae, endostome conspicuous with broad, keeled segments on a high basal membrane . . . . *6. Sematophyllum*

352

10 Leaves oval-ovate, small, not over 1.5 mm long 11
Leaves larger. . . . . 6. *Sematophyllum* p.p.

11 Leaves concave in central part, midleaf cells regularly incrassate with
fusiform lumina . . . . 6. *Sematophyllum* p.p.
Leaves flat, midleaf cells thin-walled or irregularly incrassate . . . .
. . . . 2. *Colobodontium*

1. **ACROPORIUM** Mitt., J. Linn. Soc., Bot. 10: 182. 1868.
Type: A. brevicuspidatum Mitt.

Description see Musci III in Fl. Guianas, ser. C, 1: 386. 1996.

D i s t r i b u t i o n : Pantropical; in the Guianas 1 species.

1. **Acroporium pungens** (Hedw.) Broth. in Engler, Nat. Pflanzenfam.
ed. 2, 11: 436. 1925. – *Hypnum pungens* Hedw., Sp. Musc. Frond. 1:
237. 1801. Type: Jamaica, Swartz s.n. (G).

Description and synonymy see Musci III in Fl. Guianas, ser. C, 1:
387. 1996.

D i s t r i b u t i o n : Pantropical.

E c o l o g y : Epiphytic on tree trunks and branches in undergrowth and
lower canopy of open forest types, also on logs; rather common from
lowland to higher altitudes.

2. **COLOBODONTIUM** Herzog, Memoranda Soc. Fauna Fl. Fenn.
27: 110. 1952.
Type: C. aciculare Herzog

Maguireella W.R. Buck, Mem. New York Bot. Gard. 64: 193. 1990.
Type: M. vulpina (Mont.) W.R. Buck (Neckera vulpina Mont.)

Description see Musci III in Fl. Guianas, ser. C, 1: 394. 1996, as *Potamium*.

D i s t r i b u t i o n : Tropical S America; in the Guianas 2 species.

N o t e s : In the revision of the genus *Potamium* (Florschütz-de Waard
1992) I recognized 2 species of *Potamium*: *P. vulpinum* (Mont.) Mitt. and

P. deceptivum Mitt. *P. lonchophyllum* (Mont.) Mitt. and *P. pacimoniense* Mitt. were transferred to the genus *Sematophyllum*, based on the typical *Sematophyllum* peristome (Florschütz-de Waard 1990); all other species described as *Potamium* are synonyms. I designated *P. vulpinum* as lectotype (Florschütz-de Waard 1992).
Meanwhile Buck (1990) had indicated *P. lonchophyllum* as lectotype of *Potamium*. It is not possible to ignore this lectotypification, but I still see the place of this species in *Sematophyllum*. Consequently *Potamium* should be considered as synonym of *Sematophyllum*.
Buck (1990) introduced *Maguireella vulpina* as a new combination for *Potamium vulpinum*. But an earlier genus name was found in a collection from Colombia: *Colobodontium aciculare* Herzog. Churchill & Linares (1995) concluded that this species is identical with *Potamium vulpinum* and made the combination *Colobodontium vulpinum*. It was considered as a monotypic genus with *Potamium deceptivum* as a synonym.
I have a different interpretation and recognize 2 species.

KEY TO THE SPECIES

1   Leaves with broad-acute to short-acuminate apex. Exostome teeth linear-lanceolate, gradually tapering, papillose throughout, endostome segments usually rudimentary . . . . *1. C. deceptivum*
Leaves with obtuse to rounded-acute apex. Exostome teeth lanceolate with a broad base, rather quickly narrowed to a slender, papillose upper part, often broken and then teeth short and blunt; endostome segments slender, usually persistent. . . . . *2. C. vulpinum*

1. **Colobodontium deceptivum** (Mitt.) J. Florsch., comb. nov. – *Potamium deceptivum* Mitt., J. Linn. Soc., Bot. 12: 473. 1869. Type: Brazil, Rio Negro, Spruce s.n., hb. Mitten 826 (H, NY).

Description and synonymy see Musci III in Fl. Guianas, ser. C, 1: 394. 1996, as *Potamium deceptivum*.

Distribution: Colombia, Venezuela, Brazil, Guyana, Suriname.

Ecology: On bark of trees or decaying wood in temporarily submerged areas.

Note: The new combination in *Colobodontium* was not yet made for *Potamium deceptivum* because Buck (1990) did not recognize this as a separate species and reported it as a synonym of *Maguireella vulpina*.

2. **Colobodontium vulpinum** (Mont.) Churchill & Linares, Bibliot. José Jeronimo Triana 12: 759. 1995. – *Neckera vulpina* Mont., Ann. Sci. Nat., Bot. sér. 2, 3: 204. 1835. – *Potamium vulpinum* (Mont.) Mitt., J. Linn. Soc., Bot. 12: 473. 1869. – *Maguireella vulpina* (Mont.) W.R. Buck, Mem. New York Bot. Gard. 64: 193. 1990. Type: French Guiana, Sources du Jary, Leprieur s.n. (PC).

Sematophyllum maguireorum W.R. Buck, Mem. New York Bot. Gard. 64: 194. 1990. Type: Brazil, Río Cauaburí, Maguire *et al.* 60142 (NY).

Description and synonymy see Musci III in Fl. Guianas, ser. C, 1: 397. 1996, as *Potamium vulpinum*.

Distribution: Amazonian Basin, Guianas.

Ecology: On stones, roots and overhanging branches in temporarily submerged areas; locally common.

Note: *Sematophyllum maguireorum*, as described by Buck (1990), is identical with *C. vulpinum*. The type and syntypes mentioned are all representatives of the latter species.

3. **DONNELLIA** Austin, Bull. Torrey Bot. Club 7: 5. 1880. Type: D. floridana Austin, nom. illeg. [= D. commutata (Müll. Hal.) W.R. Buck]

Meiothecium sect. *Pterogonidiopsis* Broth. in Engler & Prantl, Nat. Pflanzenfam. 1(3): 1101. 1908. *Meiotheciopsis* Broth. in Engler & Prantl, Nat. Pflanzenfam. 1(3): 1105. 1908.

Slender to medium-sized plants growing in thin mats. Stems creeping, irregularly branched, branches prostrate or ascending. Leaves concave, ovate-oblong to ovate-lanceolate, apex gradually acuminate, margins entire to serrulate, costae short and double or absent; laminal cells smooth, firm-walled, elongate-rhomboidal to long-hexagonal, alar cells rounded quadrate, in the basal row oval, little inflated. Autoicous. Seta short, smooth; capsule suberect, cylindric, operculum long-rostrate; peristome double but endostome strongly reduced and difficult to demonstrate, exostome teeth closely spaced, blunt, with prominent crosswalls, trabeculate at base. Calyptra cucullate.

Distribution: Tropical America and Africa; in the Guianas 1 species.

1. **Donnellia commutata** (Müll. Hal.) W.R. Buck, Bryologist 91: 134.
   1988. – *Neckera commutata* Müll. Hal., Bot. Zeitung (Berlin) 15:
   385. 1857. – *Meiothecium commutatum* (Müll. Hal.) Broth. in Engler
   & Prantl, Nat. Pflanzenfam. 1(3): 1101. 1908. – *Meiotheciopsis
   commutata* (Müll. Hal.) W.R. Buck, Contr. Univ. Michigan Herb. 15:
   138. 1982. Type: Brazil, near Rio de Janeiro, Beske s.n.

Description see Musci III in Fl. Guianas, ser. C, 1: 392. 1996, as
Meiothecium commutatum.

Distribution: Florida, Mexico, West Indies, C and tropical S America.

Ecology: Epiphytic on trunks and in canopy of open forests; in the
Guianas rare, only known from a few collections from Guyana and
French Guiana.

Note: Vegetatively this species is close to *Meiothecium boryanum*.
The place in the genus *Donnellia* with double peristome, as proposed
by Buck (1988), seemed doubtful; because endostome segments in
this species are scarcely visible I left it in *Meiothecium* (Musci III in
Fl. Guianas, ser. C, 1: 389. 1996). However, differences in peristome
structure made this decision unsatisfactory and therefore I accepted here
the transfer to *Donnellia*.

4. **MEIOTHECIUM** Mitt., J. Linn. Soc., Bot. 10: 185. 1868.
   Type: M. stratosum Mitt.

Description see Musci III in Fl. Guianas, ser. C, 1: 389. 1996.

Distribution: Pantropical; in the Guianas 1 species.

1. **Meiothecium boryanum** (Müll. Hal.) Mitt., J. Linn. Soc., Bot. 12:
   469. 1869. – *Neckera boryana* Müll. Hal., Syn. Musc. Frond. 2: 75.
   1851. Type: Dominican Republic, Bory s.n. (BM).

Description see Musci III in Fl. Guianas, ser. C, 1: 389. 1996.

Distribution: Mexico, West Indies, C and tropical S America.

Ecology: Epiphytic in exposed areas of the forest canopy.

5. **PTEROGONIDIUM** Müll. Hal., Bull. Herb. Boissier 5: 209. 1897. Type: P. pulchellum (Hook.) Müll. Hal. (Pterogonium pulchellum Hook.)

Description see Musci III in Fl. Guianas, ser. C, 1: 399. 1996.

Distribution: Neotropical; in the Guianas 1 species.

1. **Pterogonidium pulchellum** (Hook.) Müll. Hal. in Engler & Prantl, Nat. Pflanzenfam. 1(3): 1100. 1908. – *Pterogonium pulchellum* Hook., Musci Exot. 1: 4. 1818. Type: Colombia, Mt. Quindio near el Moral, Humboldt & Bonpland s.n. (BM).

Description and synonymy see Musci III in Fl. Guianas, ser. C, 1: 399. 1996.

Distribution: Mexico, West Indies, C and tropical S America.

Ecology: Epiphytic, in Suriname common in cultivated areas, frequently on bases of palm trees; seldom collected in Guyana and French Guiana. If temporarily inundated a growth form develops with elongate branches and flaccid leaves which are broader and rounded at apex (see also Florschütz-de Waard 1992).

6. **SEMATOPHYLLUM** Mitt., J. Linn. Soc., Bot. 8: 5. 1864. Type: S. demissum (Wilson) Mitt. (Hypnum demissum Wilson)

Potamium Mitt., J. Linn. Soc., Bot. 12: 472. 1869. Type: P. vulpinum (Mont.) Mitt. (Neckera vulpina Mont.)

Description see Musci III in Fl. Guianas, ser. C, 1: 402. 1996.

Distribution: Pantropical; in the Guianas 6 species.

KEY TO THE SPECIES

1 Leaves ovate, oval or semi-circular, 1-2 times as long as wide 2
   Leaves more elongate, over 3 times as long as wide. 4

2 Leaves 0.7-1.2(-1.5) mm long; laminal cells regularly incrassate with fusiform lumina. Plants of sun-exposed habitats . . . . . . *5. S. subpinnatum*
   Leaves 1-2 mm long; laminal cells thin-walled or irregularly incrassate. Plants of semi-aquatic habitats (if leaves smaller see *Colobodontium*) 3

3   Leaves firm, concave, in older leaves cochleariform; leaf apex rounded and apiculate or abruptly short-acuminate. Capsules on a slender seta (6-10 mm long) . . . . *1. S. cochleatum*
Leaves flaccid, not concave; leaf apex blunt or rounded acute. Capsules on a short, firm seta (2-3 mm long) . . . . *4. S. pacimoniense*

4   Leaves flaccid, oblong to linear with rounded or short-acuminate apex. Plants of semi-aquatic habitats . . . . *3. S. lonchophyllum*
Leaves firm, oblong-lanceolate with acute-acuminate apex. Plants of dryer habitats 5

5   Apex abruptly acuminate upper margins often broadly inflexed; laminal cells usually strongly incrassate . . . . *2. S. galipense*
Apex acute or gradually acuminate, upper margins plane or narrowly reflexed; laminal cells thin-walled to slightly incrassate . . . . *6. S. subsimplex*

1. **Sematophyllum cochleatum** (Broth.) Broth. in Engler, Nat. Pflanzenfam. ed. 2, 11: 433. 1925. – *Rhaphidostegium cochleatum* Broth., Bih. Kongl. Svenska Vetensk.-Akad. Handl. 21 Afd. 3(3): 51. 1895. Type: Brazil, São Paulo, Mosén 2 (H, NY).

Description and synonymy see Musci III in Fl. Guianas, ser. C, 1: 403. 1996.

Distribution: Colombia, Suriname, French Guiana, Brazil.

Ecology: Temporarily submerged on branches and stones along creeks and rivers; locally abundant.

2. **Sematophyllum galipense** (Müll. Hal.) Mitt., J. Linn. Soc., Bot. 12: 480. 1869. – *Hypnum galipense* Müll. Hal., Bot. Zeitung (Berlin) 6: 780. 1848. Type: Venezuela, Galipan, Funck & Schlim 345 (BM, G).

Description and synonymy see Musci III in Fl. Guianas, ser. C, 1: 405. 1996.

Distribution: Mexico, West Indies, C and tropical S America.

Ecology: On rocks and decaying wood, at higher altitudes; not common, not seen from French Guiana.

3. **Sematophyllum lonchophyllum** (Mont.) J. Florsch., Trop. Bryol. 3: 96. 1990. – *Hypnum lonchophyllum* Mont., Syll. Gen. Sp. Crypt. 10. 1856. – *Potamium lonchophyllum* (Mont.) Mitt., J. Linn. Soc., Bot. 12: 473. 1869. Type: French Guiana, Cayenne, Leprieur 1378 (PC).

358

Description and synonymy see Musci III in Fl. Guianas, ser. C, 1: 407. 1996.

Distribution: Colombia, Venezuela, Suriname, French Guiana, Brazil.

Ecology: On rocks, logs or terrestrial along rivers and flooded areas; not common, not known from Guyana.

Note: Buck (1990) provided a different description of the sporophyte of *S. lonchophyllum*, describing the capsule as very small, almost globular and pendent with furrowed exostome teeth. Based on the latter character he designated this species as lectotype of *Potamium*. However, the sporophytes of *S. lonchophyllum* that I have seen possess typical *Sematophyllum* capsules, characterized by exostome teeth with high transverse lamellae at inner side and a well-developped endostome with broad, keeled segments and cilia. The furrow on the outer side of the exostome teeth is a variable character. For a detailed discussion see Florschütz-de Waard (1992).

4. **Sematophyllum pacimoniense** (Mitt.) W.R. Buck, Mem. New York Bot. Gard. 64: 193. 1990. – *Potamium pacimoniense* Mitt., J. Linn. Soc., Bot. 12: 474. 1869. Lectotype (Florschütz-de Waard 1992): Venezuela, Rio Pacimoni, Uaiauca, Spruce s.n., hb. Mitten 829 (NY).

Distribution: Amazonian Brazil and Venezuela, Suriname.

Ecology: On vegetation along creek, just above water level and on wet rocks; in the Guianas rare.

5. **Sematophyllum subpinnatum** (Brid.) E. Britton, Bryologist 21: 28. 1918. – *Leskea subpinnata* Brid., Muscol. Recent. Suppl. 2: 54. 1812. Type: Hispaniola, 'ad arbores', Poiteau s.n., hb. Bridel 747 (B).

Sematophyllum allinckxiorum W.R. Buck, Mem. New York Bot Gard. 76, 3: 146. 2003. Type: French Guiana, Cr. Saint-Eloi, Buck 33106 (NY).
Sematophyllum squarrosum W.R. Buck, Mem. New York Bot. Gard. 76, 3: 146. 2003. Type: French Guiana, 1 km N of Eaux Claires, Buck 33051 (NY).

Description and synonymy see Musci III in Fl. Guianas, ser. C, 1: 411. 1996.

Distribution: Pantropical.

Ecology: Epiphytic on branches in sun-exposed habitats, on logs in dry savanna forest, occasionally on stones; common.

Note: *S. allinckxiorum* and *S. squarrosum* described from French Guiana (Buck 2003) are here considered to be forms of *S. subpinnatum*. As already commented in Musci III in Fl. Guianas, ser. C, 1: 413. 1996, this species has different phenotypes, apparently correlated with environmental conditions: in dry habitats curved branches with homomallous leaves, in more humid habitats erect branches with spreading, sometimes recurved leaves. Between these extremes all intermediates occur. The type of *S. squarrosum* (NY) is a very striking plant macroscopically by the extremely small and recurved leaves, but microscopically the plant is very similar to *S. subpinnatum*.

6. **Sematophyllum subsimplex** (Hedw.) Mitt., J. Linn. Soc., Bot. 12: 494. 1869. – *Hypnum subsimplex* Hedw., Sp. Musc. Frond.: 270. 1801. Type: India occidentalis, Swartz s.n.

Description and synonymy see Musci III in Fl. Guianas, ser. C, 1: 414. 1996.

Distribution: Mexico, West Indies, C and tropical S America, tropical Africa.

Ecology: Epiphytic on logs, tree bases and branches in understory and lower canopy of all types of rainforest; very common.

7. **TAXITHELIUM** Spruce ex Mitt., J. Linn. Soc., Bot. 12: 496. 1869. (Taxithelium Spruce, Cat. Musc. 14. 1867, nom. nud.). Lectotype (E.G. Britton 1920): T. planum (Brid.) Mitt. (Hypnum planum Brid.)

Description see Musci III in Fl. Guianas, ser. C, 1: 417. 1996.

Distribution: Pantropical; in the Guianas 3 species.

KEY TO THE SPECIES

1 Leaves lanceolate with little differentiated alar cells; midleaf cells with 3-6 papillae . . . . *3. T. pluripunctatum*
Leaves oval or ovate , with a distinct group of alar cells sometimes partly inflated; midleaf cells with 4-10 papillae 2

2   Branch leaves oval with broad-acute, rounded apex; alar cells quadrate in a conspicuous group of 10-25 cells....................*1. T. concavum*
Branch leaves ovate with acute-acuminate apex; alar cells few, irregular, in the basal row partly inflated.........................*2. T. planum*

1. **Taxithelium concavum** (Hook.) Spruce, Cat. Musc. 14. 1867. – *Hypnum concavum* Hook. in Kunth, Syn. Pl. 1: 63. 1822. Type: Venezuela, Rio Negro, San Carlos, Humboldt 34, hb. Hooker 3176 (BM).

Description and synonymy see Musci III in Fl. Guianas, ser. C, 1: 418. 1996.

D i s t r i b u t i o n : Amazon Basin, probably more widespread but then recorded as *T. planum*.

E c o l o g y : Common in wet habitats such as submerged rocks, logs and soil, and branches overhanging water.

N o t e : The very concave and blunt leaves and the conspicuous group of quadrate alar cells are distinct and constant characters to distinguish this species from *T. planum*. See also commment in Musci III in Fl. Guianas, ser. C, 1: 420. 1996.

2. **Taxithelium planum** (Brid.) Mitt., J. Linn. Soc., Bot. 12: 496. 1869. – *Hypnum planum* Brid., Muscol. Recent. Suppl. 2: 97. 1812. Type: Hispaniola, Poiteau s.n., hb. Brid. 819 (B), hb. Meyer 2 (GOET).

Description see Musci III in Fl. Guianas, ser. C, 1: 420. 1996.

D i s t r i b u t i o n : Pantropical.

E c o l o g y : Very common on all kinds of substrate in the understory of lowland rainforest.

3. **Taxithelium pluripunctatum** (Renauld & Cardot) W.R. Buck, Moscosoa 2: 60. 1983. – *Trichosteleum pluripunctatum* Renauld & Cardot, Bull. Soc. Roy. Bot. Belgique 29: 184. 1890. Type: Martinique, St. Marie, Bordaz s.n. (NY).

Description and synonymy see Musci III in Fl. Guianas, ser. C, 1: 422. 1996.

D i s t r i b u t i o n : West Indies, Brazil, Guianas.

E c o l o g y : Epiphytic in understory of lowland rainforest, also terrestrial or epilithic in moist areas.

361

8. **TRICHOSTELEUM** Mitt., J. Linn. Soc., Bot. 10: 181. 1868.
Type: T. fissum Mitt.

Description see Musci III in Fl. Guianas, ser. C, 1: 424. 1996.

Distribution: Pantropical; in the Guianas 4 species.

KEY TO THE SPECIES

1   Leaves lanceolate, apex acuminate.............................2
    Leaves ovate to elliptic, apex acute 4

2   Upper leaf margin serrate; papillae coarse, blunt, at base as broad as cell
    lumina.......................................*3. T. papillosum*
    Upper leaf margin entire or weakly serrulate; papillae small............3

3   Leaves 1.5-2 mm long, abruptly acuminate .... *1. T. bolivarense*
    Leaves to 1.5 mm long, gradually acuminate............ *2. T. intricatum*

4   Leaves ovate, 0.7-1.2 mm long; median laminal cells thin-walled, elongate-
    rhomboidal with distinct papillae...*4a. T. subdemissum* var. *subdemissum*
    Leaves oval-elliptic, 1-2 mm long; median laminal cells incrassate, flexuose-
    linear, papillae indistinct ....*4b. T. subdemissum* var. *subglabrum*

1. **Trichosteleum bolivarense** H. Robins., Acta Bot. Venez. 1: 78.
    1965. Type: Venezuela, Cerro Venamo, Steyermark & Dunsterville
    92253 (US).

Description see Musci III in Fl. Guianas, ser. C, 1: 425. 1996.

Distribution: Venezuela, the Guianas.

Ecology: Epiphytic on living trees and logs, also terrestrial on litter
over rocks; not common, only at higher altitudes.

Note: This species with inconspicuous papillae is sometimes difficult to
recognize. The large size of the concave, slenderly acuminate leaves and
the long, erect perichaetial leaves are good additional characters.

2. **Trichosteleum intricatum** (Thér.) J. Florsch., Trop. Bryol. 3: 98.
    1990. – *Acroporium intricatum* Thér., Ann. Bryol. 7: 159. 1934.
    Type: French Guiana, St. Jean de Maroni, Gouv. Rey s.n., coll.
    Galliot (NY, PC).

Description see Musci III in Fl. Guianas, ser. C, 1: 431. 1996.

Distribution: Brazil, the Guianas.

Ecology: Epiphytic in light forest, often shelving in horizontal rows from tree trunks.

3. **Trichosteleum papillosum** (Hornsch.) A. Jaeger, Ber. St. Gallischen Naturwiss. Ges. 1876-77: 419. 1878. – *Hypnum papillosum* Hornsch. in Mart., Fl. Bras. 1(2): 82. 1840. Type: Brazil, Minas Geraïs, Beyrich s.n., hb. Hooker (BM, NY).

Description and synonymy see Musci III in Fl. Guianas, ser. C, 1: 433. 1996.

Distribution: Brazil, Colombia, the Guianas. Perhaps this range will be wider if this species proves to be synonymous with the more widespread *Trichosteleum sentosum* (Sull.) A. Jaeger, a species from the West Indies with longer acuminate, sharply dentate apex and very tall papillae.

Ecology: Epiphytic on bark of trees and rotting logs; common in lowland rainforest.

4. **Trichosteleum subdemissum** (Besch.) A. Jaeger, Ber. St. Gallischen Naturwiss. Ges. 1876-77: 418. 1878. – *Rhaphidostegium subdemissum* Schimp. ex Besch., Ann. Sci. Nat., Bot. sér. 6, 3: 250. 1876. Type: Guadeloupe, L'Herminier s.n. (NY).

Trichosteleum hornschuchii (Hampe) A. Jaeger, Ber. St. Gallischen Naturwiss. Ges. 8: 418. 1878. – *Hypnum hornschuchii* Hampe, Icon. Musc. 9 (annotation). 1844, nom. illeg. – *Hypnum microcarpum* Hornsch. in Mart., Fl. Bras. 1(2): 84. 1840, nom. illeg. Type: Brazil, Pará, Martius s.n., hb. Hooker 880 (BM).

Description and synonymy see Musci III in Fl. Guianas, ser. C, 1: 427. 1996, as *Trichosteleum hornschuchii*.

Note : The name *T. subdemissum* replaces the illegitimate name *T. hornschuchii*.

4a. var. **subdemissum**

Description see Musci III in Fl. Guianas, ser. C, 1: 430. 1996, as *Trichosteleum hornschuchii* var. *hornschuchii*.

363

Distribution: Mexico, West Indies, C and tropical S America, tropical Africa.

Ecology: In the Guianas the typical variety is rather common in moist areas in mesophytic rainforest.

4b. var. **subglabrum** (J. Florsch.) J. Florsch., comb. nov. – *Trichosteleum hornschuchii* (Hampe) A. Jaeger var. *subglabrum* J. Florsch., Fl. Guianas Ser. C, 1: 430. 1996. Type: Guyana, Mabura Hill, Cornelissen & ter Steege 208 (L).

Description see Musci III in Fl. Guianas, ser. C, 1: 430. 1996, as *Trichosteleum hornschuchii* var. *subglabrum*.

Distribution: The Guianas; so far not recorded from other countries.

Ecology: This variety is restricted to temporarily submerged habitats. It could have been considered as a mere growthform, but the very different habit of the plants and the absence of intermediate forms support the decision to distinguish it as a variety.

9. **WIJKIA** H.A. Crum, Bryologist 74: 170. 1971.
Type: W. extenuata (Brid.) H.A. Crum (Hypnum extenuatum Brid.)

Acanthocladium sensu Broth.in Engler, Nat. Pflanzenfam. ed. 2, 11: 412. 1925.

Description see Musci III in Fl. Guianas, ser. C, 1: 436. 1996.

Distribution: Pantropical; in the Guianas 1 species.

1. **Wijkia costaricensis** (Dixon & E.B. Bartram) H.A. Crum, Bryologist 74: 170. 1971. – *Acanthocladium costaricense* Dixon & E.B. Bartram, J. Washington Acad. Sci. 21: 294. 1931. Type: Costa Rica, Prov. San José, Standley & Valerio 43395 (FH, NY).

Description see Musci III in Fl. Guianas, ser. C, 1: 436. 1996.

Distribution: Mexico, West Indies, C and tropical S America.

Ecology: Epiphytic in canopy and undergrowth of open forest at higher altitudes; rare in the Guianas, only known from Suriname and French Guiana.

32. STEREOPHYLLACEAE

Medium-sized plants growing in flat mats. Stems creeping, simple or irregularly branched, complanate or terete, pseudoparaphyllia filamentous or foliose. Leaves oblong, ovate or lanceolate, complanate to erect-spreading, apex obtuse to acuminate, costa single and strong or short and double; laminal cells linear to rhomboid, smooth, uni-papillose or prorate, alar cells differentiated, quadrate or rectangular, often in large areas, unequally distributed on either side of the costa. Autoicous. Seta elongate, smooth; capsule erect or pendent, ovoid to cylindric, operculum conic to long-rostrate; peristome double, exostome striolate below, papillose above, endostome with basal membrane and keeled segments, cilia usually present. Calyptra cucullate, smooth.

Distribution: Pantropical; in the Guianas 4 genera.

Notes: Previously the genera *Pilosium* and *Stereophyllum* were classified in the family PLAGIOTHECIACEAE (Musci II in Fl. Suriname 6: 351. 1986). Buck & Ireland (1985) restricted this family to the genus *Plagiothecium* and placed *Stereophyllum* in the family STEREOPHYLLACEAE together with *Entodontopsis*. *Pilosium* was transferred to the HOOKERIACEAE. In 1994 Ireland & Buck included *Pilosium* in the STEREOPHYLLACEAE, as a monogeneric subfamily PILOSIOIDEAE.

Ireland & Buck (1994) restricted the genus name *Stereophyllum* to the species with short laminal cells; the species reported for the Guianas as *Stereophyllum* (Musci II in Fl. Suriname 6: 352. 1986) were transferred to *Entodontopsis*.

LITERATURE

Buck, W.R. & R.R. Ireland. 1985. A reclassification of the Plagiotheciaceae. Nova Hedwigia 41: 89-125.

Ireland, R.R. & W.R. Buck. 1994. Stereophyllaceae. Fl. Neotr. Monogr. 65: 1-50.

KEY TO THE GENERA

1 Leaves dimorphic, dorsal leaves symmetric with little differentiated alar cells, lateral leaves cultriform with a large group of alar cells at one side of the costa; costa short or absent . . . . *3. Pilosium*
  Leaves uniform; costa strong, extending 1/3-3/4 leaf length 2

2   Laminal cells isodiametric to rhombic , smooth or unipapillose over lumina
. . . . *4. Stereophyllum*
Laminal cells elongate, smooth or prorulose by projecting cell ends. 3

3   Laminal cells linear, gradually tapering at cell ends, smooth. . . . .
. . . . *1. Entodontopsis*
Laminal cells elongate-rectangular, prorulose by projecting cell ends. . . . . .
. . . . *2. Eulacophyllum*

1. **ENTODONTOPSIS** Broth. in Engler & Prantl, Nat. Pflanzenfam. 1(3): 895. 1907.
Type: E. contorte-operculata (Müll. Hal.) Broth. (Hypnum contorte-operculatum Müll. Hal.)

Glossy plants in thin to dense mats. Stems 1-3 cm long, simple or sparingly branched, rhizoids in clusters on ventral side, pseudoparaphyllia filamentous. Stem and branch leaves ovate-lanceolate to oblong-lanceolate, sometimes lingulate, apex obtuse, rounded acute or short-acuminate, margins entire or serrulate in upper part, costa stout, extending 1/2-3/4 of leaf length; laminal cells thin-walled, linear-fusiform, smooth, alar cells quadrate or short-rectangular, collenchymatous, unequally distributed in usually triangular groups. Autoicous. Seta long, smooth; capsule erect or inclined, cylindric, operculum conic to short-rostrate, peristome with the characters of the genus.

D i s t r i b u t i o n : Pantropical; in the Guianas 3 species.

KEY TO THE SPECIES

1   Leaves oblong-lingulate, apex broad-acute to obtuse. . . . . *3. E. nitens*
Leaves lanceolate or oblong-lanceolate, apex acute-acuminate 2

2   Leaf margins serrulate in upper part, apex acute . . . *1. E. contorte-operculata*
Leaf margins entire, apex acuminate . . . . *2. E. leucostega*

1. **Entodontopsis contorte-operculata** (Müll. Hal.) Broth. in Engler & Prantl, Nat. Pflanzenfam. 1(3): 896 1907. – *Hypnum contorte-operculatum* Müll. Hal., Syn. Musc. Frond. 2: 682. 1851. – *Stereophyllum contorte-operculatum* (Müll. Hal.) Mitt., J. Linn. Soc., Bot. 12: 543. 1869. Type: Costa Rica, Feb.-Apr. 1847, Oersted s.n. (FH).

Entodontopsis rhabdodonta Cardot, Rev. Bryol. 37: 12. 1910. – *Stereophyllum rhabdodonta* (Cardot) Grout, Bryologist 48: 67. 1945. Type: Mexico, Jalisco, Pringle 15221 (NY).

Description see Musci II in Fl. Suriname 6: 356. 1986, as *Stereophyllum rhabdodonta*.

Distribution: Mexico, Costa Rica, Guatemala, Suriname.

Ecology: Terrestrial and on rock; rare, only one collection in Guyana.

2. **Entodontopsis leucostega** (Brid.) W.R. Buck & Ireland, Nova Hedwigia 41: 89. 1985. – *Leskea leucostega* Brid., Bryol. Univ. 2: 333. 1827. – *Stereophyllum leucostegum* (Brid.) Mitt., J. Linn. Soc., Bot. 12: 543. 1869. Type: Puerto Rico, Bertero s.n. (B).

Description see Musci II in Fl. Suriname 6: 354. 1986, as *Stereophyllum leucostegum*.

Distribution: Florida, Mexico, West Indies, C and tropical S America, India, Africa.

Ecology: On rock, tree roots and branches in moist savanna forest; locally common in Guyana.

3. **Entodontopsis nitens** (Mitt.) W.R. Buck & Ireland, Nova Hedwigia 41: 105. 1985. – *Stereophyllum nitens* Mitt., Trans. Linn. Soc. London 23: 51. 1860. Type: Africa, Fernando Po, Barter s.n. (NY).

Stereophyllum obtusum Mitt., J. Linn. Soc., Bot. 12: 542. 1869. Type: Peru, Tarapoto, Spruce 1313 (NY).

Description see Musci II in Fl. Suriname 6: 352. 1986, as *Stereophyllum obtusum*.

Distribution: Mexico, C and tropical S America, India, Africa.

Ecology: On wet rock, occasionally on tree bases near creek; not common.

2. **EULACOPHYLLUM** W.R. Buck & Ireland, Nova Hedwigia 41: 108. 1985.
Type: E. cultelliforme (Sull.) W.R. Buck & Ireland (Hypnum cultelliforme Sull.)

D i s t r i b u t i o n : Mexico, West Indies, C and northern S America; 1 species.

1. **Eulacophyllum cultelliforme** (Sull.) W.R. Buck & Ireland, Nova Hedwigia 41: 108. 1985. – *Hypnum cultelliforme* Sull., Proc. Amer. Acad. Arts 5: 289. 1861. – *Stereophyllum cultelliforme* (Sull.) Mitt., J. Linn. Soc., Bot. 12: 544. 1869. Type: Cuba, Wright s.n. (NY, not seen).   – Fig. 116

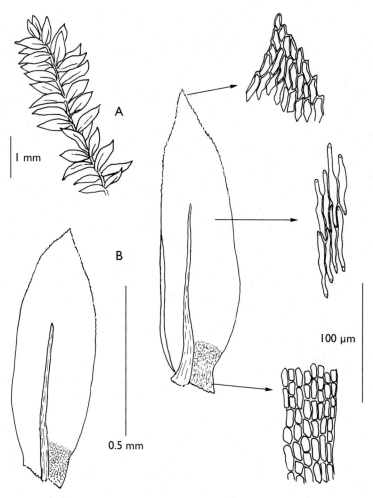

Fig. 116. *Eulacophyllum cultelliforme* (Sull.) W.R. Buck & Ireland: A, end of stem; B, leaves. (A-B, Buck 18532).

Small plants in thin mats. Stems 1-2 cm long, creeping, with clusters of rhizoids at ventral side, irregularly branched, microphyllous branches sometimes present, pseudoparaphyllia filamentous. Leaves distant, more or less complanate, ovate-oblong to oblong-lanceolate, asymmetric, sometimes cultriforme, to 1 mm long and 0.3 mm wide, apex acute, margins serrulate in upper part, costa firm, extending 1/2-2/3 of leaf length; laminal cells thin-walled, flexuose elongate-rectangular with blunt, prorulose cell ends, 35-50 μm long, shorter near apex; alar cells differentiated in a large group in the lower edge of the leaf base (on ventral side of stem), small, quadrate to rectangular, slightly incrassate. Autoicous. Seta elongate, smooth; capsule cernuous, ellipsoid to ovoid, operculum short-rostrate, oblique, peristome double, exostome teeth cross-striolate below, papillose above, endostome with keeled segments on a high basal membrane, cilia present. Calyptra cucullate. (Sporophyte description after Buck & Ireland 1985).

Distribution: Mexico, West Indies, C and northern S America.

Ecology: On rock in non-flooded moist forest; rare, only collected in French Guiana.

Specimen examined: French Guiana, Maripasoula, 6 km N of Saül, Buck 18532 (NY).

3. **PILOSIUM** (Müll. Hal.) M. Fleisch., Musc. Buitenzorg 4: 1158. 1923. – *Stereophyllum* sect. *Pilosium* Müll. Hal., Flora 83: 340. 1897. Lectotype (Grout 1945): P. chlorophyllum (Hornsch.) Müll. Hal. (Hypnum chlorophyllum Hornsch.)

Distribution: Mexico, West Indies, C and tropical S America; 1 species.

1. **Pilosium chlorophyllum** (Hornsch.) Müll. Hal., Flora 83: 340. 1897. – *Hypnum chlorophyllum* Hornsch. in Mart., Fl. Bras. 1(2): 89. 1840. Type: Brazil, prov. Minarum ad terram, sin. coll.

Description and synonymy see Musci II in Fl. Suriname 6: 358. 1986.

Distribution: Mexico, West Indies, C and tropical S America.

Ecology: On logs and tree bases, occasionally on twigs in lowland rainforest, also reported from upper stem and lower canopy; very common in the Guianas.

4. **STEREOPHYLLUM** Mitt., J. Linn. Soc., Suppl. Bot. 1: 117. 1859.
Type: S. indicum (Bél.) Mitt. (Pterygophyllum indicum Bél.)

Distribution: Worldwide, tropical and subtropical; in the Guianas
1 species.

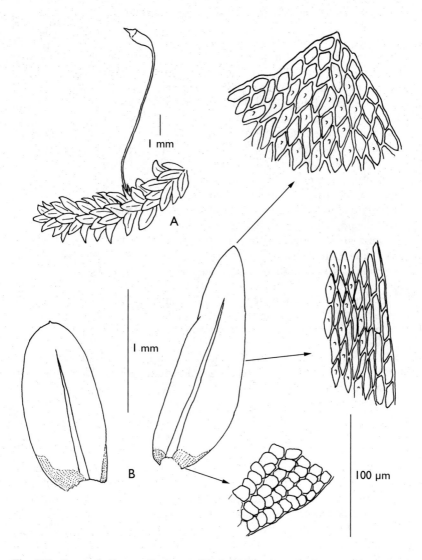

Fig. 117. *Stereophyllum radiculosum* (Hook.) Mitt.: A, end of stem with capsule;
B, leaves (A, Reese 14104; B, Lindeman & Stoffers *et al*. 387).

Note: This genus name was long used in a broader sense but is now restricted to the species with short, unipapillose laminal cells (Ireland & Buck 1994).

1. **Stereophyllum radiculosum** (Hook.) Mitt., J. Linn. Soc., Bot. 12: 542. 1869. – *Hookeria radiculosa* Hook., Musci Exot.1: 51. 1818. Type: Venezuela, Caripe, Humboldt & Bonpland s.n. (NY).
– Fig. 117

Dull-green plants in thin to dense mats. Stems creeping, 1-3 cm long, sparingly branched, rhizoids in clusters at ventral side of stem, pseudoparaphyllia filamentous. Stem and branch leaves similar, oblong-ovate to nearly lingulate, 1.5-2 mm long and 0.5-0.8 mm wide, apex obtuse to abruptly short-acuminate or apiculate, margins entire or serrulate near apex, costa stout, extending 2/3-3/4 of leaf length; laminal cells rhomboidal to elongate-hexagonal, ca. 35 µm long and 10 µm wide, often thickwalled, uni-papillose over lumen, often smooth along the margins; alar cells small, rounded-quadrate or transversely elongate, collenchymatous, forming an extensive group, at base often reaching the costa and extending far up the margins. Autoicous. Perigonia small, leaves ovate, acute, ecostate, cells rhomboidal, smooth. Perichaetia to 1.5 mm high, leaves lanceolate, acuminate, serrate at apex, costa feeble, cells flexuose-linear, smooth. Seta to 1.5 cm long; capsule inclined, ovoid, operculum rostrate; peristome with the characters of the family.

Distribution: Worldwide, tropical and subtropical.

Ecology: On rocks, logs and treebases in open savanna forest.

Specimens examined: Guyana: Makatui savanna, Lindeman & Stoffers *et al*. 387 (L); Potaro-Siparuni region, Yowan creek trail, alt. 520 m , Newton *et al*. 3645 (L, US).

33. THAMNOBRYACEAE

Primary stems creeping, secondary stems erect, pinnately branched, frondose, usually stipitate, branches often flagelliform at distal end; leaves ovate-oblong to ligulate, costa strong, extending beyond midleaf; laminal cells smooth or prorulose by projecting cell ends, rhomboidal to linear, short in apex. Dioicous. Seta elongate; capsule erect to horizontal, operculum rostrate, peristome double, exostome teeth 16, cross-striate at base, distally papillose, endostome well-developed with high basal membrane, 16 segments and cilia. Calyptra cuculate.

Distribution: Pantropical; in the Guianas 1 genus.

Note: THAMNOBRYACEAE, originally included in NECKERACEAE, are segregated as a separate family, characterized by dendroid or frondose habit, often with flagellate branches (Buck & Vitt 1986). Enroth (1994) argued for reunion of the families based on his study of the evolutionary history. New molecular data recognize the THAMNOBRYACEAE as a terrestrial clade of the NECKERACEAE. I have here followed the treatment of Sastre-De Jesús (1987).

LITERATURE

Buck, W.R. & D.H. Vitt. 1986. Suggestions for a new familial classification of Pleurocarpous Mosses. Taxon 35: 21-60.

Enroth, J. 1994. On the evolution and circumscription of the Neckeraceae. J. Hattori Bot. Lab. 76: 13-20.

Sastre-De Jesús, I. 1987. A revision of the Neckeraceae Schimp. and Thamnnobryaceae Marg. & Dur. in the Neotropics. Ph.D. dissertation, City Univ. New York.

Smith, D.K. 1985. Lectotypification and taxonomic notes for Porotrichum cobanense C.M.. Bryologist 88: 115-117.

1. **POROTRICHUM** (Brid.) Hampe, Linnaea 32: 154. 1863. – *Climacium* subgenus *Porotrichum* Brid., Bryol. Univ. 2: 275. 1827. Lectotype (M. Fleischer, Hedwigia 45: 85. 1906): P. longirostre (Hook.) Mitt. (Neckera longirostris Hook.)

Description see Musci II in Fl. Suriname 6: 283. 1986.
Additions to the description: Stipe leaves erect to squarrose-spreading; secondary stem- and branch leaves ovate-oblong to ovate-lanceolate.

Distribution: Pantropical; in the Guianas 4 species.

KEY TO THE SPECIES

1 Robust plants, irregularly branched. Leaves coarsely toothed (teeth composed of more than one cell). . . . . *3. P. stipitatum*
Smaller plants, bipinnately branched. Leaves serrulate to coarsely serrate (teeth generally composed of one cell). 2

2　Leaf cells prorulose by projecting cell ends . . . . *4. P. substriatum*
　　Leaf cells smooth 3

3　Stipe leaves spreading. Midleaf cells elongate, apical cells shorter, rhombic
　　. . . . . *1. P. korthalsianum*
　　Stipe leaves erect, appressed. Midleaf and apical cells rhombic . . . .
　　. . . . . *2. P. piniforme*

1. **Porotrichum korthalsianum** (Dozy & Molk.) Mitt., J. Linn. Soc.,
Bot. 12: 463. 1869.– *Neckera korthalsiana* Dozy & Molk., Prodr.
Fl. Bryol. Surinam. (appendix) 42: 9. 1854. Type: Venezuela, Rio
Catuche, ex. hb. Dozy & Molk. (L).

　　Porotrichum cobanense Müll. Hal., Bull. Herb. Boissier 5: 202. 1879.
　　Lectotype (Smith 1985): Guatamala, Alta Vera Paz, Türckheim s.n., ex. hb.
　　Levier (BM).

Description see Musci II in Fl. Suriname 6: 286. 1986.

D i s t r i b u t i o n : Mexico, West Indies, C and tropical S America.

E c o l o g y : Epiphytic on tree trunks and branches in the understory of
tropical rainforest at higher altitudes, occasionally on rock; in the Guianas
not common, not known from Guyana.

N o t e s : The more widely distributed *P. cobanense* proved to be identical
with *P. korthalsianum*, which name has priority (Sastre-De Jésus 1987).
Smith (1985) illustrated the great variability of the species.
The collections of *P. usagarum* Mitt. reported for French Guiana (Buck
2003) appear to me not to be different in any aspect from *P. korthalsianum*.

2. **Porotrichum piniforme** (Brid.) Mitt., J. Linn. Soc., Bot. 12: 465.
1869. – *Pilotrichum piniforme* Brid., Bryol. Univ. 2: 260. 1827. –
Pinnatella piniformis (Brid.) M. Fleisch., Hedwigia 45: 81. 1906.–
Homaliodendron piniforme (Brid.) Enroth, Nova Hedwigia 51: 55.
1990. Type: Guadeloupe, hb. Bridel 732 (B).

Description see Musci II in Fl. Suriname 6: 281. 1986, as *Pinnatella
piniformis*.

D i s t r i b u t i o n : Mexico, West Indies, C and tropical S America, tropical
Africa, Madagascar.

373

Ecology: Epiphytic on tree trunks and branches in the understory of tropical rainforest at higher altitudes; in the Guianas not common, not known from Guyana.

Note: The placement of this species in *Pinnatella* was questionable as already discussed in Musci II in Fl. Suriname 6: 281. 1986. It is here included in *Porotrichum* because erect stipe leaves, used as differentiating character for *Pinnatella*, also occur in *Porotrichum*. Buck (2003), following Enroth, placed the species in *Homaliodendron* because of the lack of a central strand in the stipe.

3. **Porotrichum stipitatum** (Mitt.) W.R. Buck, Mem. New York Bot. Gard. 76(3): 119. 2003. – *Trachyloma stipitatum* Mitt., J. Linn. Soc., Bot. 7: 156. 1863. Type: not seen.   – Fig. 118

Robust plants growing in loose tufts. Primary stem firm, short, sometimes tuberous thickened. Secondary stems erect, to 6 cm long, irregularly frondose; stipes long, stipe leaves distant, squarrose to erect-spreading, triangular, upward gradually larger; secondary stem leaves and branch leaves complanate-spreading, ovate-oblong, to 2.5 mm long and 1 mm wide, gradually shorter toward end of branches, apex rounded acute or short-acuminate, margin serrate-dentate below, coarsely toothed in upper part with teeth consisting of more than one cell, costa firm, extending ca. 2/3 of leaf length; laminal cells thin-walled or slightly thickened, prorate, elongate hexagonal to oblong-linear, 50-80 μm long and 5-7 μm wide, apical cells shorter, elongate-rhomboid, basal cells shorter and incrassate. Seta 1.5-2.6 cm long; capsule 1.5-3 mm long, exostome teeth lanceolate, cross-striolate at base, coarsely papillose above, endostome segments from a high basal membrane, keeled, perforate, as long as the teeth, cilia in groups of 1-3, papillose. (Sporophyte description after Buck 2003).

Distribution: Tropical Africa, French Guiana (first records in the New World).

Ecology: Epiphytic in understory of humid forest; rare, but locally abundant.

Specimens examined: French Guiana: Saül, alt. 300 m, Sipman 31715 (L); Mt. Galbao, alt. 450-500 m, Buck 25720, 33318 (NY).

374

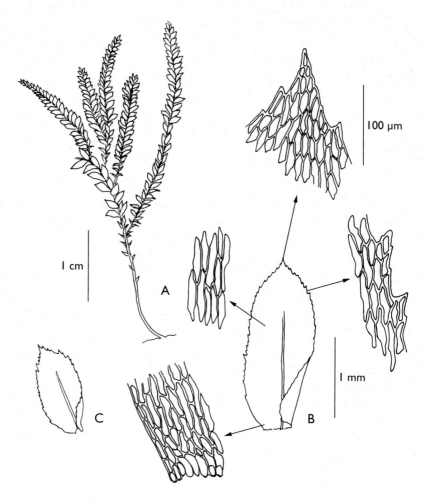

Fig. 118. *Porotrichum stipitatum* (Mitt.) W.R. Buck: A, secondary stem with branches; B, secondary stem leaf; C, upper branch leaf (A-C, Sipman 31715).

4. **Porotrichum substriatum** (Hampe) Mitt., J. Linn. Soc., Bot. 12: 463. 1869. – *Neckera substriata* Hampe, Ann. Sci. Nat., Bot. sér. 5, 5: 340. 1866. Type: Bogota, Lindig s.n. (not seen).

Porotrichum plicatulum Mitt., J. Linn. Soc., Bot. 12: 461. 1869. Type: Trinidad, Arima, Purdie s.n. (NY).

Description see Musci II in Fl. Suriname 6: 284. 1986, as *Porotrichum plicatulum*.

D i s t r i b u t i o n : Mexico, West Indies, C and tropical S America.

E c o l o g y : Epiphytic on bark of trees, occasionally on rocks; in Suriname and French Guiana rather common at higher altitudes, in Guyana collected only once.

34. THUIDIACEAE

Description see Musci III in Fl. Suriname 6: 371. 1996.

D i s t r i b u t i o n : Worldwide; in the Guianas 2 genera.

N o t e s : This family was previously divided in 2 subfamilies: THUIDIOIDEAE and CYRTOHYPNOIDEAE (Buck & Crum, 1990).
In a multi-character analysis Touw (2001a) rearranged the taxa in this family. In a taxonomic revision for tropical Asia he distinguished 7 genera, 4 dioicous and 3 monoicous (Touw 2001b). *Thuidium*, the most widespread of the dioicous genera, is characterized by the large size and strongly forked paraphyllia.
The monoicous genera are characterized by small size, simple or weakly branched paraphyllia and short setae. Of the monoicous genera only *Pelekium* is represented in the Neotropics. Touw (2001a) included *Cyrtohypnum* in the genus *Pelekium* because the type species proved to be identical with a species of *Pelekium*.

LITERATURE

Buck, W.R. & H. Crum, 1990. An evaluation of familial limits among the genera traditionally alligned with the Thuidiaceae and Leskeaceae. Contr. Univ. Michigan Herb. 17: 55-69.
Touw, A. 2001a. A review of the Thuidiaceae and a realignment of taxa traditionally accommodated in Thuidium sensu amplo (Thuidium Schimp., Thuidiopsis (Broth.) M. Fleisch., and Pelekium Mitt.), including Aequatoriella gen. nov. and Indothuidium gen. nov. J. Hattori Bot. Lab. 90: 167-209.
Touw, A. 2001b. A taxonomic revision of the Thuidiaceae of tropical Asia, the western Pacific, and Hawaii. J. Hattori Bot. Lab. 91: 1-136.

KEY TO THE GENERA

1   Plants small, first order branches 5 mm or less long. Paraphyllia simple, occasionally weakly branched. Laminal cells with small papillae on both surfaces. Autoicous, seta often rough . . . . *1. Pelekium*
Plants medium-sized, first order branches at least 5 mm long. Paraphyllia strongly branched. Laminal cells with rather large papillae on abaxial surface. Dioicous, seta smooth . . . . *2. Thuidium*

1. PELEKIUM Mitt., J. Linn. Soc., Bot. 10: 176. 1868.
Type: P. velatum Mitt.

> *Cyrto-hypnum* (Hampe) Hampe & Lorentz , Bot. Zeit. (Berlin) 27: 455. 1869.
> – *Hypnum* subgenus *Cyrto-hypnum* Hampe, Flora 50: 78. 1867. – *Thuidium* subgenus *Microthuidium* Limpr., Rabenh. Krypt.-Fl. ed. 2, 4: 822. 1895.
> Type: C. brachythecium (Hampe & Lorentz) Hampe & Lorentz (Hypnum brachythecium Hampe & Lorentz)

Slender plants, usually bipinnate, occasionally uni- or tripinnate. Stems mostly 3-5 cm long, paraphyllia unbranched, to 5 cells long, smooth to strongly papillose. Stem leaves triangular to subcordate, smooth or weakly plicate, costa strong, percurrent, margins often recurved below, crenulate to serrulate; branch leaves ovate or ovate-oblong, concave, apex acute or rounded, costa ca. 3/4 of leaf length; laminal cells pluripapillose at both surfaces, alar cells not differentiated. Autoicous. Perigonial leaves ovate-lanceolate, acuminate; perichaetial leaves ovate-oblong to triangular, long-acuminate, margins with unicellular teeth or multicellular cilia, cells elongate, smooth; seta to 2.5 cm long, smooth or mammillose, sometimes spinulose, capsule inclined, ovoid to cylindrical, peristome with the characters of the family.

D i s t r i b u t i o n : Pantropical, also subtropical in N and S America; in the Guianas 3 species.

KEY TO THE SPECIES

1   Branching unipinnate. Stem leaves not larger than branch leaves . . . .
. . . . *1. P. involvens*
Branching bi- or tripinnate. Stem leaves distinctly larger than branch leaves 2

2   Plants lax, branches catenulate (branch leaves distant, semi-circinate when dry); inner perichaetial leaves not or scarcely ciliate . . . . *2. P. scabrosulum*
Plants rigid, branch leaves loosely imbricate, incurved when dry; inner perichaetial leaves strongly ciliate . . . . *3. P. schistocalyx*

1. **Pelekium involvens** (Hedw.) Touw, J. Hattori Bot. Lab. 90: 203.
2001. – *Leskea involvens* Hedw., Sp. Musc. Frond. 218. 1801. –
Thuidium involvens (Hedw.) Mitt., J. Linn. Soc., Bot. 12: 575. 1869.
– *Cyrto-hypnum involvens* (Hedw.) W.R. Buck & H.A. Crum, Contr.
Univ. Michigan Herb. 17: 66. 1990. Type: Jamaica, Swartz s.n. (BM).

Description see Musci III in Fl. Suriname 6: 373. 1996, as *Cyrto-hypnum involvens*.

Distribution: Pantropical, extending into subtropical areas.

Ecology: Terrestrial, also on rotten logs; rare in the Guianas.

2. **Pelekium scabrosulum** (Mitt.) Touw, J. Hattori Bot. Lab. 90: 204.
2001. – *Thuidium scabrosulum* Mitt., J. Linn. Soc., Bot. 12: 574.
1869. – *Cyrto-hypnum scabrosulum* (Mitt.) W.R. Buck & H.A. Crum,
Contr. Univ. Michigan Herb. 17: 67. 1990. Type: sine loc., Humboldt
s.n., hb. Hooker 40 (BM).

Description see Musci III in Fl. Suriname 6: 375. 1996, as *Cyrto-hypnum scabrosulum*.

Distribution: West Indies, C and tropical S America.

Ecology: Epiphytic in canopy and open areas in light forest, also
terrestrial; common, seldom collected in Guyana.

3. **Pelekium schistocalyx** (Müll. Hal.) Touw, J. Hattori Bot. Lab. 90:
204. 2001. – *Hypnum schistocalyx* Müll. Hal., Syn. Musc. Frond. 2:
691. 1851. – *Thuidium schistocalyx* (Müll. Hal.) Mitt., J. Linn. Soc.,
Bot. 12: 575. 1869. – *Cyrto-hypnum schistocalyx* (Müll. Hal.) W.R.
Buck & H.A. Crum, Contr. Univ. Michigan Herb. 17: 67. 1990. Type:
Nicaragua, Matagalpa in Segovia, Oersted s.n., Dec. 1847 (BM).

Description see Musci III in Fl. Suriname 6: 377. 1996, as *Cyrto-hypnum schistocalyx*.

Distribution: Florida, Mexico, West Indies, C and tropical S America.

Ecology: Epiphytic on tree trunks in light forest, also on rocks and
logs; not common.

2. **THUIDIUM** Bruch &Schimp., Bryol. Europ. 5: 157. 1852.
   Lectotype (Grout, Moss Fl. N. Amer. 3: 174. 1932): T. tamariscinum
   (Hedw.) Schimp. (Hypnum tamariscinum Hedw.)

Description see Musci III in Fl. Suriname 6: 379. 1996.

Distribution: World-wide; in the Guianas 2 species.

KEY TO THE SPECIES

1 Laminal cells unipapillose . . . . *1. T. peruvianum*
   Laminal cells pluripapillose . . . . *2. T. tomentosum*

1. **Thuidium peruvianum** Mitt., J. Linn. Soc., Bot. 12: 578. – *Thuidium*
   delicatulum (Hedw.) Schimp. var. *peruvianum* (Mitt.) H.A. Crum,
   Bryologist 87: 211. 1984. Type: Peru, Mathews s.n. (NY).

Description see Musci III in Fl. Suriname 6: 379. 1996.

Distribution: Mexico, Guatemala, Guyana, Peru.

Ecology: On rock and soil at higher altitudes; rare, only one collection
known from Guyana (Appun, without locality).

2. **Thuidium tomentosum** Schimp., Mém. Soc. Sci. Nat. Cherbourg
   16: 237. 1872. Type: Mexico, Orizaba, F. Müller s.n. (NY).

Description and synonymy see Musci III in Fl. Suriname 6: 381. 1996.

Distribution: Mexico, West Indies, C and tropical S America.

Ecology: On stones, humus, logs and bark of trees in understory of
rainforest, rather common.

Order **Buxbaumiales**

Small plants, gametophytes often greatly reduced, sporophytes well-
developed. Perichaetia terminal, leaves strongly differentiated, capsules
immersed, large in proportion to the plant, peristome double, exostome
teeth much reduced, endostome membranous, consisting of a high 16- or
32-plicate cone.

35. DIPHYSCIACEAE

Distribution: Worldwide; 1 genus.

Note: *Diphyscium* was originally included in the family BUXBAUMIACEAE, characterized by strongly reduced, annual gametophytes and large, asymmetric-ovoid sporophytes. The relationship was mainly based on shape and structure of the capsule. The DIPHYSCIACEAE are distinct in the well-developed gametophytic plants with bistratose leaves and conspicuously differentiated perichaetial leaves and by the immersed capsules with annulus and complete peristome.

LITERATURE

Allen, B. 1996. Diphyscium pocsii an African species new to Honduras. Nova Hedwigia 62: 371-375.
Magombo, Z.L.K. 2003. Taxonomic revision of the moss family Diphysiaceae M. Fleisch. J. Hattori Bot. Lab. 94: 1-86.

1. **DIPHYSCIUM** D. Mohr, Observ. Bot. 34. 1803.
   Type: D. foliosum (Hedw.) D. Mohr (Buxbaumia foliosa Hedw.)

Small, perennial plants, growing scattered or in low mats. Stems short, erect. Leaves crowded,ovate-lanceolate, spatulate or lingulate, apex rounded-obtuse to broad-acute, margins entire or crenulate below, towards apex dentate or denticulate, costa strong, percurrent to short-excurrent. Upper laminal cells 2(-3)-stratose, rounded-quadrate or rounded-hexagonal, basal laminal cells unistratose, thin-walled, rectangular or oblong-hexagonal. Dioicous. Perigonial leaves little differentiated, perichaetial leaves conspicuous, much longer than vegetative leaves with long-excurrent costa. Capsules immersed, seta very short, capsule asymmetrically ovoid-conic, tapering to the mouth, operculum conic, annulus present, peristome with 16 short exostome teeth and a membranous 16-plicate, conic endostome. Calyptra conic, small.

Distribution: Worldwide; in the Guianas 1 species.

1. **Diphyscium longifolium** Griff., Calcutta J. Nat. Hist. 2:477. 1842. – *Webera longifolia* (Griff.) Broth. in Engler & Prantl, Nat. Pflanzenfam. 1(3): 664. 1904. Type: SE Asia, Khasia, Moosmai, on wet rocks, Griffith 771 (BM). – Fig. 119

380

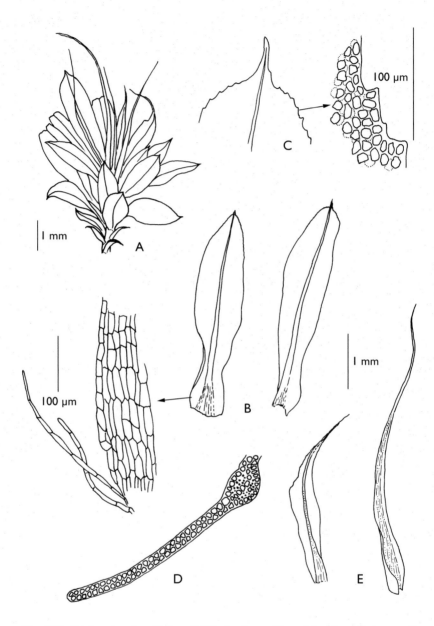

Fig. 119. *Diphyscium longifolium* Griff.: A, plant with perichaetium; B, leaves; C, leaf apex; D, cross-section upper lamina; E, perichaetial leaves (A-E, Allen 20868).

Diphyscium peruvianum Spruce ex Mitt., J. Linn. Soc., Bot. 12: 622.1869.
Type: Peru, Mt. Guayrapurina, Spruce 1492.
Diphyscium ulei Müll. Hal., Hedwigia 36: 334. 1897. Lectotype (Allen
1996): Brazil, Rio de Janeiro, Sept. 1893, Ule s.n. (H).

Small, dark green plants; stems 2-5 mm long, little divided. Leaves
crispate with incurved margins when dry, wide-spreading when moist,
oblong-spatulate, 4-6(-9) mm long, basal leaves shorter, apex broad-acute
or obtuse, mucronate or apiculate, in upper leaves often abruptly
acuminate, margins dentate in upper part, crenulate or smooth towards
base; costa distinct, percurrent or short excurrent, towards base broadened
and indistinct, in cross-section with 2 layers of stereids and a central layer
of guide cells; upper laminal cells from costa to margins bistratose,
irregularly rounded-quadrate, 10-14 µm in diameter, smooth, basal
laminal cells thin-walled, hyaline, rectangular, 36-70 µm long and
10-18 µm wide; axillary hairs abundant on abaxial side of the costa,
hyaline, septate. Dioicous. Perichaetia terminal on stem and branches,
leaves narrowly lanceolate to linear, to 6(-7) mm long, at apex often erose,
long-acuminate, costa stout, long-excurrent, flexuose and dentate at the
tip. Seta short, to 0.3 mm long, capsule oblong-ovoid, obliquely
asymmetric, 3-5 mm long, abruptly narrowed at base and mouth,
operculum obtuse, to 1 mm long, annulus consisting of 1-3 layers of
thick-walled cells; peristome with the characters of the family. Calyptra
cucullate, smooth. (Sporophyte description after Magombo 2003).

Distribution: Widespread in the tropics of C and S America, Asia and
Pacific Islands.

Ecology: On rocks and sandstone boulders or on soil in crevices
between rocks, usual at higher altitudes.

Collections examined: Guyana: Mt. Wokomung, moist tropical
forest, alt. 1200 m, Boom *et al.* 9233 (NY). Suriname: Tafelberg, on talus
boulder, alt. 530 m, Allen 20868 (MO, L).

Order **Polytrichales**

Plants robust, acrocarpous. Stems erect, usually simple; leaves long and
narrow with many longitudinal rows of green lamellae on upper surface,
costa strong. Sporophytes terminal, setae elongate, capsules symmetric,
erect or inclined; peristome single with 32 or 64 lingulate teeth joined
terminally to an epiphragm.

36. POLYTRICHACEAE

Small to robust plants. Stems simple, erect. Leaves with a unistratose sheathing base and a firm limb with numerous longitudinal, photosynthetic lamellae on upper surface, costa single, strong, narrow to nearly filling the limb, percurrent to excurrent. Dioicous. Seta elongate, capsule erect to inclined, terete or 2-4 sided, peristome single with 32 or 64 teeth, attached to an epiphragm.

Distribution: Worldwide; in the Guianas 1 genus.

LITERATURE

Smith, G.L. 1971. A conspectus of the genera of Polytrichaceae. Mem. New York Bot. Gard. 21(3): 1-83.
Merrill, G.L.S., 1994. Polytrichaceae. In Sharp, Crum & Eckel. Moss Flora of Mexico. Mem. New York Bot Gard. 69: 1070-1092.

1. **POLYTRICHUM** Hedw., Sp. Musc. Frond. 88. 1801.
Type: P. commune Hedw.

Medium-sized to large plants growing in loose tufts. Stems simple. Leaves with sheathing base and erect to wide-spreading limb, sheath obovate-oblong with hyaline margin, limb lanceolate, acute or acuminate, margin broadly infolded, entire or flat and sharply serrate, costa percurrent or excurrent. Dioicous. Perigonia terminal, leaves strongly differentiated. Perichaetia terminal, leaves with reduced limb. Seta stout, capsule erect to horizontal, 4-angled, operculum rostrate, peristome with 64 teeth attached to a thick, oval epiphragm. Calyptra cucullate, hairy.

Distribution: Worldwide; in the Guianas 1 species.

1. **Polytrichum juniperinum** Hedw., Sp. Musc. Frond. 89. 1801. Type: not designated. – Fig. 120

Medium-sized to large plants in loose tufts. Stems erect, 4-20 cm long. Leaves lanceolate, 5-11 mm long; sheath oblong, 2-3 mm long, cells elongate-rectangular to linear in basal part, 70-120 μm long and ca. 10 μm wide, narrower along margins forming a hyaline border, shorter in upper

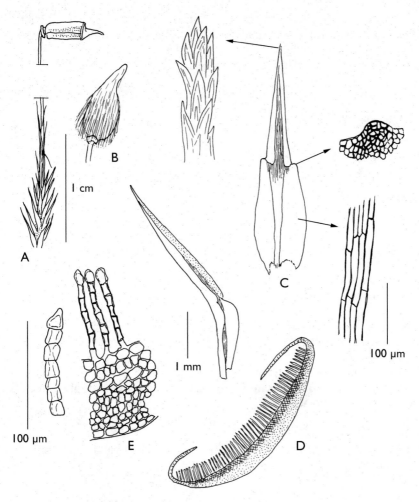

Fig. 120. *Polytrichum juniperinum* Hedw.: A, end of stem with perichaetium and capsule; B, capsule with calyptra; C, leaves; D, cross-section of canaliculate limb with lamellae; E, cross section of costa (A-B, Cleef 3335, Colombia; C-E, Gradstein 5397).

part, at shoulders roundish or transversely elongate, margins entire; costa strong, broadening towards the canaliculate limb, in limb covered with numerous longitudinal lamellae, excurrent in a strong, toothed awn; cells multilayered, small, rounded quadrate to transversely elongate, lamellae 6-7 cells high, marginal cells oval to pyriform. Dioicous. Perigonial leaves broad, forming a conspicuous rosette; perichaetial leaves hyaline with short limb and a slender awn. Seta 4-5 cm long; capsule inclined to horizontal, sharply 4-angled with distinctly delimited apophysis, operculum umbonate with a short beak, peristome with the characters of the genus. Calyptra hairy, covering the capsule.

Distribution: Worldwide.

Ecology: In the Guianas rare, only known from Guyana, at higher altitude.

Specimen examined: Guyana, N slope of Mt. Roraima, alt. 2000-2300 m, Gradstein 5397 (L).

NEW SPECIES AND TAXONOMIC CHANGES

New species:
Thamniopsis gradsteinii J. Florsch.

New combinations:
Sphagnum perichaetiale Hampe var. **tabuleirense** (Yano & H.A. Crum) H.R. Zielman
Colobodontium deceptivum (Mitt.) J. Florsch.
Trichosteleum subdemissum (Besch.) A. Jaeger var. **subglabrum** (J. Florsch.) J. Florsch.

New synonym:
Fissidens saülensis Pursell & W.R. Buck to Fissidens pellucidus Hornsch. var. pellucidus

NUMERICAL LIST OF ACCEPTED TAXA

Most of the species listed are published earlier in the Flora of Suriname (1964 and 1986) and the Flora of the Guianas (1996), sometimes in older names (placed in *italics* and between brackets). Additional species for the Guianas region are recorded here with an asterisk *.

Order SPHAGNALES

1. **Sphagnaceae**
   1-1. Sphagnum L.
       1-1-1. S. cyclophyllum Sull. & Lesq. *
       1-1-2 S. lescurii Sull. (*S. auriculatum* Schimp. var. *ovatum* Warnst.)
       1-1-3. S. magellanicum Brid.
       1-1-4. S. ornatum H.A. Crum *
       1-1-5. S. oxyphyllum Warnst. *
       1-1-6. S. perichaetiale Hampe
       1-1-6a. S. perichaetiale Hampe var. perichaetiale (*S. palustre* L.)
       1-1-6b. S. perichaetiale Hampe var. tabuleirense (Yano & H.A. Crum) H.R. Zielman *
       1-1-7. S. portoricense Hampe
       1-1-8. S. tenerum Sull. & Lesq. *
       1-1-9. S. trinitense Müll. Hal. (*S. cuspidatum* Ehrh. ex Hoffm. var. *serrulatum* (Schlieph.) Schlieph.)

Order ARCHIDIALES

2. **Archidiaceae**
   2-1. Archidium Brid.
       2-1-1. A. globiferum (Brid.) Frahm *

Order FISSIDENTALES

3. **Fissidentaceae**
   3-1. Fissidens Hedw.
       3-1-1. F. allionii Broth. *
       3-1-2. F. amazonicus Pursell *
       3-1-3. F. anguste-limbatus Mitt. (*F. pennula* Broth.)
       3-1-4. F. angustifolius Sull. *
       3-1-5. F. brevipes Besch. *
       3-1-6. F. dendrophilus Brugg.-Nann. & Pursell *
       3-1-7. F. elegans Brid. *

3-1-8.   F. flaccidus Mitt. (*F. mollis* Mitt.)
3-1-9.   F. geijskesi Florsch.
3-1-10. F. guianensis Mont.
3-1-10a. F. guianensis Mont. var. guianensis
3-1-10b. F. guianensis Mont. var. pacaas-novosensis
         Pursell & W.D. Reese *
3-1-11. F. hornschuchii Mont. *
3-1-12. F. inaequalis Mitt.
3-1-13. F. intramarginatus (Hampe) A. Jaeger *
3-1-14. F. lagenarius Mitt.
3-1-14a. F. lagenarius Mitt. var. lagenarius
3-1-14b. F. lagenarius Mitt. var. muricatulus (Spruce
         ex Mitt.) Pursell (*F. muriculatus* Spruce ex
         Mitt. var. *richardsii* (R.S. Williams) Florsch.)
3-1-15. F. leptophyllus Mont.
3-1-16. F. neglectus H.A. Crum *
3-1-17. F. oblongifolius Hook.f. & Wilson (as *F.
         asplenioides* Hedw.)
3-1-18. F. obtusissimus (Florsch.) Pursell (*F. leptophyllus*
         Mont. fo. *obtusissimus* Florsch.)
3-1-19. F. ornatus Herzog *
3-1-20. F. pallidinervis Mitt. *
3-1-21. F. palmatus Hedw. (*F. reticulosus* (Müll. Hal.)
         Mitt.)
3-1-22. F. pellucidus Hornsch.
3-1-22a. F. pellucidus Hornsch. var. pellucidus (*F.
         prionodes* Mont. fo. *flexinervis* (Mitt.) Florsch.)
3-1-22b. F. pellucidus Hornsch. var. papilliferus (Broth.)
         Pursell *
3-1-23. F. perfalcatus Broth. *
3-1-24. F. prionodes Mont. (*F. prionodes* Mont. fo.
         prionodes (Mont.) Florsch.)
3-1-25. F. radicans Mont.
3-1-26. F. ramicola Broth. *
3-1-27. F. scariosus Mitt.
3-1-28. F. serratus Müll. Hal. (*F. prionodes* Mont. fo.
         puiggari (Geh. & Hampe) Par.
3-1-29. F. submarginatus Bruch (*F. intermedius* Müll.
         Hal.)
3-1-30. F. subramicola Broth. *
3-1-31. F. weirii Mitt.
3-1-32. F. zollingeri Mont. (*F. kegelianus* Müll. Hal.)

Order DICRANALES

4. **Dicranaceae**
   4-1.   Bryohumbertia P. de la Varde & Thér.
       4-1-1.   B. filifolia (Hornsch.) Frahm (*Campylopus nanofilifolius* (Müll. Hal.) Paris)
   4-2.   Campylopus Brid.
       4-2-1.   C. angustiretis (Austin) Sull. & James *
       4-2-2.   C. arctocarpus (Hornsch.) Mitt.
       4-2-3.   C. bryotropii Frahm *
       4-2-4.   C. cubensis Sull. (*C. harrisii* (Müll. Hal.) Paris)
       4-2-5.   C. dichrostis (Müll. Hal.) Paris *
       4-2-6.   C. lamellinervis (Müll. Hal.) Mitt. *
       4-2-7.   C. luteus (Müll. Hal.) Paris *
       4-2-8.   C. pilifer Brid. *
       4-2-9.   C. richardii Brid. (*C. atratus* Broth.)
       4-2-10. C. savannarum (Müll. Hal.) Mitt.
       4-2-11. C. subcuspidatus (Hampe) A. Jaeger (*C. praealtus* (Müll. Hal.) Paris)
       4-2-12. C. surinamensis Müll. Hal.
       4-2-13. C. trachyblepharon (Müll. Hal.) Mitt.
   4-3.   Dicranella (Müll. Hal.) Schimp.
       4-3-1.   D. hilariana (Mont.) Mitt.
   4-4.   Eucamptodontopsis Broth.
       4-4-1.   E. pilifera (Mitt.) Broth.
       4-4-2.   E. tortuosa H. Rob. *
   4-5.   Holomitrium Brid.
       4-5-1.   H. arboreum Mitt.
   4-6.   Leucoloma Brid.
       4-6-1.   L. cruegerianum (Müll. Hal.) A. Jaeger
       4-6-2.   L. mariei Besch. *
       4-6-3.   L. serrulatum Brid.
       4-6-4.   L. tortellum (Mitt.) A. Jaeger

5. **Ditrichaceae**
   5-1.   Garckea Müll. Hal.
       5-1-1.   G. flexuosa (Griff.) Margad. & Nork. *

6. **Leucobryaceae**
   6-1.   Holomitriopsis H. Rob.
       6-1-1.   H. laevifolia (Broth.) H. Rob. (*Leucobryum laevifolium* Broth.)

6-2.　Leucobryum Hampe
  6-2-1.　　L. albicans (Schwägr.) Lindb. *
  6-2-2.　　L. antillarum Schimp. ex Besch. *
  6-2-3.　　L. crispum Müll. Hal.
  6-2-4.　　L. martianum (Hornsch.) Hampe
6-3.　Ochrobryum Mitt.
  6-3-1.　　O. subobtusifolium Broth. *
  6-3-2.　　O. subulatum Hampe *
6-4.　Octoblepharum Hedw.
  6-4-1.　　O. albidum Hedw.
  6-4-2.　　O. ampullaceum Mitt. *
  6-4-3.　　O. cocuiense Mitt.
  6-4-4.　　O. cylindricum Mont.
  6-4-5.　　O. erectifolium Mitt. ex R.S. Williams
  6-4-6.　　O. pulvinatum (Dozy & Molk.) Mitt.
  6-4-7.　　O. stramineum Mitt.
  6-4-8.　　O. tatei (R.S. Williams) E.B. Bartram *

7.　**Leucophanaceae**
7-1.　Leucophanes Brid.
  7-1-1.　　L. molleri Müll. Hal. (*L. mittenii* Cardot)

Order POTTIALES

8.　**Calymperaceae**
8-1.　Calymperes Sw.
  8-1-1.　　C. afzelii Sw. (*C. donnellii* Austin)
  8-1-2.　　C. erosum Müll. Hal.
  8-1-3.　　C. guildingii Hook. & Grev.
  8-1-4.　　C. levyanum Besch.
  8-1-5.　　C. lonchophyllum Schwägr.
  8-1-6.　　C. mitrafugax Florsch.
  8-1-7.　　C. nicaraguense Renauld & Cardot
  8-1-8.　　C. othmeri Herzog *
  8-1-9.　　C. palisotii Schwägr. (*C. richardii* Müll. Hal.)
  8-1-10.　　C. pallidum Mitt. (*C. uleanum* Broth.)
  8-1-11.　　C. platyloma Mitt.
  8-1-12.　　C. rubiginosum (Mitt.) Reese (*C. rufum* Herzog)
  8-1-13.　　C. smithii E.B. Bartram
  8-1-14.　　C. venezuelanum (Mitt.) Broth. ex Pittier
8-2.　Syrrhopodon Schwägr.
  8-2-1.　　S. africanus (Mitt.) Paris subsp. graminicola
     (R.S. Williams) Reese (*S. parasiticus* (Brid.)
     Besch. var. *disciformis* (Müll. Hal.) Florsch.)

8-2-2.   S. cryptocarpos Dozy & Molk.
8-2-3.   S. cymbifolius Müll. Hal. *
8-2-4.   S. elatus Mont. (*S. incompletus* Schwägr. var.
          elatus (Mont.) Florsch.)
8-2-5.   S. elongatus Sull. var. glaziovii (Hampe) Reese *
8-2-6.   S. flexifolius Mitt. (*S. parasiticus* (Brid.) Besch.
          var. *flexifolius* (Mitt.) Reese)
8-2-7.   S. gaudichaudii Mont.
8-2-8.   S. helicophyllus Mitt. *
8-2-9.   S. hornschuchii Mart. *
8-2-10.   S. incompletus Schwägr.
8-2-10a. S. incompletus Schwägr. var. incompletus
8-2-10b. S. incompletus Schwägr. var. luridus (Paris &
          Broth.) Florsch.
8-2-11.   S. lanceolatus (Hampe) Reese (*Calymperes*
          lanceolatum Hampe)
8-2-12.   S. leprieurii Mont.
8-2-13.   S. ligulatus Mont.
8-2-14.   S. lycopodioides (Brid.) Müll. Hal. *
8-2-15.   S. parasiticus (Brid.) Besch.
8-2-16.   S. prolifer Schwägr.
8-2-16a. S. prolifer Schwägr. var. prolifer
8-2-16b. S. prolifer Schwägr. var. acanthoneuros (Müll.
          Hal.) Müll. Hal.
8-2-16c. S. prolifer Schwägr. var. cincinnatus (Hampe)
          Reese
8-2-16d. S. prolifer Schwägr. var. scaber (Mitt.) Reese
8-2-17.   S. rigidus Hook. & Grev.
8-2-18.   S. simmondsii Steere
8-2-19.   S. tortilis Hampe *

9. **Pottiaceae**
    9-1. Barbula Hedw.
          9-1-1.   B. indica (Hook.) Spreng. ex Steud. (*B. cruegeri*
                   Sond. ex Müll. Hal.)
    9-2. Hyophila Brid.
          9-2-1.   H. involuta (Hook.) A. Jaeger
    9-3. Hyophiladelphus (Müll. Hal.) Zander
          9-3-1.   H. agrarius (Hedw.) Zander (*Barbula agraria*
                   Hedw.)
    9-4. Trichostomum Bruch
          9-4-1.   T. duidense E.B. Bartram *

Order FUNARIALES

10. **Ephemeraceae**
10-1. Micromitrium Austin
    10-1-1.   M. thelephorothecum (Florsch.) Crosby
        (*Nanomitrium thelephorothecum* Florsch.)

11. **Funariaceae**
11-1. Entosthodon Schwägr.
    11-1-1.   E. bonplandii (Hook.) Mitt. *
11-2. Funaria Hedw.
    11-2-1.   F. calvescens Schwägr. (*F. hygrometrica* Hedw.
        var. *calvescens* (Schwägr.) Mont.)

12. **Splachnobryaceae**
12-1. Splachnobryum Müll. Hal.
    12-1-1.   S. obtusum (Brid.) Müll. Hal.

Order BRYALES

13. **Bartramiaceae**
13-1. Breutelia (Bruch & Schimp.) Schimp.
    13-1-1.   B. tomentosa (Brid.) A. Jaeger
13-2. Leiomela (Mitt.) Broth.
    13-2-1.   L. bartramioides (Hook.) Paris *
13-3. Philonotis Brid.
    13-3-1.   P. hastata (Duby) Wijk & Margad. (*P. uncinata*
        (Schwägr.) Brid. var. *gracillima* (Ångstr.)
        Florsch.)
    13-3-2.   P. sphaerocarpa (Hedw.) Brid. *
    13-3-3.   P. uncinata (Schwägr.) Brid. (*P. uncinata*
        (Schwägr.) Brid. var. *glaucescens* (Hornsch.)
        Florsch.

14. **Bryaceae**
14-1. Brachymenium Hook. ex Schwägr.
    14-1-1.   B. klotzschii (Schwägr.) Paris *
    14-1-2.   B. speciosum (Hook. & Wilson) Steere (*B.
        sipapoense* E.B. Bartram)
    14-1-3.   B. wrightii (Sull.) Broth. (*B. wrightii* (Sull.)
        Broth. var. *mnioides* (Besch.) Florsch.)
14-2. Bryum Hedw.
    14-2-1.   B. apiculatum Schwägr. (*B. cruegerii* Hampe)
    14-2-2.   B. argenteum Hedw.

14-2-2a. B. argenteum Hedw. var. argenteum (*B. candicans* Taylor)
14-2-2b. B. argenteum Hedw. var. lanatum (P. Beauv.) Hampe *
14-2-3. B. beyrichianum (Hornsch.) Müll. Hal. (*Rhodobryum beyrichianum* (Hornsch.) Müll. Hal.)
14-2-4. B. billardieri Schwägr. (*B. truncorum* (Brid.) Brid.)
14-2-5. B. capillare Hedw. *
14-2-6. B. coronatum Schwägr.

15. **Phyllodrepaniaceae**
15-1. Mniomalia Müll. Hal.
    15-1-1.   M. viridis (Mitt.) Müll. Hal. *
15-2. Phyllodrepanium Crosby
    15-2-1.   P. falcifolium (Schwägr.) Crosby (*Drepanophyllum falcifolium* (Schwägr.) Wijk & Margad.)

16. **Rhizogoniaceae**
16-1. Pyrrhobryum Mitt.
    16-1-1.   P. spiniforme (Hedw.) Mitt. (*Rhizogonium spiniforme* (Hedw.) Bruch)

Order ORTHOTRICHALES

17. **Macromitriaceae**
17-1. Groutiella Steere
    17-1-1.   G. chimborazensis (Spruce ex Mitt.) Florsch. *
    17-1-2.   G. mucronifolia (Hook. & Grev.) H.A. Crum & Steere
    17-1-3.   G. obtusa (Mitt.) Florsch.
    17-1-4.   G. tomentosa (Hornsch.) Wijk & Margad. (*G. fragilis* (Mitt.) H.A. Crum & Steere)
17-2. Macromitrium Brid.
    17-2-1.   M. cirrosum (Hedw.) Brid. (*M. schwaneckeanum* Hampe)
    17-2-2.   M. harrisii Paris *
    17-2-3.   M. leprieurii Mont.
    17-2-4.   M. orthostichum Nees ex Schwägr. *
    17-2-5.   M. pellucidum Mitt.
    17-2-6.   M. perichaetiale (Hook. & Grev.) Müll. Hal. *
    17-2-7.   M. podocarpi Müll. Hal.

17-2-8.   M. punctatum (Hook. & Grev.) Brid. (*M. pentastichum* Müll. Hal.)

17-2-9.   M. scoparium Mitt. *

17-2-10.   M. stellulatum (Hornsch.) Brid.

17-2-11.   M. trinitense R.S. Williams

17-2-12.   M. ulophyllum Mitt. *

17-2-13.   M. xenizon B.H. Allen & W.R. Buck *

17-3. Schlotheimia Brid.

      17-3-1.   S. angustata Mitt. *

      17-3-2.   S. jamesonii (Arn.) Brid.

      17-3-3.   S. rugifolia (Hook.) Schwägr.

      17-3-4.   S. torquata (Hedw.) Brid.

Order LEUCODONTALES

18. **Leptodontaceae**

    18-1. Pseudocryphaea E. Britton

       18-1-1.   P. domingensis (Spreng.) W.R. Buck

19. **Meteoriaceae**

    19-1. Floribundaria M. Fleisch.

       19-1-1.   F. cardotii (Broth.) Broth. *

       19-1-2.   F. floribunda (Dozy & Molk.) M. Fleisch. *

    19-2. Lepyrodontopsis Broth.

       19-2-1.   L. trichophylla (Sw. ex Hedw.) Broth. *

    19-3. Meteoridium (Müll. Hal.) Manuel

       19-3-1.   M. remotifolium (Müll. Hal.) Manuel (*Meteoriopsis remotifolia* (Müll. Hal.) Broth.)

    19-4. Orthostichella Müll. Hal.

       19-4-1.   O. versicolor (Müll. Hal.) B.H. Allen & W.R. Buck (*Pilotrichella versicolor* (Müll. Hal.) A. Jaeger)

    19-5. Papillaria (Müll. Hal.) Lorentz

       19-5-1.   P. nigrescens (Sw. ex Hedw.) A. Jaeger

    19-6. Squamidium (Müll. Hal.) Broth.

       19-6-1.   S. leucotrichum (Taylor) Broth.

       19-6-2.   S. macrocarpum (Mitt.) Broth. *

    19-7. Zelometeorium Manuel

       19-7-1.   Z. patulum (Hedw.) Manuel (*Meteoriopsis patula* (Hedw.) Broth.)

20. **Neckeraceae**

    20-1. Isodrepanium (Mitt.) E. Britton

       20-1-1.   I. lentulum (Wilson) E. Britton

20-2. Neckeropsis Reichardt
    20-2-1.   N. disticha (Hedw.) Kindb.
    20-2-2.   N. undulata (Hedw.) Reichardt

21. **Phyllogoniaceae**
   21-1. Phyllogonium Brid.
    21-1-1.   P. fulgens (Hedw.) Brid.
    21-1-2.   P. viride Brid. *

22. **Pterobryaceae**
   22-1. Calyptothecium Mitt.
    22-1-1.   C. planifrons (Renauld & Paris) Argent *
   22-2. Henicodium (Müll. Hal.) Kindb.
    22-2-1.   H. geniculatum (Mitt.) W.R. Buck
             (*Leucodontopsis geniculata* (Mitt.) H.A.
             Crum & Steere)
   22-3. Hildebrandtiella Müll. Hal.
    22-3-1.   H. guyanensis (Mont.) W.R. Buck
             (*Orthostichidium guyanensis* (Mont.) Broth.)
   22-4. Jaegerina Müll. Hal.
    22-4-1.   J. scariosa (Lorentz) Arzeni *
   22-5. Orthostichopsis Broth.
    22-5-1.   O. praetermissa W.R. Buck (*O. crinita* (Sull.)
             Broth.)
    22-5-2.   O. tetragona (Hedw.) Broth.
   22-6. Pireella Cardot
    22-6-1.   P. cymbifolia (Sull.) Cardot *
    22-6-2.   P. pohlii (Schwägr.) Cardot
   22-7. Renauldia Müll. Hal.
    22-7-1.   R. paradoxica B.H. Allen *
   22-8. Spiridentopsis Broth.
    22-8-1.   S. longissima (Raddi) Broth.

23. **Racopilaceae**
   23-1. Racopilum P. Beauv.
    23-1-1.   R. tomentosum (Hedw.) Brid.

24. **Rhacocarpaceae**
   24-1. Rhacocarpus Lindb.
    24-1-1.   R. purpurascens (Brid.) Paris (*R. humboldtii*
             (Hook.) Lindb.)

394

Order HOOKERIALES

25. **Daltoniaceae**
    25-1. Daltonia Hook. & Taylor
        25-1-1.   D. longifolia Taylor *
    25-2. Leskeodon Broth.
        25-2-1.   L. cubensis (Mitt.) Thér. *

26. **Leucomiaceae**
    26-1. Leucomium Mitt.
        26-1-1.   L. steerei B. Allen & Veling
        26-1-2.   L. strumosum (Hornsch.) Mitt.

27. **Pilotrichaceae**
    27-1. Actinodontium Schwägr.
        27-1-1.   A. sprucei (Mitt.) A. Jaeger *
    27-2. Brymela Crosby & B.H. Allen
        27-2-1.   B. parkeriana (Hook. & Grev.) W.R. Buck
                (*Hookeriopsis parkeriana* (Hook. & Grev.)
                A. Jaeger)
    27-3. Callicostella (Müll. Hal.) Mitt.
        27-3-1.   C. colombica R.S. Williams (*Schizomitrium*
                grossiretis (E.B. Bartram) H.A. Crum)
        27-3-2.   C. guatemalensis (E.B. Bartram) J. Florsch.
                (*Schizomitrium guatemalensis* (E.B. Bartram)
                J. Florsch.)
        27-3-3.   C. merkelii (Hornsch.) A. Jaeger (*Schizomitrium*
                merkelii (Hornsch.) J. Florsch.)
        27-3-4.   C. pallida (Hornsch.) Ångstr. (*Schizomitrium*
                pallidum (Hornsch.) H.A. Crum & L.E.
                Anderson)
        27-3-5.   C. rivularis (Mitt.) A. Jaeger *
        27-3-6.   C. rufescens (Mitt.) A. Jaeger (*Schizomitrium*
                rufescens (Mitt.) J. Florsch.)
    27-4. Crossomitrium Müll. Hal.
        27-4-1.   C. epiphyllum (Mitt.) Müll. Hal. *
        27-4-2.   C. patrisiae (Brid.) Müll. Hal.
        27-4-3.   C. sintenisii Müll. Hal. *
    27-5. Cyclodiction Mitt.
        27-5-1.   C. varians (Sull.) O. Kuntze
    27-6. Hemiragis (Brid.) Besch.
        27-6-1.   H. aurea (Brid.) Kindb.
    27-7. Hypnella (Müll. Hal.) A. Jaeger
        27-7-1.   H. diversifolia (Mitt.) A. Jaeger (*Neohypnella*
                diversifolia (Mitt.) W.H. Welch & H.A. Crum)

27-7-2.   H. pallescens (Hook.) A. Jaeger (*H. cymbifolia* (Hampe) A. Jaeger)

27-8. Lepidopilidium (Müll. Hal.) Broth.
    27-8-1.   L. portoricense (Müll. Hal.) H.A. Crum & Steere

27-9. Lepidopilum (Brid.) Brid.
    27-9-1.   L. affine Müll. Hal.
    27-9-2.   L. cubense (Sull.) Mitt.
    27-9-3.   L. denticulatum (Thér.) F.D. Bowers *
    27-9-4.   L. polytrichoides (Hedw.) Brid.
    27-9-5.   L. purpurascens Schimp. ex Besch. *
    27-9-6.   L. radicale Mitt.
    27-9-7.   L. scabrisetum (Schwägr.) Steere (incl. *L. stolonaceum* Müll. Hal.)
    27-9-8.   L. surinamense Müll. Hal.
    27-9-9.   L. tortifolium Mitt. *

27-10. Pilotrichum P. Beauv.
    27-10-1.   P. bipinnatum (Schwägr.) Brid.
    27-10-2.   P. evanescens (Müll. Hal.) Crosby
    27-10-3.   P. fendleri Müll. Hal.

27-11. Thamniopsis (Mitt.) M. Fleisch.
    27-11-1.   T. gradsteinii J. Florsch. *
    27-11-2.   T. incurva (Hornsch.) W.R. Buck *
    27-11-3.   T. killipii (R.S. Williams) E.B. Bartram
    27-11-4.   T. undata (Hedw.) W.R. Buck (*Hookeriopsis crispa* (Müll. Hal.) A. Jaeger)

27-12. Trachyxiphium W.R. Buck
    27-12-1.   T. guadalupense (Spreng.) W.R. Buck (*Hookeriopsis falcata* (Hook.) A. Jaeger)

Order HYPNALES

28. **Fabroniaceae**
28-1. Anacamptodon Brid.
    28-1-1.   A. cubensis (Sull.) Mitt. *

29. **Hydropogonaceae**
29-1. Hydropogon Brid.
    29-1-1.   H. fontinaloides (Hook.) Brid.
29-2. Hydropogonella Cardot
    29-2-1.   H. gymnostoma (Bruch. & Schimp.) Cardot

30. **Hypnaceae**
30-1. Chryso-hypnum Hampe
    30-1-1.   C. diminutivum (Hampe) W.R. Buck

30-2. Ectropothecium Mitt.
 30-2-1. E. leptochaeton (Schwägr.) W.R. Buck
30-3. Isopterygium Mitt.
 30-3-1. I. subbrevisetum (Hampe) Broth.
 30-3-2. I. tenerum (Sw.) Mitt.
30-4. Mittenothamnium Henn.
 30-4-1. M. reptans (Hedw.) Cardot
30-5. Phyllodon Schimp.
 30-5-1. P. truncatulus (Müll. Hal.) W.R. Buck
30-6. Rhacopilopsis Renauld & Cardot
 30-6-1. R. trinitensis (Müll. Hal.) E. Britton & Dixon
30-7. Vesicularia (Müll. Hal.) Müll. Hal.
 30-7-1. V. vesicularis (Schwägr.) Broth.

31. **Sematophyllaceae**
31-1. Acroporium Mitt.
 31-1-1. A. pungens (Hedw.) Broth.
31-2. Colobodontium Herzog
 31-2-1. C. deceptivum (Mitt.) J. Florsch. (*Potamium deceptivum* Mitt.)
 31-2-2. C. vulpinum (Mont.) Churchill & Linares (*Potamium vulpinum* (Mont.) Mitt.)
31-3. Donnellia Austin
 31-3-1. D. commutata (Müll. Hal.) W.R. Buck (*Meiothecium commutatum* (Müll. Hal.) Broth.)
31-4. Meiothecium Mitt.
 31-4-1. M. boryanum (Müll. Hal.) Mitt.
31-5. Pterogonidium Müll. Hal.
 31-5-1. P. pulchellum (Hook.) Müll. Hal.
31-6. Sematophyllum Mitt.
 31-6-1. S. cochleatum (Broth.) Broth.
 31-6-2. S. galipense (Müll. Hal.) Mitt.
 31-6-3. S. lonchophyllum (Mont.) J. Florsch.
 31-6-4. S. pacimoniense (Mitt.) W.R. Buck
 31-6-5. S. subpinnatum (Brid.) E. Britton
 31-6-6. S. subsimplex (Hedw.) Mitt.
31-7. Taxithelium Spruce ex Mitt.
 31-7-1. T. concavum (Hook.) Spruce
 31-7-2. T. planum (Brid.) Mitt.
 31-7-3. T. pluripunctatum (Renauld & Cardot) W.R. Buck
31-8. Trichosteleum Mitt.
 31-8-1. T. bolivarense H. Robins.
 31-8-2. T. intricatum (Thér.) J. Florsch.

31-8-3.   T. papillosum (Hornsch.) A. Jaeger
31-8-4.   T. subdemissum (Besch.) A. Jaeger (*T.
hornschuchii* (Hampe) A. Jaeger)
31-8-4a. T. subdemissum (Besch.) A. Jaeger var.
subdemissum (*T. hornschuchii* (Hampe) A.
Jaeger var. *hornschuchii*)
31-8-4b. T. subdemissum (Besch.) A. Jaeger var.
subglabrum (J. Florsch.) J. Florsch.
(*T. hornschuchii* (Hampe) A. Jaeger var.
subglabrum J. Florsch.)
31-9. Wijkia H.A. Crum
31-9-1.   W. costaricensis (Dixon & E.B. Bartram)
H.A. Crum

32. **Stereophyllaceae**
32-1. Entodontopsis Broth.
32-1-1.   E. contorte-operculata (Müll. Hal.) Broth.
(*Stereophyllum rhabdodonta* (Cardot) Grout)
32-1-2.   E. leucostega (Brid.) W.R. Buck & Ireland
(*Stereophyllum leucostegum* (Brid.) Mitt.)
32-1-3.   E. nitens (Mitt.) W.R. Buck & Ireland
(*Stereophyllum obtusum* Mitt.)
32-2. Eulacophyllum W.R. Buck & Ireland
32-2-1.   E. cultelliforme (Sull.) W.R. Buck & Ireland *
32-3. Pilosium (Müll. Hal.) M. Fleisch.
32-3-1.   P. chlorophyllum (Hornsch.) Müll. Hal.
32-4. Stereophyllum Mitt.
32-4-1.   S. radiculosum (Hook.) Mitt. *

33. **Thamnobryaceae**
33-1. Porotrichum (Brid.) Hampe
33-1-1.   P. korthalsianum (Dozy & Molk.) Mitt.
33-1-2.   P. piniforme (Brid.) Mitt. (*Pinnatella piniformis*
(Brid.) M. Fleisch.)
33-1-3.   P. stipitatum (Mitt.) W.R. Buck *
33-1-4.   P. substriatum (Hampe) Mitt. (*P. plicatulum* Mitt.)

34. **Thuidiaceae**
34-1. Pelekium Mitt.
34-1-1.   P. involvens (Hedw.) Touw (*Cyrto-hypnum
involvens* (Hedw.) W.R. Buck & H.A. Crum)
34-1-2.   P. scabrosulum (Mitt.) Touw (*Cyrto-hypnum
scabrosulum* (Mitt.) W.R. Buck & H.A. Crum)

34-1-3.　P. schistocalyx (Müll. Hal.) Touw (*Cyrto-hypnum schistocalyx* (Müll. Hal.) W.R. Buck & H.A. Crum)
34-2. Thuidium Bruch & Schimp.
　　　34-2-1.　T. peruvianum Mitt.
　　　34-2-2.　T. tomentosum Schimp.

Order BUXBAUMIALES

35.　**Diphysciaceae**
　35-1. Diphyscium D. Mohr
　　　35-1-1.　D. longifolium Griff. *

Order POLYTRICHALES

36.　**Polytrichaceae**
　36-1. Polytrichum Hedw.
　　　36-1-1.　P. juniperinum Hedw. *

COLLECTIONS STUDIED
(Numbers in **bold** represent types)

GUYANA

Andel, T. van, 1963 (1-1-9)
Appun, C.F., **819** (8-1-11)
Aptroot, A., 15705 (8-1-9); 17050
(4-3-1); 17052 (6-4-4); 17054
(14-2-2b); 17057 (17-1-3); 15061
(19-7-1); 17063 (8-2-1); 17064
(8-2-15); 17065 (6-2-4); 17066
(4-2-10); 17068 (27-7-2); 17070
(8-2-5); 17071 (9-4-1); 17075
(27-7-2); 17078 (4-4-1); 17080
(8-1-13); 17081 (8-2-12); 17091
(6-4-8); 17092 (27-7-1); 17099
(1-1-4); 17100 (27-7-2); 17101
(27-9-5); 17112 (4-2-7); 18924
(6-2-4); 18924A (27-7-2); 18924B
(31-8-3); 18924C (27-3-6)
Boom, B.M., *et al.*, 7390 (21-1-2);
7514 (27-11-4); 7620 (4-2-6);
7708 (27-7-2); 8914 (27-11-4);
9001 (27-7-2); 9066 (4-4-2);
9067 (17-2-9); 9069 (4-4-2);
9086, 9100 (17-2-3); **9131** (27-
7-2); 9133 (27-12-1); 9135 (27-
11-4); 9146 (4-4-2); 9166, 9177,
9179 (24-1-1); 9219 (17-2-2);
9233 (35-1-1)
Cornelissen, J.H.C., *et al.*, 8 (8-1-
5); 13 (6-2-4); 36 (32-3-1); 38
(4-5-1); 40 (6-2-4); 42 (17-2-5);
49 (8-2-10b); 54 (22-5-2); 58
(8-1-5); 59 (22-5-2); 60 (19-7-
1); 63 (8-1-5); 76 (32-3-1); 78
(6-4-6); 82 (14-2-6); 86 (8-1-2);
90 (8-2-13); 90A (8-2-10a); 92
(15-2-1); 94 (3-1-24); 96 (8-2-
10b); 98 (27-3-4); 99 (8-1-8);
101 (8-1-1); 104 (4-3-1); 109
(6-2-4); 121 (17-2-8); 122 (6-
4-6); 123 (6-4-1); 125 (32-3-1);

127 (22-2-1); 129A (15-1-1);
131, 132 (32-3-1); 134 (8-1-1);
135 (27-3-4); 137 (3-1-12); 138
(32-3-1); 142 (27-3-4); 143 (8-
1-5); 147 (3-1-7, 11); 149 (4-2-
12); 151 (8-1-5); 152 (27-10-2);
153 (6-2-3); 156 (27-10-1); 161
(14-2-6); 166 (17-2-8); 177 (27-
3-4); 182 (3-1-10a); 187 (27-10-
2); 197 (3-1-22a); 201 (8-1-1);
202 (3-1-17); 204 (15-2-1); **208**
(31-8-4b); 211 (32-3-1); 212
(15-2-1); 500 (3-1-10a); 501 (3-
1-10a); 504 (8-1-2); 505, 506
(8-2-6); 507 (8-2-13); 508, 509
(8-1-2); 513 (8-2-15); 514 (8-1-
5); 515 (8-1-2); 520 (8-2-10a);
521, 522 (8-2-6); 525 (8-1-5);
524 (8-2-13); 525 (8-1-5); 529
(8-2-18); 531, 532 (8-2-15); 534
(8-2-10a); 536 (8-2-6); 537 (8-
2-3); 538 (8-2-1); 539 (8-2-6);
540 (8-2-1); 542 (8-2-15); 543
(8-2-10a); 544 (8-2-3); 545 (8-
2-15); 546 (8-2-10a); 547, 548
(8-2-10b); 549 (8-2-10a); 551,
552, 553 (8-1-5); 554 (15-1-1);
555, 556 (8-1-11); 557 (8-2-4);
559 (8-2-10a); 561 (8-2-18);
932, 933 (32-3-1); 934 (27-3-6);
935 (27-9-8); 936 (6-4-6); 937
(8-2-18); 939 (32-3-1); 941 (6-
4-6); 942 (32-3-1); 943 (8-1-2);
944 (15-1-1); 945 (17-1-3); 946
(15-1-1); 947 (22-5-2); 948 (17-
1-3); 949 (19-7-1); 950 (32-3-
1); 951 (8-2-2); 953 (8-2-6); 954
(8-1-2); 955, 956 (8-2-13); 957
(8-1-13); 958, 962 (6-4-6); 963
(17-1-3); 965 (6-4-1); 967 (17-
2-8); 968 (22-2-1); 969 (19-7-

4909 (8-1-14); 4912 (8-1-2); 4915 (32-3-1); 4928 (3-1-28); 4928A (15-1-1); 4933 (8-2-17); 4942 (17-2-1); 4951 (27-4-3); 4952 (22-2-1); 4953 (4-2-10); 4938 (4-3-1); 4945 (17-1-3); 4949 (17-2-8); 4954 (4-2-10); 4967 (3-1-22a); 4968 (3-1-2, 22a); 4969 (3-1-22a); 4975 (3-1-22a); 4981 (1-1-8); 4981A (1-1-6a); 4984 (6-2-4); 4970 (4-3-1); 4997 (8-1-6); 5007 (27-7-2); 5009 (1-1-6b); 5015 (6-4-3); 5024 (8-2-17); 5026 (27-7-2); 5034 (8-2-12); 5037 (6-2-3); 5045 (8-2-6); 5046 (15-2-1); 5047 (27-11-3); 5053 (20-1-1); 5057 (17-2-1); 5058 (15-2-1); 5061 (3-1-22a); 5064 (27-10-1); 5072 (4-3-1); 5070 (6-2-4); 5071A (22-5-1); 5073 (11-1-1); 5077 (4-2-12); 5078 (1-1-6b); 5083 (8-2-17); 5086A (6-2-4); 5089 (27-11-3); 5094 (27-7-2); 5095 (4-4-2); 5097 (6-1-1); 5102 (17-2-3); 5105 (27-7-2); 5107 (4-4-1); 5108 (27-7-2); 5109 (27-11-4); 5116 (4-4-2); 5153 (6-1-1); 5153A (6-4-3); 5154 (8-1-13); 5155 (4-4-2); 5156 (4-4-1); 5157 (4-4-2); 5157A (27-7-2); 5158 (9-4-1); 5166 (17-2-3); 5168 (17-2-6); 5170 (14-2-4); 5187 (22-7-1); 5200 (8-2-17); 5214 (9-4-1); 5216 (27-6-1); 5217 (27-11-4); 5218 (8-2-14); 5219 (4-4-2); 5221 (27-9-5); 5225 (6-2-1); 5228 (17-2-3); 5230 (17-2-6); 5231 (14-2-4); 5235 (27-7-2); 5238 (1-1-3); 5239 (4-4-2); 5240 (9-4-1); 5242 (27-9-3); **5244** (27-11-1); 5244A (27-11-4); 5245 (27-6-1); 5245A (26-1-1); 5248

(1-1-3); 5259 (6-1-1); 5261 (6-2-1); 5264 (8-2-17); 5268 (25-2-1); 5271 (27-11-1); 5284 (27-11-4); 5293 (4-2-8); 5297 (8-2-17); 5299 (27-7-1); 5306 (27-7-2); 5311 (27-7-1); 5311A (27-6-1); 5313 (13-2-1); 5314 (27-7-2); 5318 (8-2-17); 5322 (27-12-1); 5325 (8-2-16c); 5329 (8-2-14); 5336 (4-2-7); 5346 (6-1-1); 5348 (24-1-1); 5350 (4-2-11); 5358 (4-2-9); 5372 (1-1-4); 5379 (4-2-7); 5382 (4-2-4); 5383 (27-9-5); 5384 (8-2-14); 5385 (13-1-1); 5386 (27-7-2); 5392 (17-2-12); 5397 (36-1-1); 5405 (24-1-1); 5406 (17-2-11); 5408 (27-7-1); 5412A (1-1-3); 5413 (21-1-1); 5422 (27-6-1); 5423 (8-2-17); 5433 (27-7-2); 5444 (27-12-1); 5446 (8-2-14); 5447 (8-2-16c); 5450 (6-1-1); 5454 (27-9-5); 5455 (4-4-2); 5457 (27-7-2); 5458 (8-1-3); 5469 (25-1-1); 5539 (27-4-2); 5540 (6-2-3); 5542 (32-3-1); 5558 (22-5-1); 5569 (3-1-22a); 5575 (3-1-24); 5577 (8-2-13); 5578 (27-3-4); 5580 (19-7-1); 5586 (8-1-5); 5588 (6-2-3); 5591 (16-1-1); 5592 ((19-2-1); 5593(4-2-4); 5601 (4-6-1); 5605A (8-2-7). 5606 (4-2-11); 5609 (1-1-4); 5610 (4-2-4); 5611 (8-2-5); 5615 (4-2-4); 5623 (19-6-1); 5625 (1-1-6a); 5626 (4-2-11); 5632, 5633 (8-2-5); 5637, 5638 (17-2-1); 5639 (17-3-2); 5640 (17-2-8); 5641 (17-2-6); 5641A (17-3-4); 5642 (17-1-3); 5643 (6-2-2); 5653 (4-2-4); 5654 (4-2-11); 5662 (14-2-1); 5663 (6-4-4); 5671 (19-3-1); 5672 (8-2-7); 5674 (23-1-1);

5831 (9-4-1); 5834 (16-1-1); 5907 (8-2-17); 7398 (17-1-4); 7465 (1-1-6a); 7465A (1-1-3); 7466 (4-2-11); 7477 (6-2-1); 7478 (6-2-3); 7481 (4-1-1); 7575 (6-2-4); 7574 (6-4-3)

Newton, A.E., *et al.*, 3307 (19-6-1); 3309 (17-2-1); 3310 (17-2-11) 3311 (6-2-3); 3313 (6-4-1); 3314 (8-2-19); 3315, 3316 (4-2-4); 3317 (6-4-3); 3322 (27-11-3); 3325 (14-2-4); 3326 (9-4-1); 3327 (4-2-10); 3329 (32-3-1); 3330 (30-6-1); 3333 (8-1-7); 3334 (3-1-14b); 3335 (27-9-2); 3338 (14-2-1); 3340 (13-3-3); 3343 (8-2-12); 3344 (19-7-1); 3350 (8-2-5); 3355 (27-11-3); 3356 (4-2-5); 3357 (1-1-8); 3358 (1-1-6a); 3359 (4-6-3); 3361 (20-1-1); 3363 (6-2-4); 3364 (31-8-4b); 3365 (4-2-4); 3367 (3-1-25); 3368 (9-4-1); 3372 (32-3-1); 3373 (30-2-1); 3375 (27-11-3); 3377 (8-2-9); 3379 (26-1-1?); 3382 (3-1-22a); 3386 (22-5-1); 3387 (1-1-3); 3388 (8-2-12); 3389 (4-1-1); 3394 (27-10-1) 3396 (8-1-14); 3397 (22-5-1); 3399 (17-2-1); 3402 (31-1-1); 3404 (31-8-1); 3407 (17-2-8); 3408 (31-1-1); 3411 (22-5-1); 3414 (8-1-11); 3415 (17-2-1); 3416 (8-1-7); 3420 (8-2-12); 3421 (27-9-7); 3423 (6-2-3); 3425 (17-2-1); 3426 (8-1-14); 3431 (17-2-7); 3432 (17-2-1); 3440 (17-2-11); 3441 (4-4-1); 3442 (31-1-1); 3444 (4-6-1); 3448 (31-2-2); 3453 (1-1-1); 3454 (31-8-4b); 3455 (31-8-4b); 3456 (4-2-1); 3457 (6-4-3); 3460 (17-2-11); 3461 (20-1-1); 3462

(4-6-2); 3465 (27-9-8); 3466 (34-2-2); 3467 (20-1-1); 3468 (27-9-7); 3470 (27-9-8); 3473 (6-2-4); 3475 (17-1-3); 3477 (30-3-2); 3478 (16-1-1); 3479, 3480 (31-7-2); 3481 (4-6-3); 3484 (27-3-4); 3485 (31-7-1); 3487 (14-2-4); 3488 (4-2-10); 3489 (4-2-1); 3493 (31-1-1); 3501 (31-8-1); 3504 (6-2-4); 3507 (17-2-1); 3509 (4-2-13); 3510 (17-3-3); 3512 (22-4-1); 3513 (22-2-1); 3520 (8-1-11); 3521 (21-1-2); 3523 (4-6-3); 3524 (30-2-1); 3527 (22-5-1); 3528A (30-2-1); 3528B (27-11-2); 3528C (27-3-4); 3529 (32-3-1); 3532 (19-6-1); 3533 (6-2-3); 3535 (4-6-3); 3536 (19-6-1); 3540 (17-2-1); 3541 (17-3-1); 3546 (17-2-1); 3547 (30-4-1); 3548 (27-3-4); 3549 (27-11-2); 3550 (3-1-12, 19, 22a); 3552 (22-3-1); 3555 (3-1-22a); 3556 (31-6-6); 3558 (4-2-10); 3559 (4-6-3); 3561 (31-7-1); 3565 (30-7-1); 3567 (27-3-5); 3571 (3-1-27); 3573 (32-3-1); 3574 (31-6-6); 3575 (17-3-3); 3576 (31-3-1); 3577 (4-5-1); 3579 (31-6-6); 3582 (8-2-16a); 3589 (11-2-1); 3590 (34-2-2); 3591 (27-3-5); 3593 (30-7-1); 3596 (3-1-8); 3598 (30-1-1); 3601 (17-3-1); 3605 (8-2-13); 3606, 3607 (31-6-6); 3608 (34-2-2); 3609 (13-3-3); 3612, 3614 (31-6-5); 3620 (17-1-4); 3622 (30-2-1); 3623 (31-6-5); 3626 (22-6-1); 3627 (8-1-1); 3628 (20-2-1); 3629 (20-2-2); 3630 (19-3-1); 3631 (22-4-1); 3632 (8-1-5); 3635 (31-7-2); 3641 (30-1-1); 3644 (6-3-1); 3645

Florschütz, P.A., *et al*., 187 (8-1-7); 256 (3-1-3); 261 (3-1-3, 8); 262 (3-1-3); 339A (3-1-19, 22a); 343A (3-1-22a); 386 (3-1-10a); 394A (3-1-10a); 395 (17-2-13); 403 (3-1-15); 408 (3-1-10a); **417** (3-1-15, 18); 429, 466 (3-1-10a); 482 (3-1-15); 489 (3-1-3); 497, 546 (3-1-29); 690 (1-1-6a); 932 (3-1-32); 992 (3-1-29); 994 (3-1-32); 999 (3-1-12, 22a); **1105A** (3-1-30); 1172 (3-1-3); 1188 (3-1-22a); 1218A (3-1-3); 1231, 1257 (3-1-10a); 1284B (8-1-8); 1320 (17-2-13); 1322 (8-1-8); 1341 (3-1-10a, 32); 1359 (23-1-1); 1370, 1426 (27-10-2); 1476 (8-1-8); 1537 (15-1-1); 1554A (8-1-8); 1556 (27-10-2); 1584 (3-1-7, 11, 22a); 1584A (3-1-22a); 1631 (27-4-3); 1644 (3-1-24); 1674 (3-1-12, 29); 1707 (3-1-29, 12); **1707A** (3-1-12, 29); 1708 (3-1-29); 1752, 1754 (3-1-24); 1771 (3-1-10a); 1776 (3-1-22a); 1788 (3-1-24); 1789 (3-1-17); 1790 (3-1-10a); 1799 (8-1-8); 1813 (8-2-4); 1829 (3-1-7); 1831 (27-10-2); 1832 (3-1-31); 1844 (3-1-10a); 1891, 1905A (3-1-32); 2041, 2042, 2057 (3-1-22a); 2065 (3-1-32); 2073 (3-1-26); 2124, 2126 (3-1-17); 2130 (3-1-7); 2173 (3-1-3); 2196 (3-1-10a); 2200 (3-1-21); 2200A (3-1-10a); 2207 (3-1-15); 2226 (3-1-31); 2230 (17-2-13); 2231 (27-10-2); 2233 (3-1-32); 2236 (8-1-8); 2245 (3-1-29); **2253** (10-1-1); 2255 (3-1-29); 2256 (3-1-12); 2281 (27-10-2); 2288 (3-1-12, 22a); 2291 (3-1-29); 2312 (8-1-2); 2328 (8-2-10a); 2330 (22-5-2); 2354 (3-1-3); 2373 (3-1-29); 2376 (6-4-1): 2379 (8-1-2); 2382 (6-4-6); 2384 (19-7-1); 2388 (8-1-5); 2390 (8-1-1); 2438 (8-1-5); 2439 (3-1-10a); 2456 (8-1-5); 2489 (3-1-12); 2490 (3-1-29); 2491 (3-1-19, only slide found); 2492 (8-2-10a); 2493 (8-2-2); 2511 (3-1-29); 2539 (3-1-12, 21, 32); 2540 (3-1-21); 2554, 2557 (8-1-5); 2608 (17-3-3); 2608A (17-2-13); 2622, 2623 (3-1-3); 2625 (3-1-7); 2641 (3-1-10a); 2643 (7-1-1); 2644 (3-1-10a); 2645, 2646, 2647 (3-1-10a); 2648 (9-2-1); 2649 (3-1-7); 2649A (3-1-3); 2649C (3-1-7); 2649C (29-2-1); 2658 (3-1-29); 2677 (4-3-1); 2680 (13-3-3); 2681 (6-4-1); 2682 (8-1-9); 2685, 2686 (3-1-29); 2687, 2688 (4-3-1); 2693 (3-1-29); 2694, 2696 (9-2-1); 2711 (8-1-5); 2712 (3-1-29); 2714 (7-1-1); 2720 (19-7-1); 2773 (3-1-10a); 2803 (23-1-1); 2811 (6-4-6); 2812 (8-2-10a); 2837 (23-1-1); 2838 (8-1-8); 2868 (21-1-1); 2871 (19-3-1); 2876 (23-1-1); 2881 (22-5-1); 2882 (6-2-3); 2915 (19-6-1); 2916 (6-4-3); 2917 (4-1-1); 2918 (8-2-10b); 2921 (21-1-1); 2927 (6-4-3); 2928 (4-2-2); 2929 (8-1-5); 2930 (4-6-3); 2933 (19-3-1); 2962 (6-2-3); 2965 (8-2-11); 2966 (6-4-3); 2968 (17-2-6); 2992 (22-5-1); 3023 (17-2-1); 3024 (6-2-3); 3025 (17-3-2); 3027A (4-6-3); 3028 (4-1-1); 3027 (17-2-6); 3029 (8-1-11); 3032 (8-2-11); 3036 (8-2-16b); 3064 (8-1-1); 3066 (8-1-5); 3068 (3-1-17); 3076 (8-2-17); 3077 (3-1-10a);

3097 (14-2-4); 3120 (8-2-17); 3122 (8-2-16b); 3155 (17-2-1); 3157D (8-2-3); 4486 (3-1-22a); 4513 (3-1-32); 4514 (3-1-29); 4515 (17-1-4); 4516 (8-1-9); 4517 (14-2-1); 4519 (17-1-4); 4534 (3-1-29); 4540 (8-2-10a); 4541 (8-1-5); 4546 (8-2-10a); 4553 (33-1-4); 4560 (8-1-5); 4561 (8-1-9); 4564 (27-10-1); 4567 (8-1-9); 4572 (4-2-10); 4573 (17-2-8); 4573A (17-1-3); 4574 (4-2-12); 4574A (8-2-7); 4575 (6-4-4); 4578 (4-2-10); 4580 (8-2-10a); 4583 (9-3-1); 4614 (8-2-10a); 4614A (6-2-2); 4619 (8-1-2); 4631 (4-3-1); 4632 (19-1-1); 4635 (19-7-1); 4636 (8-2-15); 4636A (8-2-1); 4642 (14-2-1); 4643 (6-2-4); 4644 (9-2-1); 4648 (3-1-22a); 4649 (8-1-5); 4650 (13-3-3); 4651 (3-1-8); 4652 (4-3-1); 4653 (3-1-22a); 4654 (4-2-2); 4655 (6-2-2); 4556 (22-2-1); 4660 (4-6-1); 4661 (17-2-1); 4663 (3-1-24); 4664 (4-2-4); 4666 (4-3-1); 4668 (19-6-1);4669, 4672 (22-5-2); 4672A (22-5-1); 4674 (17-2-1); 4677 (6-2-2); 4679 (4-6-1); 4680 (6-4-6); 4682 (18-1-1); 4684 (8-2-16a); 4685 (4-6-3); 4694 (27-10-1); 4698 (3-1-7); 4723 (4-5-1); 4703 (13-3-3); 4715 (23-1-1); 4729 (19-7-1); 4729A (19-1-1); 4730 (8-1-5); 4734 (8-1-2); 4745 (19-1-1); 4751 (19-1-2); 4754 (8-1-5); 4755 (15-2-1); 4776 (1-1-6a); 4778 (9-1-1); 4780 (8-1-9); 4781A (8-2-18); 4782 (6-4-6); 4786 (8-1-2); 4788 (8-2-13); 4790 (4-2-12); 4794 (19-1-1); 4796 (8-2-15); 4796A (14-2-6);

4797 (17-1-3); 4799 (3-1-19, 22a); 4806 (3-1-7); 4807 (13-3-3); 4815 (22-5-1); 4816 (4-6-3); 4818 (17-1-4); 4825 (27-10-1); 4828 (21-1-2); 4829 (4-6-3); 4830 (19-4-1); 4834 (17-2-8); 4835 (22-5-1); 4836 (19-4-1); 4837 (14-2-4); 4838 (4-6-1); 4839 (3-1-24); 4840 (8-1-5); 4841 (3-1-17); 4842 (8-2-11); 4843 (3-1-27); 4844 (4-2-4); 4845 (4-6-3); 4846 (19-6-1); 4847A (17-2-1); 4849 (6-4-3); 4850 (3-1-28); 4851 (17-2-11); 4852 (17-2-7); 4853 (6-2-3); 4854 (6-2-4); 4861 (17-2-8); 4863 (17-2-11); 4864 (8-2-10b); 4869 (19-4-1); 4872 (19-3-1); 4873 (17-3-4); 4875 (3-1-22a); 4880 (3-1-29); 4881 (8-2-18); 4882 (6-4-3); 4883, 4885 (19-7-1); 4886 (3-1-22a); 4891 (3-1-19, 22a); 4902 (6-2-4); 4904 (8-1-5); 4907 (4-6-3); 4908 (19-6-1); 4909 (21-1-2); 4910 (17-2-1); 4918 (13-3-3); 4919 (9-1-1); 4920 (6-4-1); 4925 (19-7-1); 4929 (6-4-1); 4932, 4936 (8-1-2); 4937 (19-7-1); 4938 (17-1-1); 4998 (3-1-7)

Florschütz-de Waard, J. & H.R. Zielman, 5001 (17-1-4); 5002 (13-3-3); 5003 (19-7-1); 5006 (17-1-4); 5012A (22-2-1); 5012B (17-1-3); 5015 (17-2-5); 5016 (6-4-3); 5018 (17-1-4); 5018A (17-1-2); 5021 (23-1-1); 5026 (3-1-22a, 27); 5028 (8-2-10a); 5031 (4-6-3); 5034 (19-1-1); 5035 (6-2-2); 5035A (7-1-1); 5037 (22-5-2); 5038 (6-4-6).; 5055 (6-2-2); 5060 (8-1-9); 5062 (8-1-1); 5063 (8-1-5); 5064 (3-1-21); 5065 (3-1-25);

5070 (8-1-5); 5071 (22-5-2); 5077, 5078, 5079 (8-1-11); 5080 (8-1-5); 5084 (6-4-3); 5086 (3-1-22a); 5089 (8-1-5); 5092 (8-1-10); 5094 (6-2-4); 5099 (8-2-9); 5102 (8-1-11); 5110 (6-2-4); 5113 (6-4-3); 5114 (6-2-4); 5117 (8-1-2); 5118 (17-2-5); 5119A (17-1-3); 5120 (6-2-4); 5132 (17-2-5); 5133 (6-4-6); 5138 (8-1-5); 5146 (7-1-1); 5147 (19-7-1); 5148 (22-5-2); 5150 (8-2-10a); 5152 (6-4-6); 5153 (8-1-5); 5157 (17-1-3); 5160 (8-1-2); 5165 (6-4-1); 5166 (6-4-6); 5167 (8-1-2); 5168 (8-1-1); 5169 (8-1-5); 5170 (8-1-1); 5172 (3-1-17); 5173 (3-1-22a); 5177 (8-1-1); 5178 (8-2-3); 5182 (29-2-1); 5186 (19-7-1); 5188 (8-2-10a); 5190 (8-1-10); 5196 (6-4-6); 5202 (6-4-6); 5205 (17-1-3); 5206 (3-1-11); 5208 (8-2-10a); 5209 (8-1-5); 5216 (6-2-4); 5218 (6-4-6); 5219, 5220, 5226 (6-2-4); 5228 (3-1-12); 5230 (8-2-9); 5231 (6-4-6); 5233 (8-1-5); 5235 (3-1-10a); 5240 (15-1-1); 5241 (3-1-22a); 5242 (19-7-1); 5247 (8-1-2); 5251 (3-1-22a); 5252 (8-1-5); 5253 (6-4-3); 5255, 5257, 5261 (6-2-4); 5262 (6-4-3); 5266 (8-2-2); 5267 (6-4-3); 5272 (29-2-1); 5282 (8-1-1); 5284 (6-4-6); 5286 (22-2-1); 5287 (19-7-1); 5289 (8-1-5); 5290 (22-5-2); 5291 (3-1-24); 5293 (8-1-1); 5296 (3-1-11, 12, 22a); 5296 (9-2-1); 5297 (8-1-2); 5298 (8-1-5); 5299, 5300 (8-1-1); 5301, 5302, 5305 (8-1-2); 5309 (8-1-1); 5311 (6-2-4); 5314 (8-1-2); 5319 (8-2-10a); 5320 (6-4-6);

5322 (8-2-10a); 5323 (6-4-6); 5326 (8-1-1); 5329 (3-1-10a); 5331 (6-4-6); 5334 (8-1-1); 5336 (6-4-6); 5340 (3-1-10a); 5342 (6-4-6); 5343 (8-1-5); 5348 (22-2-1); 5352 (19-7-1); 5358 (8-1-2); 5359 (17-1-4); 5360 (22-5-2); 5361, 5363 (8-2-10a); 5365, 5366, 5367 (8-1-1); 5370 (3-1-10a); 5371 (8-1-2); 5373 (8-1-5); 5379 (6-4-6); 5384 (8-1-5); 5385 (19-7-1); 5387 (6-4-6); 5388 (8-2-10a); 5389 (6-2-4); 5392 (3-1-10a); 5393 (8-2-10a); 5395 (3-1-21); 5396, 5397, 5399 (4-3-1); 5403 (6-4-6); 5404 (6-2-4); 5405 (6-4-6); 5406 (6-2-4); 5407 (8-2-10a); 5409 (8-2-13); 5411, 5418 (6-4-6); 5419 (3-1-10a); 5422 (8-1-5); 5423 (3-1-10a); 5424 (8-1-5); 5426 (17-1-3); 5427 (6-2-4); 5428 (6-2-4); 5429 (8-2-18); 5430, 5432 (6-4-3); 5433 (6-2-4); 5435 (8-2-3); 5437 (8-1-2); 5438 (22-2-1); 5439 (17-1-3); 5441 (17-2-8); 5443, 5445 (8-1-5); 5446 (19-7-1); 5448 (17-2-5); 5449 (6-4-3); 5453 (8-1-5); 5458 (6-2-4); 5462, 5468 (6-4-6); 5473 (8-1-2); 5475 (8-1-5); 5481 (19-7-1); 5483 (4-2-10); 5484 (8-1-2); 5490 (22-2-1); 5492 (6-4-1); 5493 (8-2-1); 5497 (19-7-1); 5499 (8-2-1); 5500 (22-5-2); 5502, 5503 (8-1-5); 5505 (8-1-1); 5511, 5512 (4-2-10); 5513 (8-1-8); 5515 (14-2-5); 5517, 5518 (4-2-10); 5519 (17-1-3); 5521 (8-2-1); 5522 (8-1-9); 5526 (17-1-3); 5528 (3-1-7); 5529A (6-4-1); 5530 (22-2-1); 5532 (8-1-5); 5533 (4-3-1); 5534 (6-2-4); 5535 (3-

1829 (3-1-7); 1943 (3-1-29); 2011 (3-1-3); 2269 (3-1-17); 2285 (3-1-22a, 24); 2288 (3-1-25); 2705 (3-1-22a); 2705a (3-1-27); 2794 (3-1-17); 2795 (3-1-17); 2891 (3-1-17); 2893 (3-1-17); 6931 (3-1-22a)

LBB (Lands Bosbeheer), 15296 (3-1-1)

Lindeman J.C., *et al.*, 86C (21-1-2); 88 (23-1-1); 89 (4-3-1); 119 (6-4-1); 165A (6-4-3); 220 (21-1-2); 248B (6-2-4); 339B (21-1-2); 344 (4-6-3); 344B (6-4-6); 426B (17-3-4);435 (4-2-10); 436 (19-7-1); 455 (6-4-6); 462 (19-1-2); 462A (19-7-1); 465 (6-4-4); 588 (21-1-2); 590 (6-2-3); 603B (19-6-1); 617 (2-1-1); 4620 (1-1-9); 5515 (3-1-29, slide); 5516 (3-1-29); 6931 (3-1-22a); 6954B (8-2-6); s.n. (year 1954) (3-1-3)

Looy, C.H. van, 2 (17-1-4); 8 (4-6-1); 23 (22-5-2); 24 (6-4-1); 28 (6-2-4); 31 (8-2-10b); 32 (6-4-1); 34 (6-4-6); 35 (8-2-10b); 36 (8-2-12); 38 (9-3-1); 45 (17-1-4)

Maas, P.J.M., *et al.*, F3277 (3-1-22a); F3330 (4-2-12); F3359 (3-1-22a); F3365 (6-2-4); F3366 (3-1-12); F3366 A (3-1-22a); F3372 (13-3-3); F3374 (3-1-17); F3375 (3-1-3); F3377 (32-3-1); F3378 (3-1-22a, 29)

Maguire, B., *et al.*, 24598 (8-2-16d)

Pursell, R.A., 11623 (3-1-12); 11800 (3-1-32); 11859 (3-1-20)

Schulz, J.P., *et al.*, 10088A (4-2-10); 10314A (19-7-1); 10545 (4-2-10); 10546 (6-4-4); 10548 (4-2-12); 10572 (8-2-7); 10573 (4-2-12); 10574 (8-2-7); 10577 (6-4-4); 10585 (4-2-10); 10586A

(8-2-7); 10666 (14-2-2a); 10671 (6-4-4); 10672 (8-2-7); 10673 (4-2-12); s.n. (3-1-10a, 32)

Splitgerber, F.L., **1214** (6-4-6); s.n. (3-1-29)

Suringar, W.F.R., s.n. (3-1-10b, slide)

Teunissen, P.A., *et al.*, LBB 11925 (1-1-6a); 11927, 11928, 11929 (4-2-12); 14938, 16300 (1-1-9)

Weigelt, C., s.n. (3-1-29)

Wessels Boer, J.G., 841A (3-1-32); 871 (3-1-32)

FRENCH GUIANA

Aptroot, A., 15177 (19-7-1); 15194 (3-1-1, 19); 15229 (17-3-3); 15229A (17-1-4); 15246 (3-1-25); 15253 (22-5-2); 15265 (7-1-1); 15266 (22-5-2); 15275 (32-3-1); 15297 (20-2-2); 15298 (27-9-7); 15308 (3-1-7, 10a, 32); 15312 (20-2-2); 15315 (3-1-12, 22a); 15332 (22-5-2); 15333 (27-10-2); 15337 (17-2-5); 15395 (27-3-3); 15396 (27-10-3); 15407 (27-9-4); 15408 (27-10-1); 15434A, 15435 (8-1-1); 15452 (32-3-1); 15453 (27-3-4); 15455 (6-2-4); 15489 (17-1-2); 15541 (19-7-1); 15542 (3-1-10a); 15544 (8-1-1); 15545 (8-1-5); 15557 (27-3-4); 15558, 15561 (27-9-7); 15565 (8-1-2); 15571 (6-4-3); 15581 (27-3-4); 15585 (27-3-6); 15628 (15-2-1); 15631 (8-2-12); 15632 (8-1-7); 15630 (8-2-10b)

Bekker, J.M., 2281 (27-9-8); 2299B (32-3-1); 2305-2 (3-1-7, 31, 32); 2316A (19-3-1); 2321 (20-1-1); 2321A (21-1-1)

5674 (34-1-2); 5789 (6-2-4); 5792 (27-9-8); 5797 (17-1-2); 5801 (22-5-2); 5802-5803 (19-7-1); 5806 (27-10-1); 5810 (8-1-14); 5815 (15-2-1); 5822 (27-10-1); 5824 (19-7-1); 5826 (21-1-1); 5827 (19-7-1); 5829 (22-5-2); 5831 (8-2-12); 5843 (22-5-2); 5848 (27-9-7); 5853 (22-5-2); 5855 (6-2-4); 5857 (27-9-8); 5862-5863 (27-9-7); 5873 (33-1-4); 5876 (27-10-1); 5877 (33-1-4); 5878 (20-2-2); 5880 (8-1-5); 5881 (8-1-2); 5885 (4-6-3); 5886 (19-7-1); 5890 (6-4-3); 5891 (6-2-4); 5898 (8-2-18); 5899, 5900 (6-4-3); 5901 (8-2-10b); 5904 (19-7-1); 5907 (27-10-1); 5908 (22-5-2); 5915 (6-2-4); 5916 (4-2-4); 5917 (6-2-4); 5918 (6-4-3); 5922 (8-2-19); 5923 (6-2-4); 5925 (8-2-16d); 5926 (6-2-4); 5930 (8-2-19); 5931 (8-2-16d); 5933, 5934, 5935 (6-2-4); 5936 (27-10-1); 5937 (17-1-2); 5938 (6-4-6); 5940 (19-7-1); 5942 (27-9-7); 5966 (19-7-1); 5972 (8-2-10b); 5979 (15-2-1); 5983 (6-2-4); 5986 (8-1-5); 5987 (27-10-1); 6001 (19-7-1); 6202 (8-2-10b); 6207 (8-1-12); 6212 (6-2-4); 6213 (8-2-7); 6214 (8-2-19); 6215, 6216 (6-2-4); 6218, 6219 (6-4-3); 6222 (8-2-19); 6230 (14-2-4); 6239 (19-7-1); 6241 (19-3-1); 6256 (6-2-4); 6261 (3-1-17); 6265 (3-1-10a); 6267 (8-1-12); 6258 (19-7-1); 6272 (15-2-1); 6275 (4-5-1); 6276 (8-2-16a); 6277 (33-1-4); 6281 (6-2-4); 6282 (20-1-1); 6286 (4-6-3); 6288 (6-4-5); 6296 (20-1-1); 6305-6306 (27-9-8); 6309, 6311 (19-7-1); 6755 (6-2-4); 6758

(21-1-1); 6760 (19-3-1); 6765 (14-1-2); 6767 (6-4-3); 6778 (3-1-10a); 6778 (27-4-1); 6790 (22-5-1); 6793 (6-2-3); 6794 (6-2-1); 6796 (4-6-3); 6799 (8-1-5); 6801 (6-2-3); 6802, 6804 (19-3-1); 6806 (3-1-16); 6807 (27-3-6); 6810 (19-3-1); 6815 (17-2-1); 6817 (27-3-1); 6822 (4-6-3); 6825 (27-12-1); 6828 (27-9-4); 6834 (19-3-1); 6835 (19-4-1); 6836 (27-3-1); 6838 (27-3-4); 6841 (6-4-3); 6847 (6-4-6); 6850 (19-7-1); 6852 (6-2-4); 6857 (8-1-1); 6858 (8-2-16b); 6859, 6860 (6-4-3); 6861 (6-2-4); 6862 (4-2-12).; 6863 (6-2-4); 6865 (17-2-1); 6869 (19-6-1); 6874 (19-1-2); 6880 (19-7-1); 6883 (17-2-8); 6884 (17-1-4); 6886 (19-7-1); 6887 (17-2-5); 6895 (19-4-1); 6895C (17-1-2); 6915 (32-3-1); 7022 (6-2-4); 7024 (8-2-16a); 7025 (6-4-3); 7026 (6-2-4); 7027 (27-10-1); 7023 (8-2-16b); 7030 (19-7-1); 7117 (6-4-3); 7118 (4-2-1); 7119 (19-7-1); 7121 (17-3-3); 7122 (22-5-2); 7123 (8-2-16a); 7124, 7126 (6-2-4); 7127 (6-4-3); 7129, 7130 (6-2-4); 7132 (6-4-6); 7133 (6-2-4); 7139 (27-10-1); 7143 (20-2-2); 7144 (19-7-1); 7158 (17-1-3); 7159 (17-1-2); 7320 (17-2-1); 7596 (19-7-1); 7599 (22-5-2); 7600 (8-2-10b); 7601 (6-4-6); 7604 (27-3-4); 7606 (6-4-3); 7608 (8-2-17); 7609 (27-10-1); 7610 (8-2-17); 7612 (19-7-1); 7616 (6-4-6); 7621 (8-1-8); 7622 (6-2-4); 7624 (19-7-1); 7630 (27-10-1); 7631 (22-5-2); 7632 (7-1-1); 7634 (6-4-6); 7635 (19-7-1); 7636 (17-2-3); 7640 (21-1-2);

7637 (19-6-1); 7638 (8-2-16c); 7645 (6-4-3); 7646 (8-2-19); 7648 (6-4-5); 7650 (19-7-1); 7651, 7683 (22-5-2); 7866 (4-3-1); 8021 (19-7-1); 8022 (27-9-4); 8025 (4-6-3); 8026 (21-1-2); 8027 (27-10-1); 8028 (8-1-5); 8030 (21-1-2); 8032 (14-1-3); 8034 (22-5-1); 8043 (6-2-4); 8044 (4-6-3); 8050 (6-4-1); 8055 (4-6-3); 8346, 8350 (22-5-2); 8351 (8-2-13); 8352 (8-2-6); 8353 (6-4-3); 8354 (6-2-4); 8357 (8-1-1); 8359 (6-4-6); 8360 (8-1-1); 8362 (8-2-19); 8366 (6-4-6); 8368 (8-1-8); 8641 (3-1-29); 8643 (9-2-1); 8644 (8-1-9); 8645, 8646 (9-2-1); 8647 (8-1-9); 8745 (13-3-3); 8796 (32-3-1); 8798 (8-1-5); 8800 (22-5-2); 8802 (6-4-6); 8806 (17-2-8); 8816 (6-4-6); 8817 (8-2-2); 8820 (6-4-6); 8840 (22-5-2); 8842 (23-1-1); 8856 (6-4-6); 8857 (19-5-1); 8858 (22-6-2); 8859 (19-7-1); 8865 (20-2-2); 8903 (27-9-4); 8935 (8-2-10b); 8961 (6-4-6); 8863 (8-2-2); 8905, 8906 (19-4-1); 8952 (3-1-27, same as 'Onraedt 8952'); 8958 (27-10-3); 8960 (27-3-1); 8975 (6-4-3); 8977 (8-2-17); 8981 (22-3-1); 8983, 8984 (19-3-1); 8986 (27-9-7); 8988 (8-2-10a); 8991 (27-9-7); 8993 (8-1-5); 8994 (6-2-2); 8997 (19-1-2); 8998 (8-2-10b); 8999 (22-5-1); 9006 (8-2-10a); 9007 (23-1-1); 9008 (14-2-3); 9009 (19-3-1); 9025 (17-2-7); 9035 (19-3-1); 9036 (8-1-5); 9037 (6-2-4); 9038 (21-1-2); 9043 (19-3-1); 9057 (27-3-1); 9061(3-1-10a, 16); 9062 (4-5-1); 9065 (6-2-2); 9066 (27-3-4); 9069 (33-1-1); 9076 (8-2-16a); 9078 (27-3-4); 9080 (6-2-4); 9081 (4-2-4); 9083 (4-6-3); 9095 (19-3-1); 9103 (22-3-1); 9104A (23-1-1); 9105 (19-6-1); 9107-9108 (27-9-4); 9109 (20-2-2); 9110 (27-3-1); 9117-9118 (27-3-4); 9119 (6-2-4); 9126 (8-2-16a); 9132 (22-5-1); 9134 (33-1-2); 9135 (27-9-8); 9145 (14-2-4); 9146 (3-1-14a); 9147 (6-4-6); 9163 (6-4-3); 9165 (19-2-1); 9168 (6-2-4); 9172 (19-3-1); 9181 (17-1-4); 9189 (4-5-1); 9210 (23-1-1); 9216 (6-2-4); 9217 (4-6-1); 9219 (27-3-4); 9222 (21-1-2); 9228 (19-3-1); 9229 (23-1-1); 9238, 9241 (8-2-16a); 9248 (4-6-1); 9262 (27-3-4); 9263 (19-4-1); 9266 (8-2-10b); 9284 (19-3-1); 9293 (17-2-1); 9300, 9302 (23-1-1); 9309 (27-10-1); 9310 (27-3-4); 9312 (23-1-1); 9993, 9996 (6-4-3); 9999 (14-2-4); 10091 (6-4-6); 10288 (14-2-3); 10750 (6-4-6); 10751 (17-2-8); 10802 (32-3-1); 10956, 10961 (6-4-3); 10965 (6-4-2); 10966 (6-2-4); 10974 (17-2-1); 10976 (15-2-1); 10977 (17-2-1); 10978 (8-1-5); 10928 (15-2-1); 10986 (6-2-4); 11016 (19-6-1); 11026 (6-2-4); 11028 (19-6-1); 11033 (6-4-3); 11034 (27-3-4); 11039, 11042 (8-2-17); 11043 (6-2-4); 11052 (17-2-5); 11053, 11055 (6-2-4); 11056 (8-1-7); 11060 (27-7-2); 11064 (32-3-1); 11066 (8-2-12); 11071 (17-2-5); 11072 (8-2-16a); 11073, 11074, 11075 (6-4-3); 11094 (6-4-1); 11467 (6-4-6); 11633 (22-5-2); 11635 (8-2-10b); 11637 (19-7-1); 11639 (8-1-5); 11641

INDEX TO SYNONYMS, NAMES IN NOTES AND SOME TYPES

Acanthocladium
  costaricense Dixon & E.B. Bartram = 31-9-1
Acroporium
  brevicuspidatum Mitt., see 31-1, type
  intricatum Thér. = 31-8-2
Actinodontium
  adscendens Schwägr., see 27-1, type
  portoricense H.A. Crum & Steere = 27-1-1
ADELOTHECIACEAE, see 27, note
Adelothecium Mitt., see 27, note
Anacamptodon
  splachnoides (Brid.) Brid., see 28-1, type
Amblytropis
  denticulata Thér. = 27-9-3
Anictangium
  cirrosum Hedw. = 17-2-1
  humboldtii Hook., see 24-1, type; = 24-1-1
Archidium
  alternifolium (Dicks. ex Hedw.) Schimp., see 2-1, type
  ohioense Schimp. ex Müll. Hal. = 2-1-1
  phascoides Brid., see 2-1, type
Atractylocarpus Mitt., see 4, note
Arthrocormus
  pulvinatus Dozy & Molk. = 6-4-6
Barbula
  agraria Hedw., see 9-3, type; = 9-3-1
  cruegeri Sond. ex Müll. Hal. = 9-1-1
  sect. *Hyophiladelphus* Müll. Hal. = 9-3
  subgenus *Hyophiladelphus* (Müll. Hal.) Zander = 9-3
  unguiculata Hedw., see 9-1, type
Bartramia
  glaucescens Hornsch. = 13-3-3
  sect. *Breutelia* Bruch & Schimp. = 13-1
  sphaerocarpa (Hedw.) P. Beauv. = 13-3-2
  subsect. *Leiomela* Mitt. = 13-2
  uncinata Schwägr. = 13-3-3
Brachymenium
  mnioides Besch. = 14-1-3
  nepalense Hook. ex Schwägr., see 14-1, type
  sipapoense E.B. Bartram = 14-1-2
  wrightii (Sull.) Broth. var. *mnioides* (Besch.) Florsch. = 14-1-3

BRACHYTHECIACEAE, see 19-2, note
Breutelia
  arcuata (Sw.) Schimp., see 13-1, type
  scoparia (Schwägr.) A. Jaeger, see 13-1-1, note
Brymela
  tutezona Crosby & B.H. Allen, see 27-2, type
Bryohumbertia
  metzlerelloides P. de la Varde & Thér., see 4-1, type
Bryum
  candicans Taylor = 14-2-2a
  cruegeri Hampe = 14-2-1
  lycopodioides Brid. = 8-2-14
  parasiticum Brid. = 8-2-15
  truncorum (Brid.) Brid., see 14-2-4
Buxbaumia
  foliosa Hedw., see 35-1, type
BUXBAUMIACEAE, see 35, note
Callicosta Müll. Hal. = 27-10
  bipinnata (Schwägr.) Müll. Hal. = 27-10-1
  evanescens Müll. Hal. = 27-10-2
  fendleri (Müll. Hal.) Crosby = 27-10-3
Callicostella
  grossiretis E.B. Bartram = 27-3-1
  papillata (Mont.) Mitt., see 27-3, type
Calymperes
  androgynum Mont. = 8-2-17
  disciforme Müll. Hal. = 8-2-1
  donnellii Austin = 8-1-1
  gardneri Hook., see 8-2, type
  lanceolatum Hampe = 8-2-11
  palisotii Schwägr. subsp. *richardii* (Müll. Hal.) S.R. Edwards = 8-1-9
  richardii Müll. Hal. = 8-1-9
  rufum Herzog = 8-1-12
  rupicola P.W. Richards = 8-1-8
  uleanum Broth. = 8-1-10
Calymperopsis
  martinicensis (Broth.) Broth. = 8-2-15
Calyptothecium
  praelongum Mitt., see 22-1, type
Campylopus
  arenicola (Müll. Hal.) Mitt. = 4-2-13
  atratus Broth. = 4-2-9
  bartlettii E.B. Bartram = 4-2-10
  filifolius (Hornsch.) Mitt., = 4-1-1

flavicoma Müll. Hal., see 4-1, type
flexuosus (Hedw.) Brid., see 4-2, type
gracilicaulis Mitt. = 4-2-12
harrisii (Müll. Hal.) Paris = 4-2-4
lamellinervis (Müll. Hal.) Mitt. var. exaltatus (Müll. Hal.) Frahm, see
    4-2-6, note
nanofilifolius (Müll. Hal.) Paris = 4-1-1
praealtus (Müll. Hal.) Paris = 4-2-11
savannarum (Müll. Hal.) Mitt. subsp. *bartlettii* (E.B. Bartram) Florsch.
    = 4-2-10
surinamensis Müll. Hal. var. *angustiretis* (Austin) Frahm = 4-2-1
Carinafolium
    tatei R.S. Williams = 6-4-8
Chaetophora
    incurva Hornsch. = 27-11-2
Chryso-hypnum
    patens Hampe, see 30-1, type
Climacium
    subgenus *Porotrichum* Brid. = 33-1
Colobodontium
    aciculare Herzog, see 31-2, type and note
Conomitrium
    intramarginatum Hampe = 3-1-13
    puiggari Geh. & Hampe = 3-1-28
    reticulosum Müll. Hal. = 3-1-21
Crossomitrium
    portoricense Müll. Hal. = 27-8-1
    rotundifolium Herzog = 27-4-3
Cryptangium
    schomburgkii Müll. Hal. = 29-2-1
Cyathophorum
    dupuisii Renauld & Cardot, see 30-6, type
Cyclodiction
    laetevirens (Hook. & Taylor) Mitt., see 27-5, type
Cyrto-hypnum (Hampe) Hampe & Lorentz = 34-1
    brachythecium (Hampe & Lorentz) Hampe & Lorentz, see 34-1, type
    involvens (Hedw.) W.R. Buck & H.A. Crum = 34-1-1
    scabrosulum (Mitt.) W.R. Buck & H.A. Crum = 34-1-2
    schistocalyx (Müll. Hal.) W.R. Buck & H.A. Crum = 34-1-3
Daltonia
    splachnoides (Sm.) Hook & Taylor, see 25-1, type
Dicranodontium Bruch, see 4, note
Dicranella
    grevilliana (Brid.) Schimp., see 4-3, type

Dicranoloma
  brittonae E.B. Bartram, see 4-4, note
Dicranum
  albicans Schwägr. = 6-2-1
  angustirete Austin = 4-2-1
  arctocarpus Hornsch. = 4-2-2
  arenicola Müll. Hal. = 4-2-13
  cruegerianum Müll. Hal. = 4-6-1
  dichroste Müll. Hal. = 4-2-5
  filifolium Hornsch. = 4-1-1
  flexuosum Hedw., see 4-2, type
  glaucum Hedw., see 6-2, type
  harrisii Müll. Hal. = 4-2-4
  hilarianum Mont. = 4-3-1
  lamellinerve Müll. Hal. = 4-2-6
  lycopodioides (Brid.) Sw. = 8-2-14
  martianum Hornsch. = 6-2-4
  nanofilifolium Müll. Hal. = 4-1-1
  praealtum Müll. Hal. = 4-2-11
  savannarum Müll. Hal. = 4-2-10
  schreberi Sw. var. *grevillianum* Brid., see 4-3, type
  subcuspidatum Hampe = 4-2-11
  trachyblepharon Müll. Hal. = 4-2-13
Didymodon
  klotzschii Schwägr. = 14-1-1
Diphyscium
  foliosum (Hedw.) D. Mohr, see 35-1, type
  peruvianum Spruce ex Mitt. = 35-1-1
  ulei Müll. Hal. = 35-1-1
Distichophyllum
  cubense Mitt. = 25-2-1
Donnellia
  floridana Austin, see 31-3, type
Drepanophyllum Schwägr. = 15-2
  duidense R.S. Williams = 15-2-1
  falcifolium (Schwägr.) Wijk & Margad. = 15-2-1
  semilimbatum Mitt., see 15-1, type
  viride Mitt. = 15-1-1
Ectropothecium
  trinitense (Müll. Hal.) Mitt. = 30-6-1
  tutuilum (Sull.) Mitt., see 30-2, type
Entodontopsis
  rhabdodonta Cardot = 32-1-1

Entosthodon
    templetonii (Sm.) Schwägr., see 11-1, type
Eucamptodon
    pilifera Mitt. = 4-4-1
Fabronia
    cubensis Sull. = 28-1-1
Fissidens
    asplenioides Hedw., see 3-1-17, note
    austro-americanus Pursell & W.D. Reese = 3-1 30
    bryoides Hedw., see 3-1, type
    densiretis Sull. var. *latifolius* Grout = 3-1-7
    diplodus Mitt. = 3-1-14b
    diplodus Mitt. var. *richardsii* (R.S. Williams) Pursell = 3-1-14b
    falcifolius Schwägr. = 15-2-1
    flexinervis Mitt. = 3-1-22a
    guadelupensis Schimp. ex Besch., see 3-1-17, note
    intermedius Müll. Hal. = 3-1-29
    intromarginatus (Hampe) Mitt. = 3-1-13
    kegelianus Müll. Hal. = 3-1-32
    leptophyllus Mont. fo. *obtusissimus* Florsch. = 3-1-18
    leptopodus Cardot = 3-1-7
    marmellensis Broth. = 3-1-24
    mollis Mitt. = 3-1-8
    muriculatus Spruce ex Mitt. = 3-1-14b
    muriculatus Spruce ex Mitt. var. *richardsii* (R.S. Williams) Florsch. =
        3-1-14b
    papilliferus Broth. = 3-1-22b
    pauperculus M. Howe var. *surinamensis* Florsch. = 3-1-12
    pennula Broth. = 3-1-3
    prionodes Mont. fo. *flexinervis* (Mitt.) Florsch. = 3-1-22a
    prionodes Mont. fo. *prionodes* (Mont.) Florsch. = 3-1-24
    prionodes Mont. fo. *puiggari* (Geh. & Hampe) Florsch. = 3-1-28
    puiggari (Geh. & Hampe) Paris = 3-1-28
    ravenelii Sull. = 3-1-7
    reticulosus (Müll. Hal.) Mitt. = 3-1-21
    saülensis Pursell & W.R. Buck = 3-1-22a
    similiretis Sull. = 3-1-17
    weirii Mitt. var. *bistratosus* W.R. Buck = 3-1-31
Floribundaria
    flaccida (Mitt.) Broth., see 19-1-2, note
Fontinalis
    gymnostoma Bruch. & Schimp. = 29-2-1
Funaria
    hygrometrica Hedw., see 11-2, type

422

hygrometrica Hedw. var. *calvescens* (Schwägr.) Mont. = 11-2-1
subgenus *Entosthodon* (Schwägr.) Lindb. = 11-1
templetonii Sm., see 11-1, type
Garckea
   phascoides Müll. Hal., see 5-1, type
Garovaglia
   planifrons Renauld & Paris = 22-1-1
Glossadelphus M. Fleisch. = 30-5
   truncatulus (Müll. Hal.) M. Fleisch. =30-5-1
Grimmia
   flexuosa Griff. = 5-1-1
   fontinaloides Hook. = 29-1-1
Groutiella
   apiculata (Hook.) H.A. Crum & Steere, see 17-1-2, note
   fragilis (Mitt.) H.A. Crum & Steere = 17-1-4
   undosa (Cardot) H.A. Crum & Steere = 17-1-1
   wagneriana (Müll. Hal.) H.A. Crum & Steere, see 17-1, note
Gymnostomum
   bonplandii Hook. = 11-1-1
   involutum Hook. = 9-2-1
   javanicum Nees & Blume, see 9-2, type
   tortula Schwägr. = 9-2-1
Harpophyllum
   aureum (Brid.) Spruce = 27-6-1
Harrisonia Spreng. = 24-1
HEDWIGIACEAE, see 24, note
Hemiragis
   striata (Schwaegr.) Besch., see 27-6, type
Henicodium
   niam-niamiae (Müll. Hal.) Kindb., see 22-2, type
Hildebrandtiella
   endotrichelloides Müll. Hal., see 22-3, type
   perpinnata Broth., see 22-3, type
Holomitrium
   olfersianum Hornsch., see 4-5-1, note
   perichaetiale (Hook.) Brid., see 4-5, type
   williamsii E.B. Bartram, see 4-5-1, note
Homalia
   lentula Wilson = 20-1-1
Homaliodendron
   piniforme (Brid.) Enroth = 33-1-2
Hookeria
   crispa Müll. Hal. = 27-11-4
   cubensis Sull. = 27-9-2

cymbifolia Hampe = 27-7-2
debilis Sull., see 26-1, type
diversifolia Mitt. = 27-7-1
falcata Hook. = 27-12-1
laetevirens Hook. & Taylor, see 27-5, type
leptorrhyncha Hook. & Grev., see 27-7, type
martiana Hornsch., see 27-3, type
merkelii Hornsch. = 27-3-3
pallescens Hook. = 27-7-2
pallida Hornsch. = 27-3-4
papillata Mont., see 27-3, type
parkeriana Hook. & Grev. = 27-2-1
pendula Hook., see 27-11, type
radiculosa Hook. = 32-4-1
rivularis Mitt. = 27-3-5
rufescens Mitt. = 27-3-6
sect. *Callicostella* Müll. Hal. = 27-3
sect. *Callicostella* Müll. Hal. subsect. *Thamniopsis* Mitt. = 27-11
sect. *Hypnella* Müll. Hal. = 27-7
sect. *Lepidopilidium* Müll. Hal. = 27-8
strumosum Hornsch. = 26-1-2
subgenus *Hemiragis* (Brid.) Mitt. = 27-6
surinamensis (Müll. Hal.) Müll. Hal. = 27-9-8
tenuiseta Müll. Hal.), see 27-8, type
varians Sull. = 27-5-1
HOOKERIACEAE, see 27, note
Hookeriopsis (Besch.) A. Jaeger, see 27-11, note
crispa (Müll. Hal.) A. Jaeger = 27-11-4
falcata (Hook.) A. Jaeger = 27-12-1
guadalupense (Spreng.) A. Jaeger = 27-12-1
guatemalensis E.B. Bartram = 27-3-2
incurva (Hornsch.) Broth. = 27-11-2
killipii R.S. Williams = 27-11-3
parkeriana (Hook. & Grev.) A. Jaeger = 27-2-1
Hydropogonella
    gymnostoma (Bruch. & Schimp.) Cardot fo. *obtusifolia* P.W. Richards
        = 29-2-1
Hyophila
    javanica (Nees & Blume) Brid., see 9-2, type
    tortula (Schwägr.) Hampe = 9-2-1
Hyophiladelphus
    agrarius (Hedw.) Zander, see 9-3, type
Hypnella
    cymbifolia (Hampe) A. Jaeger = 27-7-2

guayanensis B.H. Allen & W.R. Buck = 27-7-2
leptorrhyncha (Hook. & Grev.) A. Jaeger, see 27-7, type
Hypnum
  aureum Lam. ex Brid. = 27-6-1
  bifidum Brid., see 4-6, type
  brachythecium Hampe & Lorentz, see 34-1, type
  chlorophyllum Hornsch. = 32-3-1
  concavum Hook. = 31-7-1
  contorte-operculatum Müll. Hal. = 32-1-1
  cultelliforme Sull. = 32-2-1
  demissum Wilson, see 31-6, type
  diminutivum Hampe = 30-1-1
  extenuatum Brid., see 31-9, type
  galipense Müll. Hal. = 31-6-2
  guadalupense Spreng. = 27-12-1
  hastatum Duby = 13-3-1
  hornschuchii Hampe = 31-8-4
  leptochaeton Schwägr. = 30-2-1
  leucotrichum Taylor = 19-6-1
  longissimum Raddi = 22-8-1
  lonchophyllum Mont. = 31-6-3
  microcarpum Hornsch. = 31-8-4
  niam-niamiae Müll. Hal., see 22-2, type
  nigrescens Sw. ex Hedw. = 19-5-1
  papillosum Hornsch. = 31-8-3
  patrisiae Brid. = 27-4-2
  patulum Hedw. = 19-7-1
  planum Brid. = 31-7-2
  polytrichoides Hedw. = 27-9-4
  pungens Hedw. = 31-1-1
  purpurascens Brid. = 24-1-1
  reptans Hedw. = 30-4-1
  schistocalyx Müll. Hal. = 34-1-3
  sect. *Henicodium* Müll. Hal. = 22-2
  spiniforme Hedw. = 16-1-1
  strumosum (Hornsch.) Müll. Hal. = 26-1-2
  subbrevisetum Hampe = 30-3-1
  subgenus *Cyrto-hypnum* Hampe = 34-1
  subsect. *Vesicularia* Müll. Hal. = 30-7
  subsimplex Hedw. = 31-6-6
  tamariscinum Hedw., see 34-2, type
  tenerum Sw. = 30-3-2
  tetragonum Hedw. = 22-5-2
  tomentosum Hedw. = 23-1-1

torquatum Hedw. = 17-3-4
trichophyllum Sw. ex Hedw. = 19-2-1
trinitense Müll. Hal. = 30-6-1
truncatulum Müll. Hal. = 30-5-1
tutuilum Sull., see 30-2, type
vesiculare Schwägr. = 30-7-1
HYPOPTERYGIACEAE, see 27, note
Isopterygium
    tenerifolium Mitt., see 30-3-2, note
Jaegerina
    stolonifera (Müll. Hal.) Müll. Hal., see 22-4, type
Jaegerinopsis
    scariosa (Lorentz) Broth. = 22-4-1
    squarrosa E. Britton = 22-4-1
Lepidopilidium
    tenuisetum (Müll. Hal.) Broth., see 27-8, type
Lepidopilum
    amplirete (Sull.) Mitt., see 27-9-2, note
    brevipes Mitt. = 27-9-7
    epiphyllum Mitt. = 27-4-1
    sect. *Isodrepanium* Mitt. = 20-1
    sprucei Mitt. = 27-1-1
    stolonaceum Müll. Hal. = 27-9-7
    subenerve Brid., see 27-9, type
Leptodontium Müll. Hal., see 9, note
Leptotheca
    speciosa Hook. & Wilson = 14-1-2
    wrightii Sull. = 14-1-3
LEPYRODONTOPSIDACEAE, see 19-2, note
Leskea
    floribunda Dozy & Molk. = 19-1-2
    involvens Hedw. = 34-1-1
    leucostega Brid. = 32-1-2
    remotifolia Müll. Hal. = 19-3-1
    striata Schwaegr., see 27-6, type
    subgenus *Hemiragis* Brid. = 27-6
    subpinnata Brid. = 31-6-5
    undata Hedw. = 27-11-4
Leskeodon
    auratus (Müll. Hal.) Broth., see 25-2, type
Leucobryum
    giganteum Müll. Hal., see 6-2-1, note
    glaucum (Hedw.) Ångstr., see 4-6, type
    laevifolium Broth. = 6-1-1

subobtusifolium (Broth.) B.H. Allen = 6-3-1
vulgare Hampe, see 6-2, type
Leucodon
  bartramioides Hook. = 13-2-1
  geniculatus Mitt. = 22-2-1
  pohlii Schwägr. = 22-6-2
LEUCODONTACEAE, see 18, note
Leucodontopsis Renauld & Cardot = 22-2
  geniculata (Mitt.) H.A. Crum & Steere = 22-2-1
  plicata Renauld & Cardot, see 22-2, type
Leucoloma
  bifidum (Brid.) Brid., see 4-6, type
Leucomium
  debile (Sull.) Mitt., see 26-1, type
Leucophanes
  gardneri Müll. Hal., see 6-3, type
  mittenii Cardot = 7-1-1
  octoblepharioides Brid., see 7-1, type
Macromitrium
  aciculare Brid., see 17-2, type
  chimborazense Spruce ex Mitt. = 17-1-1
  dubium Schimp. ex Müll. Hal., see 17-2-3, note
  fragile Mitt. = 17-1-4
  obtusum Mitt. = 17-1-3
  pallidum (P. Beauv.) Wijk & Margad., see 17-2, type
  pentastichum Müll. Hal. = 17-2-8
  portoricense R.S. Williams = 17-2-7; see 17-2-13, note
  schwaneckeanum Hampe = 17-2-1
  tomentosum Hornsch. = 17-1-4
  undosum Cardot = 17-1-1
Maguireella W.R. Buck = 31-2
  vulpina (Mont.) W.R. Buck, see 31-2, type; = 31-2-2
Meiotheciopsis Broth. = 31-3
  commutata (Müll. Hal.) W.R. Buck = 31-3-1
Meiothecium
  commutatum (Müll. Hal.) Broth. = 31-3-1
  sect. *Pterogonidiopsis* Broth. = 31-3
  stratosum Mitt., see 31-4, type
Meteoriopsis
  patula (Hedw.) Broth. = 19-7-1
  remotifolia (Müll. Hal.) Broth. = 19-3-1
  sect. *Meteoridium* (Müll. Hal.) Broth. = 19-3
  sect. *Squarridium* Broth. = 19-7

Meteorium
　lorentzii Müll. Hal., see 19-6, type
　macrocarpum Mitt. = 19-6-2
　s*cariosum* Lorentz = 22-4-1
Micromitrium
　austinii Sull. ex Austin,see 10-1, type
Microthamnium Mitt. = 30-4
Mittenothamnium
　diminutivum (Hampe) E. Britton = 30-1-1
Mniadelphus
　auratus Müll. Hal., see 25-2, type
Mniomalia
　bernouillii Müll. Hal. = 15-1-1
　semilimbata (Mitt.) Müll. Hal., see 15-1, type
Mnium
　arcuatum Sw., see 13-1, type
　beyrichianum Hornsch. = 14-2-3
　fontanum Hedw., see 13-3, type
　lanatum P. Beauv. = 14-2-2b
　sphaerocarpon Hedw. = 13-3-2
　tomentosum (Brid.) Sw. ex Brid. = 13-1-1
Moenkemeyera
　richardsii R.S. Williams = 3-1-14b
MYRINIACEAE, see 28, note
Nanomitrium Lindb. = 10-1
　thelephorothecum Florsch. = 10-1-1
Neckera
　bipinnata Schwägr. = 27-10-1
　boryana Müll. Hal. = 31-4-1
　commutata Müll. Hal. = 31-3-1
　disticha Hedw. = 20-2-1
　domingensis Spreng. = 18-1-1
　guyanensis Mont. = 22-3-1
　hexasticha (Schwägr.) Müll. Hal., see 19-4, type
　korthalsiana Dozy & Molk. = 33-1-1
　longirostris Hook., see 33-1, type
　scabriseta Schwägr. = 27-9-7
　sect. *Pseudopilotrichum* subsect. *Orthostichella* Müll. Hal. = 19-4
　splachnoides Sm., see 25-1, type
　subsect. *Meteoridium* Müll. Hal. = 19-3
　subsect. *Papillaria* Müll. Hal. = 19-5
　substriata Hampe = 33-1-4
　undulata Hedw. = 20-2-2

Pilotrichum
  bipinnatum (Schwägr.) Brid. var. *evanescens* (Müll. Hal.) Müll. Hal. =
    27-10-2
  cymbifolium Sull. = 22-6-1
  flagelliferum Brid. = 18-1-1
  stoloniferum Müll. Hal., see 22-4, type
  subgenus *Lepidopilum* Brid. = 27-9
Pilotrichum Brid. = 27-10
Pinnatella
  piniformis (Brid.) M. Fleisch. = 33-1-2
Pirea
  mariae Cardot, see 22-6, type
Pireella
  mariae (Cardot) Cardot, see 22-6, type
PLAGIOTHECIACEAE, see 32, note
Pleuridium
  globiferum Brid. = 2-1-1
Poecilophyllum
  tortellum Mitt. = 4-6-4
Pohlia
  apiculata (Schwägr.) H.A. Crum & L.E. Anderson = 14-2-1
Polytrichum
  commune Hedw., see 36-1, type
Porotrichum
  cobanense Müll. Hal. = 33-1-1
  longirostre (Hook.) Mitt., see 33-1, type
  piniforme Brid. = 33-1-2
  plicatulum Mitt. = 33-1-4
  usagarum Mitt., see 33-1-1, note
Potamium Mitt. = 31-6
  deceptivum Mitt. = 31-2-1
  lonchophyllum (Mont.) Mitt., see 31-2, note; = 31-6-3
  pacimoniense Mitt., see 31-2, note; = 31-6-4
  vulpinum (Mont.) Mitt. = 31-2-2
Pseudocryphaea
  flagellifera (Brid.) E. Britton = 18-1-1
Pseudohypnella
  guianensis P.W. Richards = 27-7-2
Pseudopilotrichum (Müll. Hal.) W.R. Buck & B.H. Allen = 19-4
Pterygophyllum
  indicum Bél., see 32-4, type
Pterigynandrum
  fulgens Hedw. = 21-1-1

Pterogonium
  pulchellum Hook. = 31-5-1
Renauldia
  hildebrandtielloides Müll. Hal., see 22-7, type
Rhacocarpus
  humboldtii (Hook.) Lindb., see 24-1, type; = 24-1-1
Rhacopilopsis
  dupuisii (Renauld & Cardot) Renauld & Cardot, see 30-6, type
Rhaphidostegium
  cochleatum Broth. = 31-6-1
  subdemissum Schimp. ex Besch. = 31-8-4
Rhizogonium
  lindigii (Hampe) Mitt., see 16-1, note
  sect. *Pyrrhobryum* (Mitt.) Mitt. = 16-1
  spiniforme (Hedw.) = 16-1-1
Rhodobryum
  beyrichianum (Hornsch.) Müll. Hal. = 14-2-3
Rosulabryum
  billardierei (Schwägr.) Spence, see 14-2-4, note
SAULOMATACEAE, see 27, note
SCHIMPEROBRYACEAE, see 27, note
Schizomitrium Schimp. = 27-3
  colombicum (R.S. Williams) W.R. Buck & Steere = 27-3-1
  grossiretis (E.B. Bartram) H.A. Crum = 27-3-1
  guatemalensis (E.B. Bartram) J. Florsch. = 27-3-2
  martinianum (Hornsch.) Crosby, see 27-3, type
  merkelii (Hornsch.) J. Florsch. = 27-3-3
  pallidum (Hornsch.) H.A. Crum & L.E. Anderson = 27-3-4
  rivulare (Mitt.) H.A. Crum = 27-3-5
  rufescens (Mitt.) J. Florsch. = 27-3-6
Schlotheimia
  stellulata Hornsch. = 17-2-10
Sematophyllum
  allinckxiorum W.R. Buck = 31-6-5
  demissum (Wilson) Mitt., see 31-6, type
  maguireorum W.R. Buck = 31-2-2
  squarrosum W.R. Buck = 31-6-5
Sphagnum
  auriculatum Schimp. var. *ovatum* Warnst., see 1-1-2, note
  cuspidatum Ehrh. ex Hoffm. var. *serrulatum* (Schlieph.) Schlieph. =
    1-1-9
  palustre L., see 1-1, type; see 1-1-6a, note
  tabuleirense Yano & H.A. Crum = 1-1-6b

Squamidium
  lorentzii (Müll. Hal.) Broth., see 19-6, type
Stereophyllum
  contorte-operculatum (Müll. Hal.) Mitt. = 32-1-1
  cultelliforme (Sull.) Mitt. = 32-2-1
  indicum (Bél.) Mitt., see 32-4, type
  leucostegum (Brid.) Mitt. = 32-1-2
  nitens Mitt. = 32-1-3
  obtusum Mitt. = 32-1-3
  rhabdodonta (Cardot) Grout = 32-1-1
  sect. *Pilosium* Müll. Hal. = 32-3
Syrrhopodon
  acanthoneuros Müll. Hal. = 8-2-16b
  androgynus (Mont.) Besch. = 8-2-17
  brevisetus Florsch. = 8-2-18
  cincinnatus Hampe = 8-2-16c
  disciformis Dusén, see 8-2 and 8-2-1, note
  gardneri (Hook.) Schwägr., see 8-2, type
  glaziovii Hampe = 8-2-5
  graminicola R.S. Williams = 8-2-1
  incompletus Schwägr. var. *elatus* (Mont.) Florsch. = 8-2-4
  incompletus Schwägr. var. *lanceolatus* (Hampe) Reese = 8-2-11
  incompletus Schwägr. var. *perangustifolius* Reese, see 8-2-11, note
  longisetaceus Müll. Hal. = 8-2-17
  luridus Paris & Broth. = 8-2-10b
  martinicensis Broth. = 8-2-15
  miquelianus Müll. Hal. = 8-2-9
  papillosus Müll. Hal. = 8-2-16d
  parasiticus (Brid.) Besch. var. *disciformis* (Müll. Hal.) Florsch. = 8-2-1
  parasiticus (Brid.) Besch. var. *flexifolius* (Mitt.) Reese = 8-2-6
  prolifer Schwägr. var. *papillosus* (Müll. Hal.) Reese = 8-2-16d
  rubiginosus Mitt. = 8-1-12
  rufus Hornsch. = 8-2-9
  scaber Mitt. = 8-2-16d
  surinamensis Dozy & Molk. = 8-2-9
  venezuelanus Mitt. = 8-1-14
Thamniopsis
  pendula (Hook.) M. Fleisch., see 27-11, type
Thuidium
  delicatulum (Hedw.) Schimp. var. *peruvianum* (Mitt.) H.A. Crum =
    34-2-1
  involvens (Hedw.) Mitt = 34-1-1
  scabrosulum Mitt. = 34-1-2

schistocalyx (Müll. Hal.) Mitt. = 34-1-3
subgenus *Microthuidium* Limpr. = 34-1
tamariscinum (Hedw.) Schimp., see 34-2, type
Thysanomitrium
  luteum Müll. Hal. = 4-2-7
Tortula
  indica Hook. = 9-1-1
Trachyloma
  stipitatum Mitt. = 33-1-3
Trichosteleum
  fissum Mitt., see 31-8, type
  hornschuchii (Hampe) A. Jaeger = 31-8-4

  hornschuchii (Hampe) A. Jaeger var. *subglabrum* J. Florsch. = 31-8-4b
  pluripunctatum Renauld & Cardot = 31-7-3
Trichostomum
  brachydontium Bruch, see 9-4, type and 9-4-1, note
  perichaetiale Hook., see 4-5, type
Vesicularia
  eligianum W.R. Buck = 26-1-2, see 30-7-1, note
  vesicularis (Schwägr.) Broth., see 30-7, type
Webera
  longifolia (Griff.) Broth. = 35-1-1
Weisia
  obtusa Brid. = 12-1-1
Wijkia
  extenuata (Brid.) H.A. Crum, see 31-9, type
Zygodon Hook. & Taylor, see 17, note

THE GUIANAS
yana, Suriname, French Guiana